# Econodynamics

# New Economic Windows



More information about this series at http://www.springer.com/series/6901

Vladimir N. Pokrovskii

# Econodynamics

## The Theory of Social Production

Third Edition

Vladimir N. Pokrovskii
Institute of Chemical Physics
Russian Academy of Sciences
Moscow
Russia

ISSN 2039-411X ISSN 2039-4128 (electronic)
New Economic Windows
ISBN 978-3-319-89138-5 ISBN 978-3-319-72074-6 (eBook)
https://doi.org/10.1007/978-3-319-72074-6


Softcover re-print of the Hardcover 3rd edition 2018


Printed on acid-free paper

This Springer imprint is published by Springer Nature
The registered company is Springer International Publishing AG
The registered company address is: Gewerbestrasse 11, 6330 Cham, Switzerland

# Preface to the Third Edition

The third edition of the monograph is based on the text of the second edition and keeps its main structure, though some corrections and additions, that specially are not discussed, are made in the text. Besides, the three new Chaps. 8, 12 and 13 are added. In Chap. 8, the ability of the theory to describe a real situation is illustrated, using, as an example, the historical (1960–2016) statistical data for the Russian national economy. Chapter 12 contains the application of the theory, developed in the monograph, to the reconstruction of global production activity on the Earth from the ancient times. And, finally, Chap. 13 contains discussion of the principles of organization of social production; a special attention is given to the rules of distribution of social product, which, eventually, determine social relations and structure of the society. The contents of the Chap. 3 and Appendix A are essentially changed.

The timeliness and, perhaps, even an indispensability of the publication of a new edition is justified by greater interest to the problems of economy and sociology from representatives of natural sciences (to whom I belong). The monograph proposes a new view on the economic growth theory from the point of view of a physicist. I have tried also to comprehend and consistently to state the conventional economic principles and theories, hoping, that the proposed monograph could serve as a reference book to the researchers of social problems. The experience of physicists has appeared and could appear in the future useful for investigation of economic problems.

The author will be grateful for all responses and comments.

Moscow, Russia
October 2017

Vladimir N. Pokrovskii

# Preface to the Second Edition

While studying and teaching *Methods of Mathematical Modelling of Economic Processes*, I have been confused about some discrepancies between various parts of the economic theory. There was an impression that the economic theory exists in independent fragments. Especially upsetting for me, a person who began the study economic theory with *Das Kapital*, was the fact that Marx' theory seems to have no concern in mainstream economics.

I realised later that I was not the sole person to feel a deep dissatisfaction with the situation with the economic theory and its ability to describe reality. To say nothing of the numerous papers, there are many books devoted to a critique of mainstream economics (Nelson and Winter, 1982; Kornai, 1975; Beaudreau, 1998; Keen, 2001). There is a special online *Real-World Economics Review* (http://www.paecon.net/) opposing the mainstream theories. The people—who are engaged in ecology—are traditionally confronting the conventional economic thinking and are looking for physical terms to explain the phenomena of production (Costanza, 1980; Odum, 1996). Some physicists are trying to find new approaches to the analysis of economic situations (Mantegna and Stanley, 1999).

This book contains no critique of any theories. It is devoted to understanding the principles of production and contains a consecutive exposition of the technological theory of social production, which can also be understood as the theory of production of *value*. In the foundation of the theory, the achievements of classical political economy are laid. The labour theory of value is completed, after Marx's hints in *Das Kapital*, with *the law of substitution*. The latter states that, when interpreting value, one has to ensure that the workers' efforts in production of things are substituted with the work of production equipment. A new important concept of *substitutive work, as a value-creating production factor*, was introduced and used to formulate the appropriate theory. The adequacy of the theory has been tested by using the historical data for the U.S. economy.

The book is written by a physicist for the scientifically literate reader, who wishes to understand the principles of the functioning of a national economy. The book contains a discussion of conventional models (Leontief's input–output model, the classical Walras market theory and others) and can be considered as a textbook for students of various specialties who have the necessary preparation in physics and mathematics and a desire to study economic problems. I think the monograph could be interesting for energy specialists, who are engaged in planning and analysing the production and consumption of energy carriers, and for economists, who want to know how energy and technology are affecting economic growth.

The appropriate formulation of the theory has a long history. This monograph was launched, in fact, as a revision and enlargement of my book *Physical Principles in the Theory of Economic Growth*, issued by Ashgate Publishing in 1999. However, it appears that the proper description of the theory has required the text to be completely rewritten and new material to be added, so that I have the opportunity to present a new book with a new title. I have used this edition to clarify the concepts and methods of the theory as far as it was possible for me at the moment.

I am grateful to many people—who support and encourage me in my work. I especially would like to separate a few persons, with whom I had the opportunity to discuss many relevant topics: Robert Ayres, Bernard Beaudreau, Sergio Ulgiati, Andre Maisseu, Michail Gelvanovskii, Grigorii Zuev and Irina Kiselyeva. Some issues became clearer for me after a discussion on the *generalized labour theory of value* with members of the *Socintegrum forum* (http://socintegrum.ru/); I am thankful especially to Valerii Kalyuzhnyi and Grigorii Pushnoi. Finally, I would like to express special thanks to my editors Maria Bellantone and Mieke van der Fluit at *Springer*.

Moscow, Russia
May 2011

Vladimir N. Pokrovskii

## References

Beaudreau, B. C.: Energy and Organization: Growth and Distribution Reexamined, 1st edn. Greenwood Press, Westport (1998)

Costanza, R.: Embodied Energy and Economic Valuation, Science, vol. 210, pp. 1219–1224 (1980)

Keen, S.: Debunking Economics: The Naked Emperor of the Social Sciences. Pluto Press, Sydney (2001)

Kornai, J.: Anti-Equilibrium. On Economic Systems Theory and the Tasks of Research, 2nd ed. North-Holland Publishing, Amsterdam and Oxford (1975)

Mantegna, R. N., Stanley, H. E.: An Introduction to Econophysics: Correlations and Complexity in Finance. Cambridge University Press, Cambridge (1999)

Nelson, R. R., Winter, S. A.: An Evolutionary Theory of Economic Change. Belknap Press of Harvard University Press, Cambridge (1982)

Odum, H. T.: Environmental Accounting. Emergy and Environmental Decision Making. Wiley, New York (1996)

# Contents

# Notations and Conventions

| | |
|---|---|
| A | Input-output matrix with components $a_i^j$; |
| $A = \frac{Y}{L}$ | Labour productivity; |
| B | Capital-output matrix with components $b_i^j$; |
| $E$ | Primary energy used in production; |
| $E_P$ | Primary substitutive work used in production; |
| $I$ | Gross investment in production system; |
| $I_j$ | Gross investment of product $j$; |
| $I^i$ | Gross investment in sector $i$; |
| $I_j^i$ | Gross investment of product $j$ in sector $i$; |
| $K$ | Value of production equipment in production system; |
| $K_j$ | Value of production equipment of kind $j$ in production system; |
| $K^i$ | Value of production equipment in sector $i$; |
| $K_j^i$ | Value of production equipment of kind $j$ in sector $i$; |
| $L$ | Workers' efforts in production system; |
| $L^i$ | Workers' efforts in sector $i$; |
| $M$ | Amount of circulating (paper and credit) money; |
| $M_0$ | Amount of circulating paper money; |
| $N$ | Number of population; |
| $p$ | Price of substitutive work as a production factor; |
| $p_j$ | Price of product $j$; |
| $P$ | Substitutive work |
| $P^j$ | Substitutive work in sector labeled $j$; |
| $Q_j$ | Quantity of product $j$ in natural units; |
| $R_j$ | Value of stock of fundamental product $j$; |
| $S$ | Entropy; |
| $t$ | Time; |
| $U(\cdot)$ | Utility function, welfare function; |
| $u(\cdot)$ | Subjective utility function; |
| $W$ | Value of national wealth; |

| | |
|---|---|
| $W_j$ | Value of national wealth of the kind *j*; |
| $w$ | Price of workers' efforts, wage; |
| $X_j$ | Gross output of product *j*; |
| $Y$ | Final output, gross domestic product; |
| $Y_j$ | Final output of product *j*; |
| $Z^i$ | Production of value in sector *i*; |
| $\alpha$ | Technological index; |
| $\alpha^i$ | Technological index in sector *i*; |
| $\beta = \frac{\Delta Y}{\Delta L}$ | Marginal productivity of workers' efforts at $P = const$; |
| $\beta_i$ | Marginal productivity of workers' efforts in sector *i*; |
| $\gamma = \frac{\Delta Y}{\Delta P}$ | Marginal productivity of substitutive work at $L = const$; |
| $\gamma_i$ | Marginal productivity of substitutive work in sector *i*; |
| $\delta = \frac{1}{K}\frac{dK}{dt}$ | Rate of real growth of capital stock; |
| $\tilde{\delta}$ | Rate of potential growth of capital stock; |
| $\varepsilon$ | Substitutive work requirement; |
| $\overline{\varepsilon} = \varepsilon\,\frac{K}{P}$ | Non-dimensional technological variable; |
| $\varepsilon^i$ | Substitutive work requirement in sector *i*; |
| $\overline{\varepsilon}^i = \varepsilon^i\,\frac{K^i}{P^i}$ | Non-dimensional technological variable for sector *i*; |
| $\eta$ | Rate of real (effective) growth of substitutive work; |
| $\tilde{\eta}$ | Rate of potential growth of substitutive work; |
| $\lambda$ | Labour requirement; |
| $\overline{\lambda} = \lambda\,\frac{K}{L}$ | Non-dimensional technological variable; |
| $\lambda^i$ | Labour requirement in sector *i*; |
| $\overline{\lambda}^i = \lambda^i\,\frac{K^i}{L^i}$ | Non-dimensional technological variable for sector *i*; |
| $\mu$ | Rate of capital depreciation; |
| $\nu$ | Rate of real (effective) growth of labour; |
| $\tilde{\nu}$ | Rate of potential growth of labour; |
| $\xi = \frac{\Delta Y}{\Delta K}$ | Marginal productivity of capital; |
| $\xi^i = \frac{\Delta Z}{\Delta K^i}$ | Sector marginal productivity; |
| $\Xi$ | Marginal productivities tensor with components $\xi^i_j = \frac{\Delta Y_j}{\Delta K^i}$; |
| $\rho$ | Price index; |
| $\tau$ | Time of technological rearrangement |

Latin suffixes take values $1, 2, \ldots, n$ and numerate products and sectors. As a rule, the upper suffix numerates sectors, the lower suffix numerates products. The rule about summation with respect to twice repeated suffixes is sometimes used.

The chapter number and the number of a formula in the chapter are shown in references to formulae.

# Abstract

The monograph represents one of the main problems—why do economies grow?—from a physicist's point of view. Econodynamics itself is regarded as a science with its own concepts and principles—a science which investigates processes of appearing, movement and disappearing of value, being hardly interested in its material carrier. To reconsider the theory of economic growth, the author analyses the concepts of value and utility and their relationship to the thermodynamic concepts. The approach allows the author to include characteristics of technology into description of development and to formulate phenomenological (macroeconomic, no price fluctuations are discussed) theory of production as a set of evolutionary equations in one-sector and multi-sector approximations. The monograph presents the topics in a compact and consistent form that makes it a suitable introductory text for the students of various specialties, who are studying economic problems. The professional researchers could find the monograph to be useful, if they would be ready to get rid of some of neo-classical prejudices.

Vladimir N. Pokrovskii
Doctor of Science (Physics and Mathematics)
Professor of Applied Mathematics
Moscow State University of Economics, Statistics and Informatics
Moscow, Russia

# Chapter 1
# Introduction: The Value-Creating Factors

**Abstract** It is enough to look at the contents of economic courses, to become easily convinced, that the common thing for all of them is ‘a substance’ of value. It is convenient to use the name – economic dynamics (econodynamics) – for the discipline. It investigates the processes of emergence, motion and disappearing of value, just as hydrodynamics investigates processes of motion of liquids, electrodynamics – those of changing electric and magnetic fields, thermodynamics – processes connected with the motion and conversion of heat.... In this chapter, the concept of value is reviewed and the role of basic production equipment, as a set of sophisticated devices, which allow human beings to attract energy from natural sources for the production of useful things, is discussed.

## 1.1 A National Economy at a Glance

The enormous growth of the human population through the centuries is connected with special features of the population. In contrast to any other biological population inhabiting the earth, humans have invented the highly sophisticated artificial means of supporting their own existence while developing a great level of cooperation of members of their society. Since Paleolithic times, clothing, shelter and fuel have become necessities of life almost as fundamental as the food itself. The initial motive power of production is the demand of people, their desire to consume and, consequently, to produce things. As an economic system, a society produces everything that is necessary for a survival: means for the maintenance of human existence, and tools to provide such support. Since Paleolithic times the organization of human society has also been progressing. Modern society presents itself as a huge hierarchal organization, including the government, firms, banks, colleges, libraries and so on. It is a very complex organization, and every one of the members of the society, in some way, is included in the system.

A huge amount of artificial things are accumulated by societies: buildings, transport ways, bridges, production equipment, energy supply systems, sanitation systems



V. N. Pokrovskii, *Econodynamics*, New Economic Windows,
https://doi.org/10.1007/978-3-319-72074-6_1

and so on. Aside from the tangible things, a society accumulated a great amount of intangible objects: knowledge of the laws of nature, principles of organization of society, items of the literature and arts and so on. Both the tangible and intangible constituents of *the wealth of the society* are equally important for maintaining the existence of human communities. To create and maintain national wealth, that is, things that are useful for human beings, *a social production system* was invented and maintained by humans, and this is just what distinguishes human populations from other biological populations. The society, as an economic system, produces everything that is needed for the survival of the community: both the means for supporting the human existence and the means for generating such support.

To design a theory of social production–distribution, it is necessary to keep in mind some heuristic model, which could be imagined due to the remarkable achievements of the classical political economy and neoclassical economics [1]. Some main constituents of the production–distribution system are presented in Fig. 1.1. The central position in the model is the production block, which consists of many production units: the enterprises, factories, firms, research institutes and other organizations that create everything what the man needs. From a material point of view, the production system takes minerals and ores from the environment, transforms natural substances into finished and semi-finished things, the latter istransformed into other things and so on, until all this is finally consumed, and the substances are returned into the environment as waste. This is the material side of production. Human beings always consume the products, so the products always have to be created.

The interaction between various units of the production system and the population is realized by means of the money exchange, so it is possible to present the production system and population as being immersed in a money environment, as shown schematically in Fig. 1.2. The money medium is created by the government, the central bank and many commercial banks. The central bank issues the banknotes and coins – the primary money – that is distributed over commercial banks. The mechanism of issuing assumes that all paper money is circulating among economic subjects: practically no paper money is contained in the commercial banks. The central bank also provides the commercial banks with credits, and the commercial banks can provide their customers with credit money. The records on the accounts of customers are non-paper money, which are created by the commercial banks. The central bank and commercial banks introduce an uncertain amount of the circulating money in coins, banknotes and deposits in the system consisting of the government and the many customers of the commercial banks.

The member of the society exchanges his services for an intermediate product – *money*, and then exchanges the money for products he wants. Therefore, simultaneously with the motion of products, one discovers the motion of money, which has to be considered as a separate, special artifact. The money is circulating in the economy, providing the exchange of products. Modern money is nothing more than a certificate that its owner has a right to get a certain set of products. Modern money is paper money and records on the accounts in the central and commercial banks and, thus, is inherently useless. The value of modern money derives only from the fact that it can be exchanged for the product.

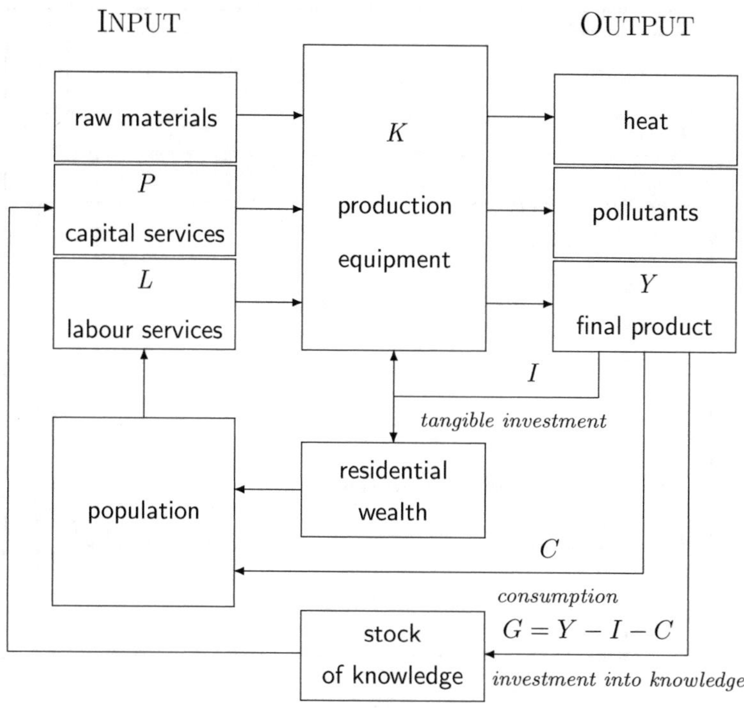


**Fig. 1.1** Fluxes in the production-consumption system. To produce a thing or a service, apart from production equipment $K$, one needs raw materials (ores, water, air, energy carriers and so on), worker efforts $L$ and some factor which can be conventionally called capital services $P$. The last factor is closely connected with production equipment – capital stock $K$, but different from it. Though capital services $P$ can be considered formally as an independent production factor, it is hardly possible to find for it any other interpretation that is different from the amount of work of production equipment, which is done with the help of external energy sources instead of the workers' efforts. The output of the production process is a multitude of things and services, which are measured by their total value $Y$. A part $C$ of final product $Y$ is directly consumed by human beings, and a part $I$ goes to the enhancement of the production system through an increase in the stock of production equipment, so that the production system itself is a subject of evolution. The production processes are accompanied by the emergence of heat and pollutant fluxes, but this is another side of the problem to which we shall not pay much attention to the monograph.

The real production and *the money system* are intervened with each other, thus one can think that an appropriate description can be achieved when these phenomena are studied together. But one ought to consider the real production as a basis of the whole system and it is possible to begin studying of the national economy with the production system, considering the money system as a neutral intermediator. The architecture of the production system appears complex, but in the simple approach the production system can be considered as a set of the interacting pure sectors; in the

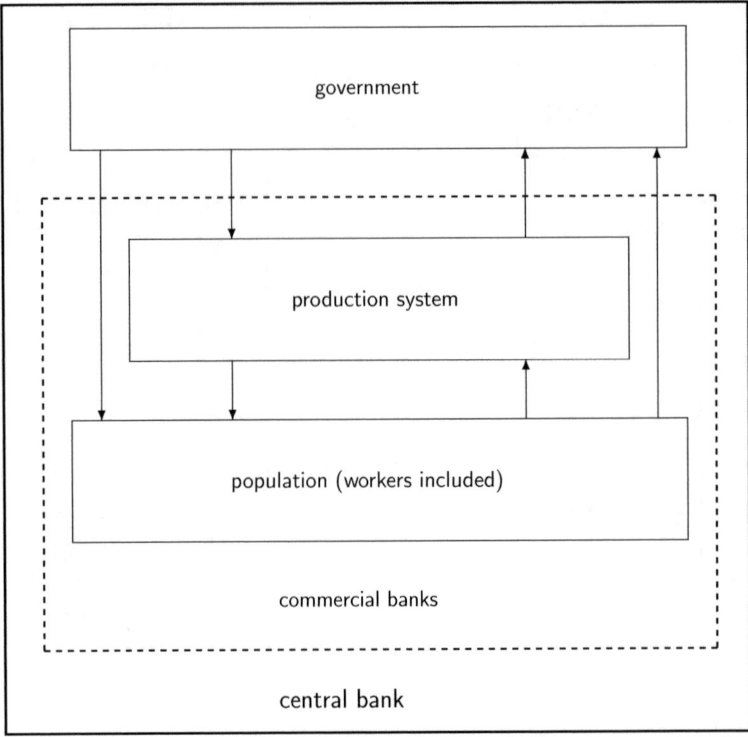


**Fig. 1.2** The architecture of a national economy. The central bank and commercial banks create a money medium for the activity of economic agents. The production system creates all products and generates the fluxes of products to workers and between production units. The fluxes of money, depicted in the picture, are moving in opposite direction. Households are buying products, and money is returning to the producers. The government receives its part of the produced value in the form of taxes, which, in different amounts, are returning to the economic agents. Each flux of money is a result of negotiation and agreement between corresponding agents.

most elementary case, the production system can be considered as the only sector. This heuristic model of the society allows us to develop the theory of the production system in a simple, so-called macroeconomic approach. The investigation of the laws of production is one of the central issues of econodynamics.

## 1.2 The Concept of Value

The notion of the *product* appears to be one of the fundamental concepts of economic theory. It can be defined as something that is produced to be consumed. It does not matter whether the moment of consumption coincides with the moment of production as, for example, in the case of transport services, or does not coincide. In the latter case, the product exists for some time in its material or non-material form. Also, it is insignificant, whether the product is intended to satisfy the needs of the producer or is prepared for sale.[1]

According to the statements of the researchers,[2] the product can be considered as the unity of *use-value* and *production-value*, which allows products to participate in the processes of exchange. In the exchange, the products oppose each other, and the use-value of one product stands against the use-value of another. Products with various use-values can be compared due to the fact that the production-values of all products differ only in quantity, not in quality. Thus, the property that allows the

[1]Let us pay attention to the distinction of the concepts *a product* and *a commodity*. The latter is defined as something that is made for sale that is for an exchange at which value is disposed. From here some people wrongly conclude that the thing made for the producer's consumption does not possess value. This statement has been rejected by Marx ([2], Chap. 1, Sect. 4): "Since Robinson Crusoe's experiences are a favourite theme with political economists, let us take a look at him on his island. Moderate though he be, yet some few wants he has to satisfy, and must, therefore do a little useful work of various sorts, such as making tools and furniture, taming goats, fishing and hunting. Of his prayers and the like we take no account, since they are a source of pleasure to him, and he looks upon them as so much recreation. In spite of the variety of his work, he knows that his labour, whatever its form, is but the activity of one and the same Robinson, and consequently, that it consists of nothing but different modes of human labour. Necessity itself compels him to apportion his time accurately between his different kinds of work. Whether one kind occupies a greater space in his general activity than another, depends on the difficulties, greater or less as the case may be, to be overcome in attaining the useful effect aimed at. This our friend Robinson soon learns by experience, and having rescued a watch, ledger, and pen and ink from the wreck, commences, like a true-born Briton, to keep a set of books. His stock-book contains a list of the objects of utility that belong to him, of the operations necessary for their production; and lastly, of the labour time that definite quantities of those objects have, on an average, cost him. All the relations between Robinson and the objects that form this wealth of his own creation, are here so simple and clear as to be intelligible without exertion, even to Mr. Sedley Taylor. And yet, those relations contain all that is essential to the determination of value".

[2]Still Aristotle, analysing the exchange of various things, wrote '... all things that are exchanged must be somehow comparable' ([3], Book 5, Sect. 5). Marx ([2], p. 14) wrote: '... when commodities are exchanged, their exchange value manifests itself as something totally independent of their use value. But if we abstract from their use value, there remains their value as defined above. Therefore, the common substance that manifests itself in the exchange value of commodities, whenever they are exchanged, is their value'. The brief history and the analysis of concept of value are exposed, for example, by A.N. Usoff in a work 'What is value' (http://www.usoff.narod.ru/Us4.htm, in Russian). Having begun with concepts of use-value and production-value, Usoff has shown, how it is necessary to introduce the concept of value, free from the pre-prepared interpretations. Everyone, who was studying in a higher educational institution in the USSR until 1990, knows the statement that 'value is the expenses of labour'. However, there is no indispensability to reduce the concept of value to expenses of labour in advance. Factorial theories of value, that is the reduction of value to labour, capital and other universal factors of production, are considered in the following section.

products to be compared and exchanged is their *exchange value* or just *value*.[3] Value is an attribute of a product, just as mass is an attribute of matter.

One believes that, irrespective of this or that factorial interpretation of value, the products are exchanged on average according to their values. This is an axiom, which gives a relative measure of value, and allows one to ascribe a certain quantity of value to the products and to estimate the value of a set of products. Value is measured in conventional money units, which are set, when the recognised means of circulation (money) are introduced into the economic system (see Chap. 3). Due to the overall exchange with the help of the money, all commodities can be evaluated, and this is considered as an estimation of their *value* in arbitrary money units (dollars, pound sterlings, euros *et cetera*). One can estimate, for example, a multitude of services and things produced by a nation for a year. This quantity is called the Gross Domestic Product (GDP).

The mechanism of exchange has been scrutinized. Some scholars emphasized the demand side of the phenomenon and argued, that there is no value without utility, so that value ought to be considered as a market estimate of the utility of a thing. Other scholars argued that there are some things (water and air, for example), which have utility without market value, and thus, to understand the meaning of value, one has to refer to the supply side and take into account the production costs of things. It was accepted later (the contributions of Walras [5] and Marshall [6] used to be especially stressed) that both the cost of production (supply) and utility (demand) were interdependent and mutually determinant of value of things. It had appeared to be fruitless to argue whether demand or supply determines value, as, in Marshall's words, 'we might as reasonably dispute whether it is the upper or under blade of a pair of scissors that cuts a piece of paper, as whether value is governed by utility or costs of production'.

The motion and transformations of products in an economy can be described as fluxes of value, which appears at the first touching substances of nature with the hand of a human being, moves together with material substance of a product, leaving its material form, transfers into other substances, and disappears at final consumption. The study of these processes is a subject of an empirical science that can be called *economic dynamics (econodynamics),* which can be defined as a science which investigates the processes of emerging, moving and disappearing of value, being hardly interested in its material carriers. The concept of value in econodynamics is as important as the concepts of energy and entropy in physics. Now we have

[3]Marx distinguished concepts *exchange value* and *value*, the last was identified by him as expenses of labour. It is possible to find the statement in the forth volume of 'Capital' (The Theories of Political Economy, Chap. 8, Sect. 5), that '... commodities do not exchange according to their values, but at average prices, which differ from their values... ', named in other places as 'prices of production'. Marx's observation should be interpreted in such a way, that '... commodities do not exchange according to *expenses of labour*, but to the differing from the expenses of labour quantities, which are usually interpreted as their *value*...'. The quantity *value* and *expenses of labour* are different and can be estimated separately, which has been stated already long ago [4]. Further everywhere, the term value is understood as *exchange value*. .

fragments of this science only, and one of the fragments – the theory of production – is described in this monograph.

The concept of value allows to unify the description and to speak both about the production of things, and about production of value. Thus it is possible to say, that the production system creates things, but the value of things is determined by functioning of all system of production–distribution. Note, that there is some strangeness with the modern use of concept of value. In their daily work economists estimate streams of value between people, enterprises and countries, but in theoretical constructions, economists try to avoid this concept; the concept of utility is used instead. The political economy of the nineteenth century has turned into the *economics* of our days, which is defined as '... the study of how societies use scarce resources to produce valuable commodities and distribute them among different groups' ([7], p. 5). But economic sciences cannot exist without the concept of value. Consideration of streams of value (in any monetary units) in a national economy makes possible the general descriptive schemes of production and distribution.

## 1.3 Factor Theories of Value

Over the centuries researchers have tried to understand how things get value, or, in other words, to find a certain universal source of wealth, to reduce production of value $Y$ in monetary units with quantities of some universal creating value factors, which have the special name *of production factors*. A formulated in this way *law of production of value* plays a fundamental role in economic theories and, consequently, ought to be considered as *the basic economic law*.

### *1.3.1 Classic Labour Theory of Value*

Benjamin Franklin, known for his works on electricity, was one of the first to formulate the statement that a measure of value is work spent by laborers [8]. This idea was especially developed in works by Adam Smith [9], David Ricardo and Karl Marx [2] and appears to be central in the political economy of the beginning of the nineteenth century. In classic political economy, labour was considered as the only factor creating value. According to Smith, 'value of any commodity...to the person who processes it and who means not to use or consume it himself, but to exchange it for other commodities, is equal to the quantity of labour which enables him to purchase or command'. According to Marx, 'all commodities are only definite masses of congealed labour time'. *The labour theory of value* states that value of the created product is equivalent to socially necessary expenses of labour needed for its production. According to this theory, output of production system $Y$ is defined as function of expenditure of labour $L$

$$Y = Y(L).$$

When considering the dynamics, the output is usually measured by monetary units of constant purchasing capacity (see greater details in Sect. 2.2.3), which excludes inflationary phenomena, when the quantity $Y$ creeps upwards owing to the depreciation of the monetary unit. The quantity represents a 'physical' content of the results of production. Considering the quantity $Y$ proportional to the quantity of labour, that was spent during the production of this goods and present, according to Marx, true labour value, one can formally write

$$Y = A\,L. \tag{1.1}$$

On the simple assumption, one can easily find from this relation that output changes, when the used technology of production has been modified, according to the law

$$Y = A_0 e^{\psi t}\,L,$$

where $A_0$ – value of labor productivity in initial point in time, $\psi$ – the reduction rate of consumption of work at the introduction of technological improvements.

As a rule, the growth rate of output exceeds the growth rate of expenditure of labor. Growth of labour productivity $Y/L$ is connected with an increase of productive force of labour, which, according to Marx (Capital, Volume 1, Part 1, Sect. 1), '... is determined by various circumstances, amongst others, by the average amount of skill of the workmen, the state of science, and the degree of its practical application, the social organization of production, the extent and capabilities of the means of production, and by physical conditions'. However, many efforts and time have been required to understand 'various circumstances' and to introduce variables, which would allow to formalize their influence on labor productivity $Y/L$. While the reduction of value to one factor – expenditures of labour $L$ – did not explain all phenomena of economic growth, other production factors, in line with expenditures of labour $L$, such as land and capital, have been introduced into the law of production of value [10].

### 1.3.2 *Role of Production Equipment*

The important element of production is equipment, about which we think as of assembly of different machines, buildings, roads, harbours, pipelines and similar objects, including animals and their pastures. Installation and operation of the production equipment leads to variation of output[4] at the constant expenditures of labour $L$, and to an increase of labor productivity $Y/L$; it would be strange to not consider

[4]Considering production in approximation of several branches, Marx in the third volume of 'Capital' has noticed, that profit distribution on branches conforms to distribution of expenses of work in branches then only, when the ratio of the constant capital to variable one, that is, in Marx's terms, an organic structure of the capital, or, in other words, the ratio of value of the production equipment to expenses for employment of labour in branches, is constant; otherwise conformity it is not observed.

this effect. And indeed, it has been introduced into the theory of value [10] as an additional variable – a monetary assessment of the production equipment (the basic production capital) $K$, as an essential production factor. On this assumption, output $Y$ is recorded as a function of two variables

$$Y = Y(K, L). \tag{1.2}$$

In relation (1.2) capital $K$ and labour $L$ are regarded as perfect substitutes for one another, that is, a given output can be achieved by any combination of the two factors, though, of course, there is a most efficient combination, depending on the prices of production factors.

It has been suggested (see, for example, [11]) many particular forms of function (1.2), but one of them, proposed by Cobb and Douglas [12], has appeared the most common

$$Y = Y_0 \frac{L}{L_0} \left( \frac{L_0}{L} \frac{K}{K_0} \right)^{\alpha}, \quad 0 < \alpha < 1. \tag{1.3}$$

Function (1.3) has the advantage of not depending on the initial values of production factors and is often used for interpretation of phenomena of economic development. The index $\alpha$ ought to be considered as a characteristic of the production system itself.

The other tradition [13–16], in accordance with empirical facts, considers the output as a linear function of capital (or generalised capital)

$$Y = AK. \tag{1.4}$$

The 'capital' productivity $A$, due to empirical evidence, does not depend on production factors. It is easy to see that the laws (1.3) and (1.4) are compatible only at the value $\alpha = 1$, which leads to the exception of the expenditure of labour in the law of production of value.

Another paradox, described Solow [17], exists: the theory based on the neoclassic production function (1.2) in any form does not include technological changes. Nevertheless, there has been a clear belief that, in recent centuries, technological progress was ultimately the source of economic growth in developed countries and it should be incorporated into the theory of economic growth.

To avoid the specified difficulties, it was suggested [17] to modify concepts of labour $L$ and capital $K$. The arguments of function (1.2) must be considered, not as capital and expenditures of labour, but as services of the capital $K' = A_{\mathrm{K}}(t)K$ and labour $L' = A_{\mathrm{L}}(t)L$, which are connected with measured quantities of capital stock $K$ and labour $L$, but are somewhat different from them, so that

$$Y = Y(K', L'). \tag{1.5}$$

---

It specified an indispensability of taking into account a role of the production equipment in the theory of production of value.

An extra time dependence of function (1.2) (the so-called exogenous technological progress) has been introduced, and this time dependence can be found empirically [18].

Multipliers $A_K(t)$ and $A_L(t)$ can be chosen variously, and the production function can be written in a variety of ways [17]. In one of the elementary (but often used) cases, for example, production function is recorded in the form

$$Y = Y_0 A(t) \frac{L}{L_0} \left( \frac{L_0}{L} \frac{K}{K_0} \right)^{\alpha}, \quad 0 < \alpha < 1, \tag{1.6}$$

where the only time-dependent multiplier $A(t)$ appears to describe the influence of technical progress.

The concepts of labour and capital services appeared to be necessary and very useful to explain the observed growth of output [19]. However, the problem of endogenous inclusions of technical progress in the theory remains unsolved, or, taking advantage of Solow's words [20], there remains a question: *'whether one has anything useful to say about the progress, in a form that can be made part of an aggregative growth model.'*

In past decades, there have been some attempts to improve the neoclassical theory by including in the production function new variables, such as technology, or human capital or stock of knowledge $H$ [21–25], or consumed energy $E$ [26–28]. It was assumed that output $Y$ can be written as a function of three, or more, variables

$$Y = Y(K, L, H, E, ...).$$

There was a belief that the only thing one needs to solve the problem is to find a sufficient number of appropriate variables. However, the econometric investigations of over 90 different variables, proposed as potential growth determinants, did not give a definite result [29]. A review of the latest development of the neoclassical approach can be found in a book by Aghion and Howitt [30].

The neoclassical tradition attaches much importance to the production equipment, considering its passive presence, as to some set of objects with a monetary assessment $K$, in the theory, ignoring its active role. The production equipment is set to carry out the certain actions, to facilitate the certain work, and some characteristic of activity of the existing capital stock must be introduced. Apparently, capital as an assembly of equipment, is not 'productive' in physical sense, it is dead [31]. Indeed, it is not so important, how many equipment we have, as what advantage from the installed equipment we have.

### 1.3.3 The Law of Substitution

The task of the production system is to change forms of matter, that is, to transform, for example, ores of different chemical elements into an aircraft, which can fly. To produce a good or a service, some specific work[5] must be done. Modern technologies assume that this work can be done by a human being himself and/or by some external sources, one can say by energy sources, simultaneously. To grind corn into flour, for example, one can use a hand mill, a water mill, a wind mill or a steam mill. In these cases, as in many others, production equipment is some means of attracting external sources of energy (water, wind, coal, oil, etc.) to the production of things; the workers' efforts are substituted by the work of falling water, or wind or heat. No matter who or what does the work, all of the work must be done to obtain the final result which should be compared with the consumed energy and the workers' efforts.

It is possible that the first person to write about the functional role of machinery in production was Galileo Galilei. He realised that all machines transmitted and applied force as special cases of the lever and fulcrum principle. A prominent historian of science and technology Donald Cardwell [32] wrote that Galileo in his notes *On Motion* and *On Mechanics* recognised that 'the function of a machine is to deploy and use the powers that nature makes available in the best possible way for man's purposes... the criterion is the amount of work done – however that is evaluated – and not a subjective assessment of the effort put into accomplishing it' (pp. 38–39). The advantage of machines is to harness cheap sources of energy because 'the fall of a river costs little or nothing'.

The relevance of machinery to economic performance was clearly recognised by Marx [2], who described the functional role of machinery in production processes in Chapter XV *Machinery and Modern Industry* of *Das Kapital* as follows:

> On a closer examination of the working machine proper, we find in it, as a general rule, though often, no doubt, under very altered forms, the apparatus and tools used by the handicraftsmen or manufacturing workman: with this difference that instead of being human implements, they are the implements of a mechanism, or mechanical implements (pp. 181–182). The machine proper is, therefore, a mechanism that, after being set in motion performs with its tools the same operations that were formerly done by the workman with similar tools. Whether the motive power is derived from man or from some other machine, makes no difference in this respect (p. 182). The implements of labour, in the form of machinery, necessitate the substitution of natural forces for human force, and the conscious application of science instead of rule of thumb (p. 188). After making allowance, both in the case of the machine and of the tool, for their average daily cost, that is, for the value they transmit to the product by their average daily wear and tear, and for their consumption of auxiliary substances, such as oil, coal and so on, they each do their work gratuitously, just like the forces furnished by nature without the help of man (p. 189).

These examples illustrate, that both physicists and political economists recognised the important role of machinery in production processes as having to do with the *substitution of workers' efforts by the work of machines moved by external sources of energy*, while the extent of this substitution depends on the technology per se. It is important to keep in mind that while capital is a necessary factor input, work can only be replaced by work, or put differently, work cannot be replaced by capital.

---

[5]One can understand work as a process of conversion of energy in technological processes from one form to another, for example, from mechanical into thermal form.

### 1.3.4 Generalised Labour Theory of Value

Every economist would agree that labour is the most important factor of production, but the situation appears to be more complicated. The production-value, generally speaking, does not reduce the expenses of labour; something else should be added into the theory. One can guess that the 'something' that is needed in the theory is the Marx's phenomenon of 'the substitution of natural forces for human force'. Indeed, after understanding this phenomenon, Marx could suggest that it affects the mechanism of production of value. To understand how gratuitous work influences the value of the products, he could analyse the performance of two similar enterprises. He could suggest that the first of the enterprises use a technology which requires some amounts of labour $L$ and substitution work $P$, and, to produce the same quantity of the same product, the second enterprise uses a technology with the quantities $L - \Delta L$ and $P + \Delta P$ for production factors. So far as the products are considered to be identical, the exchange values of the products of either enterprise on the market are equal, despite the difference in labour consumption. Therefore, Marx could continue to argue, value cannot be determined by labour only, but the properly accounted work of natural forces ought to be considered.[6] To produce the same quantity of value, the decrease in workers' efforts ought to be compensated by an increase in work of external sources, so that one can write the relation

$$-\beta\,\Delta L + \gamma\,\Delta P = 0,$$

where productivities $\beta$ and $\gamma$ of the corresponding production factors are introduced. Thus, equally with human efforts, the work of natural forces appears to be an important production factor. It is easy to see, that the quantity $\beta/\gamma$ determines the amount of gratuitous work of external sources which is needed to substitute for the unit of human work to get an equal effect in production of value. As the work of external forces, replacing efforts of the person, is impossible without the production equipment, this phenomenon has been apprehended and described as substitution of labour by capital. However, to describe substitution correctly, *it is necessary to introduce and consider a new production factor – true substitution work of the production equipment $P$.*

In the general case, the work performed by labour $L$ and substitutive work $P$ has to correspond to a set of products, which has the exchange value $Y$, and one can write, assuming that the production system itself remains unchanged, the relation between differentials of the quantities

$$\mathrm{d}Y = \beta\,\mathrm{d}L + \gamma\,\mathrm{d}P. \tag{1.7}$$

---

[6]In fact, Marx had encountered such a problem (see footnote 4 on p. 9), but stated opposite; he frequently repeats, that only work is a source of value, but, as argues, for example, Yatskevich [33, Chap. 6] this statement is true, if the concept 'work' is understood as 'social abstract work' defined by both actual efforts of working people and all set of machine technologies and methods of work. Taking into account the effect of substitution, a measure of abstract work is the quantity $L + P$.

The coefficients $\beta > 0$ and $\gamma > 0$ correspond to the value produced by the addition of the unit of labour input at constant pure substitutive energy consumption and by the addition of the unit of work of production equipment at constant labour input, respectively; in line with the existing practice, these quantities can be labeled as marginal productivities of the corresponding production factors. The meaning and functional structure of the marginal productivities depends on the method of measuring the quantities $Y$, $L$ and $P$. When the output is estimated by a constant hypothetical measure of value, one can expect, that the marginal productivities $\beta$ and $\gamma$ should be constants. However, such unit of measure does not exist; the possibility of introduction of such unit is discussed in the eleventh chapter (Sect. 11.3). For the comparison of value of the sets of products produced in various years, monetary units of constant purchasing capacity are used. If output is estimated by such monetary scale, the marginal productivities appear to be some functions of production factors.

The two production factors, the workers' efforts and the work of external sources of energy, can substitute for each other and, in this sense, be equivalent, so that labour is eventually, using Adam Smith's words, is 'the only universal, as well as the only accurate measure of value, or the only standard by which we can compare the values of different commodities at all times, and at all places'.

The discussed mechanism of substitution completes the Marx's labour theory of value. Really, by substitution of a labourer's work by forces of nature, that is, by substitution of efforts of people by the work of external forces of nature using production equipment, work operates in a complex of workers' efforts plus work of the equipment. Thus, the work of machines can be appreciated only so far as this work does what people wish, replacing their efforts. Consequently, a measure of value, certainly, can be the labourers' work only. It is possible to say also, according to Marx, that only labourers' work creates value, but Marx, unfortunately, did not complete the theory of substitution to the logical end. Taking into account the effect of substitution, one can say that the only universal and accurate measure of value is the work of labourers or other agents used for production.

### 1.3.5 The Modified Law of Production of Value

The law of production of value, considered in the monograph, is a special case of neoclassical description (see formula 1.5), when service of capital is defined as the substitution work $K' = P$ that is defined as an independent production factor, alongside with traditional factors of production: capital stock $K$ and expenditures of labour $L$. The production of value $Y$ is considered as function of three production factors

$$Y = Y(K, L, P).$$

In this elementary approximation the production system can be presented as *capital stock* (a collection of the equipment measured in its value $K$), getting capacity to operate through *work and services of the capital*, that is by efforts of people $L$

and work of machines that use natural energy sources (wind, water, coal...) $P$. Due to the definition of the production factors, capital stock $K$ and a certain combination of services $L$ and $P$ are complements to each other while capital services (substitutive work) $P$ and labour inputs $L$ act as substitutes for each other. The properties of the production factors allow one to specify (see details in Chap. 6) the production function for output $Y$ in the form of the two alternative lines

$$Y = \begin{cases} \xi K \\ Y_0 \dfrac{L}{L_0} \left( \dfrac{L_0}{L} \dfrac{P}{P_0} \right)^{\alpha} \end{cases} \tag{1.8}$$

The complementary descriptions of production of value can be traced back. The first line in the above formula reminds us of the Harrod–Domar approach [13–16] while the function in the second line coincides with the Cobb–Douglas production function (1.3), in which substitutive work $P$ stands in the place of capital stock $K$. The productivity of capital stock $\xi$ is an internal characteristic of the production system itself and is the 'sum' of the marginal productivities of labour and productive energy, so that capital productivity eventually determines the efficiency of 'transformation' of performed work into value.

Introduction of substitutive work as production factor solves a problem of 'endogenizing' of technical progress [34].[7] Indeed, the comparison of Eq. (1.6) with the second line of the Eq. (1.8) allows us to define the multiplier introduced in the neoclassical theory for description of influence of technical progress in the form

$$A(t) = \left( \frac{K_0}{K} \frac{P}{P_0} \right)^{\alpha}, \quad 0 < \alpha < 1. \tag{1.9}$$

The multiplier appears to be a function of the time-dependent ratio of production factors $P/K$.

For simplicity, we assumed here, that indexes in the Eqs. (1.6) and (1.8) coincide. In other cases, a situation can be more complex, but the ratio $P/K$, irrespective of assumptions made in neoclassical theories, can be considered as a measure of technical progress. The dimensionless ratio of substitutive work to an assessment of labour efforts $P/L$ defines a number of 'mechanical workers' for the only 'alive worker', and, consequently, can be a convenient characteristic of technological progress.

[7] We have concentrated our attention to processes of substitution of alive work by work of machines with attracting of natural forces, but there are also processes of substitution of other kind, to which it is necessary to pay attention. The person who is equipped by more perfect tool or is using more suitable material, can make more products with the same amount of work. For example, if the worker is equipped with a sharp iron axe instead of blunt stone one, he manages to cut more fire wood with the equivalent amounts of efforts in unit of time. At such substitution, efficiency of process of substitution of efforts of machine work increases, that is described by an increase of an index $\alpha$ in the relation (1.8), so that the endogenizing of technical progress by introducing the true work of production equipment $P$ is not complete. However, the introduction of this production factor eliminates some discrepancies in the theory of economic growth.

The theory considered in the monograph that has been designed to consider the phenomenon of production of value, keeps the main attributes of the neoclassical approach, that is, the concept of value created by production factors (donor value) and the concept of production factors themselves; it can be regarded as a generalisation and extension of the conventional neoclassical approach, while the roles of production factors are revised. In the conventional, neoclassical theory, capital as a variable played two distinctive roles: capital stock as value of production equipment and capital service as a substitute for labour. In the described theory, capital service is considered as an independent production factor, whereas capital stock is considered to be the means of attracting labour and energy services to the production. Human effort and the work of external energy sources are regarded as the true sources of value.

## 1.4 Universal Role of Energy in Production

All approaches to the inclusion of energy into the theory of production are known as *the energy theory of value*, which has a long history [35, 36], but, nevertheless, does not have an accurate and complete formulation. Reviewing the development of the discussion, Mirovski ([37], p. 816) concluded that "... the energy theory of value was never developed with any seriousness or concerted effort by any of the groups...' Despite further arguments and investigations performed in later years [38–48], up to now there are no conventional rules according to which one could calculate" the energy content of a money unit" and test the hypothesis.

The relationship between economic growth and energy consumed by the production system[8] is one of the most dramatic issues in economics. The spectrum of opinions on the relationship of energy with value is very broad. The majority of economists, who believe in the productive force of capital, consider energy (or more correctly: energy carriers) to be an ordinary intermediate product that contributes to the value of produced commodities by adding its cost to the price; in other words, consumption of energy is not a source of value. However, one can find many words and arguments in the literature in favour of the universal role of energy in economic processes (see, for example, [28, 35–50]). These researchers have long argued that energy must also be considered as a value-creating factor that must be introduced into the list of production factors in line with other production factors. Moreover, some biophysicists are arguing that energy must be considered as the only source and measure of value [49, 50], and the concept of value itself can be reduced to the concept of energy.

---

[8]It is customary to speak about energy consumption, though, for the sake of precision, the word *consumption* should be replaced by the word *conversion*. Energy cannot be *used up* in a production process. It can only be converted into other forms: chemical energy into heat energy, heat energy into mechanical energy, mechanical energy into heat energy and so on. The measure of potentially converted energy (work) is exergy.

Of course, energy carriers (primary energy[9]) are quite similar to other intermediate products participating in the production process. Nobody can distinguish energy carriers, which are used in the production of aluminium, metallurgical operations and some chemical processes, from other intermediate products. In all these cases, the cost of energy is included in the cost of the final products. But in some cases, apart from being a product, energy from external sources is used to substitute for labour in the technological processes. Energy-driven equipment works in the place of manual labour; genuine work done by the production equipment acquires all the properties of a value-creating production factor, including the property to produce surplus value. The work that is a part of consumed energy carries – it is convenient to have a special name for it: *substitutive work* or *productive energy* – cannot be considered as an intermediate product only, but must be considered as a value-creating factor which has to be introduced into the list of production factors equally with other production factors.[10] The substitutive work has to be interpreted as genuine work done by production equipment with help of external sources of energy instead of workers. Substitutive work gets all features of a creating value production factor, including property to make a surplus value. This quantity also can be considered as capital service provided by capital stock.

At the cost of introducing of the third production factor – substitutive work or productive energy, the discussed theory allows one to unravel the proper role of energy in production of value, on one side, and to eliminate the contradictions of conventional neoclassical theory, on the other side.[11] I think that the monograph proposes some reconciliation of contrasting points of view on the role of energy in the production of value.

---

[9]Primary energy is the name for the amount of primary energy carriers (oil, coal, running water, wind and so on) measured in energy units. It is convenient to measure huge amounts of energy in a special unit *quad* (1 quad $= 10^{15}$ Btu $\approx 10^{18}$ J), which is usually used by the U.S. Department of Energy.

[10]It was realized before the year 2000 that primary energy or total consumption of energy (or exergy) cannot be a proper production factor. In my book ([43], pp. 62–63) I refer to a production factor called *final energy*, which, by its definition, is primary energy input times a coefficient of efficiency. The definition is completely equivalent to that of *useful work*, used by Ayres [51]. The growth rate of *final energy* differs from that of *primary energy*, which for the U.S. economy is 1.5%, it is the growth rate of efficiency of usage of primary energy. Later, I realised [44] that it is substitutive work (not useful work) that has to be exploited as the production factor.

[11]Introduction of energy can also be justified from a thermodynamic point of view. In terms of modern thermodynamics [52] all the artificial things, as well as all biological organisms and natural structures, ought to be considered as deviations from equilibrium in our environment, the latter being reasonably considered a thermodynamic system, and the process of production of useful things is the process of creation of far-from-equilibrium objects (the dissipative structures) as is explained by Prigogine with collaborators [52, 53] (see also Chap. 11). To create and support these structures in our environment as in any thermodynamic system, the matter and energy fluxes must run through the system [52, 54]. In our case, energy comes in the form of human effort and the work of external sources which can be obtained by using the appropriate equipment. The system of a social production is the mechanism that involves a huge quantity of energy to transform 'wild' substances into useful things. The production of useful things can be connected with an establishment of the order (complexity) in the environment by human activity.

## 1.5 Organisation of the Monograph

The monograph presents a general technological theory of production, which is based on the conventional terms and concepts of classical political economy and neoclassical economics, and on conventional physical principles and methods. Some main fundamentals and concepts of modern economics, illustrated with historical (1900–2000) data for the U.S. economy, are described in Chaps. 2 and 3 to introduce the scientifically literate reader, who has not study economics, into the language and problems of the economic theory and to facilitate him eventually to understand the contents of the book. The next chapters contain consecutive derivations of the theory of production, which is the main topic of this book.

The core of the formal theory itself is contained in Chaps. 5 and 6.[12] In one-sector approximation, the list of production factors contains two production factors of conventional neoclassical economics: capital $K$ and labour $L$ and a new production factor: capital service or substitutive work $P$. Capital stock is considered to be the means of attracting labour and substitutive work to production while human efforts and the work of external energy sources are considered as true sources of value. The third factor of production $P$, representing true substitutive work or productive energy, allows to include technological variations in consideration and to explain consistently the phenomena of economic growth, as will be considered in the monograph.

To complete the theory, one needs equations for the dynamics of the production factors, which allow one to investigate trajectories of development. The equations for the growth rate of capital $K$, labour $L$ and substitutive work $P$ are formulated in Chap. 5, and some characteristics of technology, namely, labour and energy requirements, $\lambda$ and $\varepsilon$, that is, amounts of labour and substitutive work needed for unit of production equipment to be launched in action, are introduced. These quantities are combined to create the above introduced index $\alpha$, thus, connecting it with characteristics of applied technology. The index $\alpha$ appears to be a technological index, which can also be interpreted as a share of capital services in the total expenses for maintenance of production factors.

Chapters 4 and 9 contain a generalisation of the theory for a many-sector system. The basis of the development is the well-known *linear input–output model*, described in Chap. 4. The model represents the production system as a set of coupled sectors, and each of them creates its own specific product. The applicability of the linear input–output model to dynamic situations (Chap. 9) is extended in proposed work, so the restrictions imposed by production factors (labour and substitutive work) and the evolution of the production system itself, that is, structural and technological changes are taken into account. In fact, a phenomenological version of evolutionary theory of production system is formulated in these chapters.

---

[12]The principles of the theory were discussed earlier in the author's monograph [43], though some issues have been reformulated here. In particular, the concept of substitutive work was not clearly defined, and the important contribution to production of value from technological and structural changes was erroneously omitted. The correct version is given in the author's article [44].

In Chaps. 6, 8 and 9, the ability of the theory to describe a real situation is illustrated for the example of historical (1900–2000) statistical data for the U.S. economy and more modern (1960–2016) data for the Russian national economy. To identify the considered model, one needs in empirical time-series of output $Y$, capital $K$ and labour $L$. A method of separating substitutive work $P$ from the total primary consumption of energy, as well as a method of calculating the technological index $\alpha$, appears to be an organic part of the theory. Besides, at the given time-series for investment, one can estimate the technological characteristics of the production system. The comparison shows the consistency of the theory and its correspondence to empirical facts. The proposed theory can explain facts of economic growth, especially, the main fact of recent development, that output expansion has outpaced population growth in the 200 years since the industrial revolution. The theory has the means to describe the difference in productivity growth for different countries. Within empirical accuracy, the consistency is perfect, so that one can acquire a feeling that substitutive work or, more generally, capital service is the only missing production factor in the conventional two-factor theory of economic growth and no other production factors, aside from capital, labour and substitutive work, are needed to describe the path of growth quantitatively. Perhaps, the substitutive work is the same production factor that the scholars of the modern endogenous theory of economic growth have been seeking.

Chapters 10 and 11 represent an attempt to understand and interpret the very concept of value, which is a unique specific concept and as frequently used and important in economics as the concepts of 'energy' and 'entropy' in physics. The relationships among the thermodynamic concepts and economic concepts of value and utility are analysed. The analysis of a relationship between thermodynamic concepts and economic concepts of value and utility shows, that value is a close relative of entropy. Reconciliation of the two points of view on the phenomenon of production leads to a unified picture that enables us to relate some aspects of our observations of economic phenomena to physical principles.

Chapter 12 contains, as an example of the application of the theory established in the monograph, a description of the reconstruction of the global production activity on the earth from the ancient times. The chapter represents the general picture of influence of technological progress on progress of mankind. The growth of the human population on the earth, as connected with the growth of production and well-being, is considered also. The future of a human population on the earth is discussed.

And the final Chap. 13 contains discussion of the principles of organization of the social production. A special attention is given to the rules of distribution of social product, which, eventually, determine social relations and structure of a society. As consequence of the concepts, developed in the monograph about sources of production of value, the reasonable rules of formation of the public fund, which stimulate increase of efficiency of use of social resources, are proposed.

Concluding the description, the monograph investigates one of the main problems of economics – why do economies grow – and reconsiders the theory of production from a physicist's point of view. The monograph contains quantitative description of production as a social mechanism, embedded in the environment. The approach allows us to include characteristics of technology into the description and to formulate a phenomenological (macroeconomic, no price fluctuations are discussed) theory of production as a set of evolutionary equations in one-sector and multi-sector approximations.

## References

1. Blaug, M.: Economic Theory in Retrospect, 5th edn. Cambridge University Press, Cambridge (1997)
2. Marx K.: Capital. Encyclopaedia Britannica, Chicago (1952). English translation of Karl Marx, Das Kapital: Kritik der politischen Oekonomie, Otto Meissner, Hamburg (1867)
3. Aristotle: Nicomachean Ethics. (about 350 B.C.E). http://classics.mit.edu/Aristotle/nicomachaen.5.v.html. Accessed 17 Aug 2017
4. Böhm-Bawerk E.: Karl Marx and the Close of His System. T.F. Unwin, London (1898). English translation of Eugen Böhm-Bawerk, Zum Abschluss des Marxschen Systems. Berlin (1896)
5. Walras, L.: Elements d'economie Politique Pure ou Theorie de la Richesse Sociale. Corbaz, Lausanne (1874)
6. Marshall, A.: Principle of Economics, 8th edn. Macmillan, London (1920)
7. Samuelson, P., Nordhaus, W.: Economics, 13th edn. McGrow-Hill Book Company, New York (1989)
8. Franklin B.: A modest enquiry into the nature and necessity of a paper-currency. Philadelphia. (1729). Electronic Text Center, University of Virginia Library. http://xtf.lib.virginia.edu/xtf/view?docId=legacy/uvaBook/tei/FraMode.xml. Accessed 17 Aug 2017
9. Smith, A.: An Inquiry into the Nature and Couses of the Wealth of Nations. Clarendon Press, Oxford (1976). In two volumes
10. Wicksteed, PhH: Essay on the Coordination of the Laws of Distribution. MacMillan & Co, London (1894)
11. Ferguson, C.E.: The Neo-classical Theory of Production and Distribution. Cambridge University Press, Cambridge (1969)
12. Cobb, G.W., Douglas, P.N.: A theory of production. Am. Econ. Rev. Suppl. **18**(1), 139–165 (1928)
13. Harrod, R.F.: An essay in dynamic theory. Econ. J. **49**, 14–23 (1939)
14. Harrod, R.F.: Towards a Dynamic Economics. Macmillan, London (1948)
15. Domar, E.D.: Capital expansion, rate of growth and employment. Econometrica **14**, 137–147 (1946)
16. Domar, E.D.: Expansion and employment. Am. Econ. Rev. **37**, 343–355 (1947)
17. Solow, R.: Technical change and the aggregate production function. Rev. Econ. Stud. **39**, 312–330 (1957)
18. Brown, M.: On the Theory and Measurement of Technological Change. Cambridge University Press, Cambridge (1966)
19. Jorgenson, D.W., Stiroh, K.: Raising the speed limit: U.S. economic growth in the information age. Brookings Pap. Econ. Act. **1**, 125–211 (2000)
20. Solow, R.: Perspective on growth theory. J. Econ. Perspect. **8**, 45–54 (1994)
21. Griliches, Z.: Issues in assessing the contribution of research and development to productivity growth. Bell J. Econ. **10**, 92–116 (1979)
22. Griliches, Z.: Productivity puzzles and R and D: Another nonexplanation. J. Econ. Perspect. **2**, 9–21 (1988)
23. Romer, P.M.: Increasing returns and long-run growth. J. Polit. Econ. **94**, 1002–1037 (1986)
24. Romer, P.M.: Endogenous technological change. J. Polit. Econ. **98**, 71–102 (1990)
25. Lucas, R.E.: On the mechanics of economic development. J. Monet. Econ. **22**, 3–42 (1988)

26. Hudson, E.A., Jorgenson, D.W.: U.S. energy policy and economic growth, 1975–2000. Bell J. Econ. Manag. Sci. **5**, 461–514 (1974)
27. Berndt, E.R., Wood, D.O.: Engineering and econometric interpretations of energy - capital complementarity. Am. Econ. Rev. **69**, 342–354 (1979)
28. Kümmel, R.: The impact of energy on industrial growth. Energy **7**, 189–203 (1982)
29. Durlauf, S., Quah, D.: The new empirics of economic growth. In: Taylor, J., Woodford, M. (eds.) Handbook of Macroeconomics. North-Holland, Amsterdam (1999)
30. Aghion, P., Howitt, P.: The Economics of Growth. MIT Press, Cambridge (2009)
31. Beaudreau, B.C.: Energy and Organization: Growth and Distribution Reexamined, 2nd edn. Greenwood Press, New York (2008)
32. Cardwell, D.S.L.: Turning Points in Western Technology. Science History Publications, New York (1972)
33. Yatskevich V.V.: Refleksiya nad Marksizmom (The Reflection on Marxism). Kiev (2005). http://philprob.narod.ru/philosophy/philosophy.html. Accessed 19 Aug 2017
34. Pokrovskii V.N.: Endogenous technical progress in the theory of economic growth. Hindawi Publishing Corporation, ISRN Economics **2014**, Article ID 928121 (2014). http://dx.doi.org/10.1155/2014/928121. Accessed 19 Aug 2017
35. Scott, H.: Introduction to Technocracy. The John Day Company, New York (1933)
36. Soddy, F.: Cartesian Economics: The Bearing of Physical Sciences upon State Stewardship. Hendersons, London (1924)
37. Mirowski, P.: Energy and energetics in economic theory: A review essay. J. Econ. Issues **22**, 811–830 (1988)
38. Cottrell, W.F.: Energy and Society: The Relation between Energy. Social change and economic development. McGraw-Hill, New York (1955)
39. Maïsseu, A., Voss, A.: Energy, entropy and sustainable development. Int. J. Global Energy Issues **8**, 201–220 (1995)
40. Odum, H.T.: Environmental Accounting. Emergy and environmental decision making. Wiley, New York (1996)
41. Valero, A.: Thermoeconomics as a conceptual basis for energy-ecological analysis. In: Ulgiati, S. (ed.) Advances in Energy Studies Workshop: Energy Flows in Ecology and Economy (Porto Venere, Italy 1998), pp. 415–444. MUSIS, Rome (1998)
42. Sciubba, E.: On the possibility of establishing a univocal and direct correlation between monetary price and physical value: The concept of extended exergy accounting. In: Ulgiati, S. (ed.) Advances in Energy Studies Workshop: Exploring Supplies, Constraints, and Strategies (Porto Venere, Italy 2000), pp. 617–633. Servizi Grafici Editoriali, Padova (2001)
43. Pokrovski, V.N.: Physical Principles in the Theory of Economic Growth. Ashgate Publishing, Aldershot (1999)
44. Pokrovski, V.N.: Energy in the theory of production. Energy **28**, 769–788 (2003)
45. Beaudreau, B.C.: Energy Rent: A Scientific Theory of Income Distribution. New York; iUniverse, Shanghai, Lincoln (2005)
46. Hall, C., Lindenberger, D., Kümmel, R., Kroeger, T., Eichhorn, W.: The need to reintegrate the natural sciences with economics. Bioscience **51**, 663–673 (2001)
47. Ayres, R.U., Ayres, L.W., Warr, B.: Exergy, power and work in the U.S.* economy, 1900–1998. Energy **28**, 219–273 (2003)
48. Ayres, R.U., Ayres, L.W., Pokrovski, V.N.: On the efficiency of electricity usage since 1900. Energy **30**, 1092–1145 (2005)
49. Costanza, R.: Embodied energy and economic valuation. Science **210**, 1219–1224 (1980)
50. Cleveland, C.J., Costanza, R., Hall, C.A.S., Kaufmann, R.: Energy and the U.S. economy: A biophysical perspective. Science **225**, 890–897 (1984)
51. Ayres, R.U.: Sustainability economics: Where do we stand? Ecol. Econ. **67**, 281–310 (2008)
52. Prigogine, I.: From Being to Becoming. Time and complexity in the physical sciences. Freeman & Company, New York (1980)

53. Nicolis, G., Prigogine, I.: Self-Organisation in Non-Equilibrium Systems: From Dissipative Structures to Order through Fluctuations. Wiley, New York (1977)
54. Morowitz, H.J.: Energy Flow in Biology: Biological Organisation as a Problem in Thermal Physics. Academic Press, New York (1968)

# Chapter 2
# Empirical Foundation of Econodynamics

**Abstract** In this chapter, the terms and notions, which were introduced by many researchers of the phenomenon of social production and economic growth during the long period of development of economic theory, are reminded and discussed. The main chain of definitions is as follows: *product–output–investment–stock of production equipment.* The latter is a set of the real means of production: the collection of tools and all energy conversion machines, including information processing equipment, plus ancillary structures to contain and move them. The term *capital stock* is applied for the value of the stock of production equipment (the means of production). In this and the following chapters, time series of some quantities for the U.S. economy, which are collected in Appendix B, is used for illustration.

## 2.1 On the Classification of Products

To be able to have a possibility to describe the internal processes in economy in some detail, we need to focus on a variety of production units, and we also need in some classification of products, according to which the whole production system can be broken into sectors [1–4]. The division of the economy into sectors can vary, the number of sectors depends on the aims one is pursuing. For the current description and planning, the economy can be divided into no more than a few hundred sectors. An example of working classification can be found in Appendix A. The nomenclature list of products and sectors should cover, on-possibility, all activities. Nevertheless, in any society, beyond the official account, despite the tendency to expansion of areas of the account, there are the shadow (invisible) economic activities, including inalienable economic activities (work for home: cooking, cleaning of the house, repairing of home appliances and so on) and illegitimate (including, criminal) activity.

For research aims, the economy can be divided into a few sectors only [2, 3]. It appears to be convenient to apply a three-branch model that generalises the scheme of production with two divisions (branches) used in the Soviet Union. The first of the branches created the means of production, the second—the goods for the immediate consumption. Following Marx [5], it was accepted in the Soviet Union that activity of teachers, doctors, lawyers, financiers, officials and many other members of the


V. N. Pokrovskii, *Econodynamics*, New Economic Windows,
https://doi.org/10.1007/978-3-319-72074-6_2

society is not productive, that does not create value, and there was no indispensability of inclusion of such activity in the scheme of production. Now researchers have come to believe that for fuller description of functioning of production system, it is necessary to take into account the activity creating not only tangible but also intangible products,[1] and, consequently, the elementary model of production system ought to be expanded as follows.

1. The first sector devours natural resources and uses its own products and products of the second sector to produce the means of production and all material products necessary for production in all sectors. The sector includes industries of extraction of raw material (ore, stone, coal, oil, etc.), construction, transportation, manufacturing of cars, appliances for homework and furniture, etc.
2. The second sector creates the products necessary for the current production and future development, in that amount non-material information products, i.e. general knowledge and various instructions on how to organise the matter for humans use. The sector includes formation of principles of the organisation: scientific researches, design and experimental works, education, art, management, the financial system, the computer software and many instructions, which are partially embodied in the organisation of work of production system; other part exists in the non-material form of the postponed messages, forming the huge collection of *information resources*, which include results of activity of scientific and project institutes. It is necessary to note that non-material production, i.e. principles of organisation, software and results of research works, should be connected with material production, as they are useless if they are not consumed.
3. The third sector produces the things that human beings need directly. This sector includes the food processing industry, agriculture, retail, restaurants, hotels, health care and so on. We can say that one needs in the first two sectors, only to keep the third sector in action. Strictly speaking, human beings do not need the products of the first two sectors directly.

We can pay attention that the desire to distribute the products and industries, listed by the official statistics (see, for example, Appendix A) over the three branches of

[1] Let us notice, that the formal indispensability of introduction of the third sector in Marx's two-sector scheme has been recognised by Tougan-Baranovsky [6] and by Bortkiewicz [7] who suggested, that the additional sector creates luxury goods for pleasure of capitalists. However, confirming the indispensability of introduction of the third sector (see Sect. 2.2.2), it is necessary to give other interpretation of additional sector. The essential fraction of products of the second branch in our model is information products (projects, scientific results, qualification of workers, works of art), necessary for current production and the future development of the society and 'human capital' which means the worker shaping: education, health care and vocational training. Note that the first and the third sectors in our scheme correspond to the divisions of production, which were introduced by Marx [5] (and used in the USSR), as the sector of the means of production and the sector of production of commodities.

production meets with some difficulties. So, not each sector from the list of Appendix A can be completely included in this or that division. Probably, the sectors with numbers 5 and 6 (Appendix A, ferrous and nonferrous metallurgy) can be attributed to the first division, sectors 20–22 (healthcare, education, science, finance, management)—to the second division and sectors 11–12 (easy and food processing)—to the third one. The activity of the greater part of sectors, such as construction or transport, is necessary to distribute on two or three divisions of the three-sector scheme. The attribution of the actual content to that or this division of the three-branch scheme of production appears not too simple (see also discussion in Sect. 9.5), however, for the description to be complete, all production sectors should be included in one of these three sectors, and although, one can see that it is difficult to locate some of the activities listed in Appendix A, this scheme appears to be convenient and productive in the theoretical analysis.

Further, we shall use the assumption that the output of the production units can be divided into $n$ classes, which allows us to consider $n$ products, circulating in a national economy [1–4]. Following this tradition, one can assume that all enterprises of the production system can be divided into $n$ classes as well. Therefore, we imagine, following Leontief [2], that the production system of economy consists of $n$ production sectors, or, more accurately, of *$n$ pure production sectors*, so as some enterprises of production system could not be included completely in the only one production sector [2].

Let us note that, in addition to the sector classification, some groups of products also can be selected according to the aims and modes of their consumption. Some products can be used to produce other products [4]. If things are used for production many times, as, for example, instruments and tools, machinery, means of transport, agricultural land and so on, one speaks of *fixed production assets*. One speaks of *intermediate production consumption*, if products, for example, coal, oil and ore, are disappearing in the production processes. The products for *final consumption* by human beings comprise products, which are used as final products many times, e.g. residential buildings, furniture and so on (*residential assets*), and products which disappear at consumption, like food, for example. Sometimes, it is difficult to decide whether a product (for example, roads and buildings) ought to be classified as production assets or as residential wealth.

## 2.2 Motion of Products

Consider an economy as consisting of the production sectors, each of them creating its own product. The important characteristic of the sector is its *output* that is the amount of product created by the sector in a time unit

$$dQ_i, \quad i = 1, 2, \ldots, n.$$

These quantities are measured in natural units such as tonnes, metres, pieces and so on. We do not discuss here the difficulties that appear, when many primary natural products are aggregated in the only product of the sector.

To compare the quantities of different products, an empirical estimation of *value of product* is used. Value is measured with conditional monetary units, such as the ruble, dollar and others.[2] Neglecting fluctuations, which are the accidental deviations of quantity from some mean value, one defines the value of a unit of a product in arbitrarily chosen units, that is its *price*. We assume, that the prices, as empirical estimates of value, for all products are known

$$p_i, \quad i = 1, 2, \ldots, n.$$

The price of a product is not an intrinsic characteristic of the product. The price depends on the quantities of all products that are in existence at the moment. As a rule, the price decreases if the quantity of the product increases, though the situation can be more complicated. Note that there are coupled sets of products, such that an increase in the quantity of one product in a couple is followed by an increase (in case of couple of complementary products) or a decrease (in case of couple of substituting products) in price of the other product of the couple. Therefore, one ought to consider the price of a product to be a function of quantities of, generally speaking, all existing products

$$p_i = p_i(Q_1, Q_2, \ldots, Q_n). \tag{2.1}$$

One can define the *gross output* of the sector $i$ as the value of the product created by the sector labelled $i$ for a unit of time

$$X_i = p_i dQ_i, \tag{2.2}$$

so that the gross output of the economy appears to be a vector with $n$ components

$$\mathsf{X} = \begin{Vmatrix} X_1 \\ X_2 \\ \cdot \\ \cdot \\ X_n \end{Vmatrix}.$$

[2]The assessment and comparison of value of the various products existing in various points in time are complicated by the lack of a constant scale of value. As known financier Lietaer ([8], p. 254) writes: 'The world has been living without an international standard of value for decades, a situation which should be considered as inefficient as operating without standard of length or weight'. The absence of constant scale of value is a headache for both experts and analysts.

### 2.2.1 *Balance Equations*

#### 2.2.1.1 Closed Systems

To create the product of a sector, apart from fixed production capital, it is necessary to use the products of, generally speaking, all sectors. For example, to produce bread, apart from an oven, it is necessary to have flour, yeast, fuel and so on. Therefore, the gross output of each sector is distributed among the others

$$X_i = \sum_{j=1}^{n} X_i^j + Y_i, \quad i = 1, 2, \ldots, n, \tag{2.3}$$

where $X_i^j$ is an amount of the product labelled $i$ used for production of the product labelled $j$. The *intermediate production consumption* of the products is determined by the existing technology and does not include consumption of the basic production assets. The residue $Y_i$ is called the *final output*, which is the value of the products used for productive and non-productive consumption beyond the current production processes. It will be discussed later.

On the other hand, the value of the output of the sector $i$ is the sum of the values of products consumed in the production and an additive term

$$X^i = \sum_{j=1}^{n} X_j^i + Z^i, \quad i = 1, 2, \ldots, n. \tag{2.4}$$

This relation defines the quantify $Z^i$ which is called the *production of value* in sector $i$. One can consider that every sector creates value. The first terms on the right-hand sides of relations (2.3) and (2.4) represent products which are swallowed up by the acting production sectors.

The final output of the sectors $Y_i$ characterises production achievements of the society. For this purpose, it is convenient to use also the sum

$$Y = \sum_{j=1}^{n} Y_j. \tag{2.5}$$

This is the value of all the material and non-material products created by a society per unit of time (year). One calls it *the gross domestic product (GDP)*, if one is considering a national economy.

One can sum relations (2.3) and (2.4) over the suffixes and compare the results to obtain

$$Y = \sum_{j=1}^{n} Z^j. \tag{2.6}$$

**Table 2.1** Balance of products

| Gross output | $X^1\ X^2 \cdots X^n$ | Final output |
|---|---|---|
| $X_1$ | $X_1^1\ X_1^2 \cdots X_1^n$ | $Y_1$ |
| $X_2$ | $X_2^1\ X_2^2 \cdots X_2^n$ | $Y_2$ |
| ... | ... ... ... ... | ... |
| $X_n$ | $X_n^1\ X_n^2 \cdots X_n^n$ | $Y_n$ |
| Production of value | $Z^1\ Z^2 \cdots Z^n$ | $Y/Z$ |

It means that the GDP is equal to the production of value in all production sectors of the economy.

Let us pay attention that value of the sector product $Y_j$ not necessarily coincides with production of value in this sector $Z^j$ and, accordingly, the sector profit does not coincide with the sector surplus value that was revealed still by Marx, who, having defined in the first volume of 'Capital' value as expenses of work required for production of commodities, had met with contradictions considering the sector scheme of production in the third volume.[3]

The quantities incorporated in formulae (2.3)–(2.6) can be conventionally represented by a balance table (Table 2.1). All quantities in the table should be replaced by numbers in order for a real economy to be analysed.

[3]The analysis of the problem was completed by Samuelson [9], who writes: '...the valid competitive "prices" for the goods, made with the maximal intensity of work, are rather low, and the prices for the goods, made with the minimal intensity of work (or, using Marx's terminology, with the highest "organic structure of the capital"), are the highest'. Marx compared the final sector output $Y$ and the payment to workers $V$ in the sector (we use the symbols of Table 2.2) and had detected that the reality mismatched his concepts. Marx had collided with the dilemma: to keep statements of the first volume and to find a suitable explanation to the facts, or to think up something another. As Samuelson writes, '...neither Marx, nor Engels did not suppose it at all (to decline the statements of the first volume - my (VNP) comment); accepting the point of view that Marxist concept of "value" is (1) philosophically useful; (2) interesting and essential from sociological and historical points of view; (3) extremely important for the fundamental understanding of essence of capitalist operation and laws of movement of capitalist progress; practically providing you with a brand of an accessory to theorists of Marxism'. Marx has preferred to state that products exchange not for value (taking in mind an expenditure of work for creation of the product), but for some *prices of production*. Further, naturally, there appeared a problem of transformation of 'the prices of production' into 'values' and the problem has no other solution, besides, that for an explanation of production of value, alongside with expenditures of labour, the source of value connected with the production equipment (see Sect. 1.3 of Chap. 1) is necessary to account also.

#### 2.2.1.2 Open Systems

When we take into account that some products can be objects of international trade with other countries (import and export), the production balance changes a little. In this case, it is necessary to subtract the export part from the gross product of each sector $T_i^{\uparrow}$ and to add the import quantity of the product $T_i^{\downarrow}$, so that the balance parity (2.3) is recorded in the modified form

$$X_i + T_i^{\downarrow} - T_i^{\uparrow} = \sum_{j=1}^{n} X_i^j + Y_i, \quad i = 1, 2, \ldots, n, \tag{2.7}$$

where $X_i^j$ is the part of the product with index $i$ which is used for production in sector $j$. The difference between import and export can be used both for intermediate production consumption and for final consumption. The residual $Y_i$, called the final product, presents the value of the products used beyond current processes for productive and non-productive consumption.

On the other hand, the value of a product $X_i$ can be presented as the sum of value of the products consumed by production, and some additive term, which is presented by Eq. (2.4). This parity defines the quantity of value $Z^i$, created in sector $i$. One can suppose that each sector creates value, as the production equipment takes part in the production.

Summing up relations (2.4) and (2.7) on indexes $i$ and comparing the results, one obtains, instead of (2.6),

$$Z = Y + T^{\uparrow} - T^{\downarrow}. \tag{2.8}$$

In this case, the value created by the production system $Z = \sum_{j=1}^{n} Z_j$ is referred to as the GDP, which is used for productive and non-productive consumption $Y = \sum_{j=1}^{n} Y_j$ and pure export $T^{\uparrow} - T^{\downarrow}$.

### 2.2.2 *Distribution of the Social Product*

To move further, it is necessary to consider the main constituents of both the value of the final products created in sectors $Y_i$, and the production of value in sectors $Z^i$. The final product $Y_j$, defined by the balance equation (2.3), is used both for direct consumption and for maintenance and expansion of the production system of an economy. Consequently, we can present a vector of the final product as the sum of three vectors

$$Y_j = I_j + G_j + C_j, \quad j = 1, 2, \ldots, n, \tag{2.9}$$

where $C_j$ stands for the value of products which are consumed by people directly and immediately (one-time consumption), $G_j$ designates the value of intermediate products (material and non-material) not consumed and not used in production and $I_j$ designates gross investments (with inclusion of amortisation expenses) in the basic equipment (fixed production capital). It is assumed that all quantities are estimates of values of actual fluxes of products. Certainly, some of the components of fluxes $I_j$, $G_j$ and $C_j$ can be set equal to zero.

The production of value in sectors $Z^i$ was considered by Marx, who called it social product that, in each sector, can be broken, according to Marx, into wages $V^j$, surplus product $M^j$ and value of the production assets disappearing in the process of production $A^j$; consequently

$$Z^j = V^j + M^j + A^j, \quad j = 1, 2, \ldots, n. \tag{2.10}$$

One of the major characteristics of the sector performance is the index of profit, defined as

$$\frac{M^j}{A^j + V^j}, \quad j = 1, 2, \ldots, n.$$

According to Marx, because of aspirations of separate manufacturers to get the higher profit, these quantities tend to accept identical value; however, actually alignments of indexes of profit are not observed.

#### 2.2.2.1 Three-Sector Scheme of Production

The situation becomes simpler if we refer to the three-sector model described in Sect. 2.1. When parities (2.9) have been summed, the total final output $Y$, that is equal to production of value in sectors $Z$, can be represented as a sum of the three components

$$Y = I + G + C. \tag{2.11}$$

The quantities in the right part of the parity can be naturally interpreted as components of the gross domestic product $Y$ that are created by the earlier described (in Sect. 2.1) three production sectors

$$Y_1 = I, \quad Y_2 = G, \quad Y_3 = C, \tag{2.12}$$

where $I$ is the gross investments into the production equipment, $G$ is the governmental expenses for the defense, investments into the stock of knowledge and into the individuals (projects, scientific results, qualification working, works of art) and so on, $C$ is the direct consumption by the population.

The production of value in each sector includes three components

$$Z^j = V^j + M^j + A^j, \quad j = 1, 2, 3. \tag{2.13}$$

**Table 2.2** Balance of three-sector system

| Gross output | $X^1$ $X^2$ $X^3$ | Final output | Gross investment | Investment in social wealth | Private consumption |
|---|---|---|---|---|---|
| $X_1$ | $X_1^1$ $X_1^2$ $X_1^3$ | $Y_1$ | $I$ | 0 | 0 |
| $X_2$ | $X_2^1$ $X_2^2$ $X_2^3$ | $Y_2$ | 0 | $G$ | 0 |
| $X_3$ | 0 0 0 | $Y_3$ | 0 | 0 | $C$ |
| Production of value | $Z^1$ $Z^2$ $Z^3$ | $Z = Y$ | | | |
| Consumption of basic capital | $A^1$ $A^2$ $A^3$ | | $A \leq I$ | | |
| Surplus value | $M^1$ $M^2$ $M^3$ | | | $M \geq G$ | |
| Workers' wages | $V^1$ $V^2$ $V^3$ | | | | $V = C$ |

Both the wages $V^j$, and the surplus product $M^j$ can be used for direct consumption or for the further development of production.

Table 2.2 contains the symbols of all quantities used in Eqs. (2.12) and (2.13), and the relations between the quantities are also shown in the diagonal cells. The sign of equality in parities $A^1 + A^2 + A^3 \leq I$ and $M^1 + M^2 + M^3 \geq G$ refers to the case of simple reproduction. Generally, the surplus value is materialised in production and non-productive accumulation

$$M^1 + M^2 + M^3 = I - A + G.$$

Let us note that the shortage of consideration of two divisions and an indispensability of introduction additional (the second in our classification) branch becomes obvious by consideration of simple reproduction, when all product of the first branch goes on the restoring of the production equipment in all branches

$$A^1 + A^2 + A^3 = I^1 + I^2 + I^3.$$

When Eqs. (2.12) and (2.13) have been summed and combine, we find the expression for the created value which, certainly, is equal to value of all created products

$$Z = Y_1 + M^1 + M^2 + M^3 + Y^3 = Y.$$

From this equation follows, that in the case, if no second branch exists, $M^1 + M^2 = 0$ that means there is no surplus product, which is unsuitable for capitalist way of production. Apparently, such reasons have compelled Tougan-Baranovsky [6] and Bortkiewicz [7] to introduce additional branch which makes (at simple reproduction) the final product

$$Y_2 = M^1 + M^2 + M^3.$$

Tougan-Baranovsky and Bortkiewicz believed that this branch makes luxury goods and pleasures for capitalists, and the product of this branch is used by proprietors of the enterprises. However, now one can give another interpretation of the additional branch. It is necessary to consider that additional, second branch in our model, among other things, creates the information product (projects, scientific results, qualification working, works of art), necessary for current production and future development of the society.

### *2.2.3 Gross Domestic Product*

GDP represents a measure of current achievements of an economy as a whole—a measure of a multitude of fluxes of products. The equations recorded in the previous section show the various methods of calculating of the GDP, which can be estimated as the results of production, that is, the value of created products (Eq. 2.5), or by the account of the use of products (Eqs. 2.5 and 2.11), or by the contribution of separate components of the created value (Eqs. 2.6, 2.8 and 2.9). Using a similar foundation, methods of an assessment of GDP, based on the system of national accounts,[4] have been developed under the patronage of the United Nations.[5]

When an arbitrary monetary unit of value is chosen, the GDP can be estimated for a given point in time in an uncontested way. However, due to possible changes of the money units, there is a question of how to compare the GDPs for a variety of years. Assuming that values of equivalent sets of products for various years are identical, one finds a parity between monetary scales at various points in time [11]. The monetary unit, established in this way, possesses a property to have constant purchasing capacity, but has nothing to do with estimates of value at various points in time. When a monetary unit of constant purchasing capacity is used, inflation is excluded, but with variation of productivity, the value content of the monetary unit changes in due course.

As an illustration, the GDP of the U.S. economy measured in different scales of value is shown in Fig. 2.1. The direct assessment of the progress of a social production

---

[4]The System of National Accounts 1993 http://unstats.un.org/unsd/sna1993/introduction.asp.

[5]An interesting description of history of approaches to the estimation of the GDP for the various nations was given by Studenski [10].

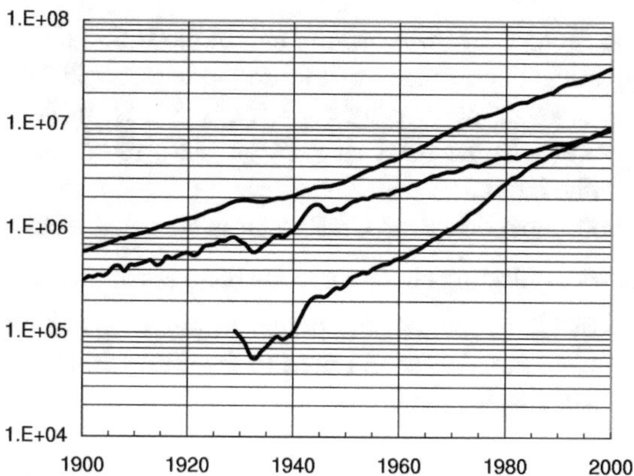


**Fig. 2.1** Production of value in the U.S. economy. The lower curve depicts GNP in millions of current dollars, the middle one—in millions of dollars of year 1996. The latter curve shows real income of the society in money units of constant purchasing power and can be approximated by exponential function (2.14). The upper curve presents values of GNP measured in millions of energy units, taken as 50000 J (see Sect. 11.3)

is made in current monetary units; for the U.S. economy, the dependence of the directly estimated total product in current monetary units can be approximated by the exponential function

$$\hat{Y} = 19.965 \times 10^9 \cdot e^{0.0518\,t} \ \text{dollar/year}.$$

Here, time $t$ is measured in years, beginning ($t = 0$) at 1900. After some tedious procedures [11], the directly estimated quantity $\hat{Y}$ can be transformed into an assessment of GDP in the monetary scale of constant purchasing capacity $Y$. In this case, the time dependence of GDP (the middle curve of Fig. 2.1) can be approximated by the exponential function

$$Y = 1.69 \times 10^{12} \cdot e^{0.0326\,t} \ \text{dollar(1996)/year}. \tag{2.14}$$

Time $t$ is measured in years, and $t = 0$ corresponds to year 1950. The upper curve of Fig. 2.1 depicts the real change of production of value with a constant money scale, which is introduced in Chap. 11 (Sect. 11.3).

The ratio of the output in the current money units to the output in the constant purchasing power money units defines *the price index*

$$\rho(t) = \hat{Y}/Y.$$

The actual price index is a pulsating quantity but with the above assessments, it is possible to see that the average price index for the U.S. has increased (since 1950) as

$$\rho \sim e^{0.0192\,t}.$$

The purchasing capacity of the monetary unit of the U.S.—the dollar—decreases as an inverse quantity. Each holder of the dollar in 1950–2000 has been losing annually nearly 2 % of its purchasing capacity, which is, in fact, an implicit tax in favour of an emitter. In the third chapter, we shall return to the discussion of money units and price index.

Though the time dependence of GDP is smooth, consideration of the rate of growth $\frac{1}{Y}\frac{dY}{dt}$ shows a pulsating character in the progress of production. On the chart of Fig. 2.2, it is possible to see that the period of pulsations of the rate of growth of GDP takes about 4 years. We shall return to the discussion of the reason for the pulsations in the sixth chapter (Sect. 6.6.2).

## 2.2.4 *Constituents of Gross Domestic Product*

### 2.2.4.1 Investments in the Production Equipment

One recognises a set of products as investments, both material and non-material, if the products are not intended for immediate consumption and are kept for use in production. In the material form, the investments are buildings, cars and the various equipment sets in various sectors. A part of a sector output is distributed over sectors, so it is possible to define quantity $I_j^i$—as a part of a product $j$ invested in sector $i$ and to consider investments as a matrix with components $I_j^i$

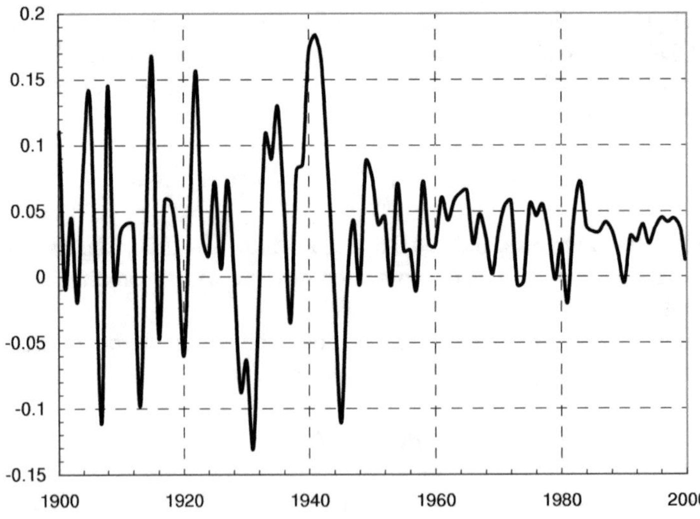


**Fig. 2.2** The rate of growth of the U.S. GDP. The rate of growth of the GDP for the U.S. economy shows a pulsating character of production

$$
\mathsf{I} = \left\| \begin{matrix} I_1^1 & I_1^2 & \dots & I_1^n \\ I_2^1 & I_2^2 & \dots & I_2^n \\ \dots & \dots & \dots & \dots \\ I_n^1 & I_n^2 & \dots & I_n^n \end{matrix} \right\| . \tag{2.15}
$$

The quantities $I_j^i$ apparently cannot be chosen arbitrarily, and the society works out the mechanisms of the choice of investments. When the development of an economy is planned, which is facilitated in the case where all means of production basically belong to the state, the choice has a directive character: the special state body centrally makes decisions about investment that define the future assortment and volumes of goods and services. When the market economy is reined, and the means of production belong to various proprietors, including the state, each proprietor itself defines the investment decision, and therefore the future production is determined spontaneously.

One can define investment of type $j$ in all sectors as

$$
I_j = \sum_{i=1}^{n} I_j^i, \quad j = 1, 2, \dots, n.
$$

Quite similarly, we can calculate the gross investment of all products in sector $i$ as

$$
I^i = \sum_{j=1}^{n} I_j^i, \quad i = 1, 2, \dots, n.
$$

The gross investment in the entire production system is now defined as

$$
I = \sum_{i=1}^{n} I^i = \sum_{j=1}^{n} I_j = \sum_{i,j=1}^{n} I_j^i. \tag{2.16}
$$

One can find very good estimates of investment $I$ for the U.S. economy (see Appendix B). The time dependence of the gross investment for the entire economy is shown in Fig. 2.3.

#### 2.2.4.2 Personal Consumption

The consumption $C$ is defined as the value of the products which are consumed by humans immediately (one-time consumption). Perhaps, a proper estimate of this quantity could be the minimum amount of products which are needed in order for the humans to subsist. To characterise the necessary consumption, it is convenient to use the poverty threshold used in the U.S. statistics. The estimates of this quantity for a person in different family situations since year 1959 can be found on the U.S.

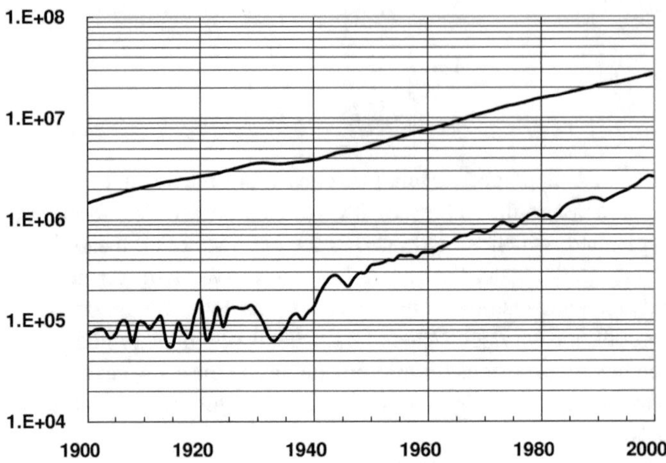


**Fig. 2.3** Production investment and capital in the U.S. economy. Estimates of value of material national wealth $K$ (upper curve) and value of investment $I$ (lower curve) are given, according to Appendix B, in million dollars for year 1996. The growth of capital can be approximated by the exponential function (2.28)

Census Bureau website.[6] One can consider the poverty threshold per person in a one-person family to give a realistic estimate of the current consumption. For year 1996, for example, this quantity is estimated as 7995 dollars per person per year. This quantity is ought to be multiplied by the number of population to get the lower estimate of the consumption in year 1996 as $C = 2,120$ billion dollars. The time dependence of the personal consumption is depicted in Fig. 2.4. On the other hand, one can use Eq. (6.32) for the cost of labour and the estimated (in Sect. 7.1.2) values of the technological index to calculate the personal consumption. The results for the U.S. in the twentieth century are shown in Fig. 2.4 by the dashed line.

One can consider consumption as the most important part of the GDP. 'Every man is rich or poor according to the degree in which he can afford to enjoy the necessaries, conveniences, and amusements of human life' ([12], p. 47).

#### 2.2.4.3 Fundamental Investment

In the recent times, one can easily get an estimate of the material and non-material stored product $G$ from formula (2.13). For example, one has estimates for year 1996: GDP $Y = 7,813$, material investment $I = 2,054$ and the current consumption $C = 2,120$ billion for 1996 dollars. Thus, one can get the estimate for the fundamental investment product as $G = Y - I - C = 3,638$ billion 1996 dollars which is about 47% of the GNP. The fundamental investment $G$ for the U.S. in various years is depicted in Fig. 2.5.

[6] http://www.census.gov/hhes/poverty/histrov/hstpov1.htm.

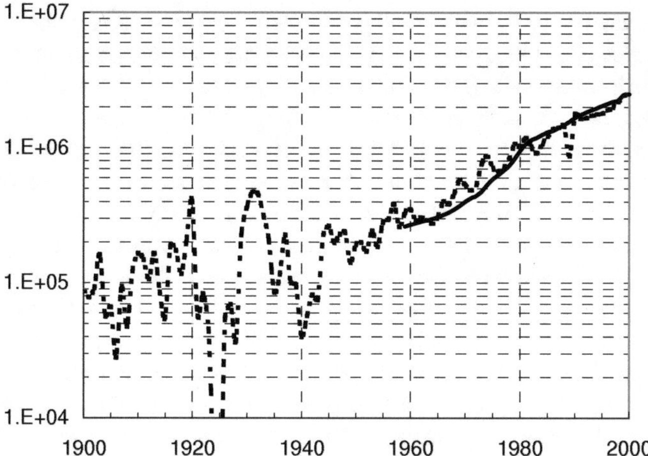


**Fig. 2.4** Personal Consumption in the U.S. economy. Estimates of value of personal consumption $C = cN$ are given in millions of dollars for year 1996. The solid line is based on direct estimates of the poverty threshold by the U.S. Census Bureau; the dashed line presents the results of calculation due to Eq. (6.32)

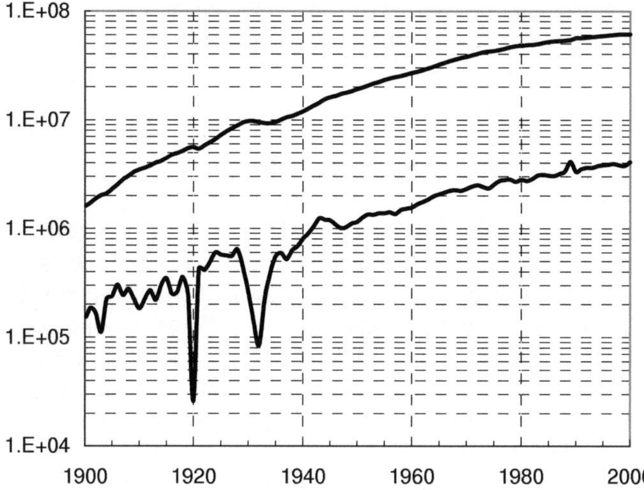


**Fig. 2.5** Fundamental investment in the U.S. economy. Values of fundamental investment $G = Y - I - C$ (lower curve) are calculated from known values of output $Y$, investment $I$ and averaged values of consumption $C$ (see Figs. 2.1, 2.3, 2.4). Values of the fundamental part of the national wealth $R$ (upper curve) are calculated according to Eq. (2.29), whereas depreciation coefficient is assumed to take the same values as for material products. All quantities are given in million dollars for year 1996

The estimated quantity includes all government's expenses (defense, social health, and so on). Besides, it includes an estimate of so-called information products. There are some businesses, such as education, science and R&D, publishers, theatres, TV, cinema, post, law services, statistics, consulting companies and so on, for which the main activity is the creation and distribution of different messages. The product of these businesses is a great amount of messages, which could be informative or not; it depends on the recipient. Therefore, one cannot say that the product of the information sectors is information. Some messages are never read; they are waiting for recipients in depositories such as libraries. Some of the messages are received by many recipients and for some of them, the messages carry no information. Some messages certainly carry valuable information for the recipients, e.g. instructions on how to use the energy of running water as a work horse, or instructions on how to organise matter to be used as a transport vehicle or an appliance. Some of the messages lose their value, and some disappear but for many years, society has stored a great deal of messages—information resources.

The total amount of produced services on the creation and distribution different messages in the U.S. economy was estimated by Machlup [13] as 29% of the GNP for year 1959 and as 46% of the GNP for year 1967. The value of the achievements of science, research and projects is essential and cannot be ignored. The information products are considered to be important for society (because much effort is spent to produce them) and the share of the information products in the GNP apparently does not decrease.

#### 2.2.4.4 Principle of Distribution of Products

All three parts of the final product for the U.S.: production investment, personal consumption and fundamental investment are comparable, and it seems possible that the final output of any society is distributed among three parts in approximately equal fractions. The distribution certainly experiences some operating influences from the society, and it would be interesting to determine whether there exists a principle which governs such division. One of the main questions to understand is: what are the rules to determine a splitting of the final output into three parts.

The future level of production and consumption depends on today's production and fundamental investments. At any moment of time, a society has to decide what part of the final product ought to be consumed and what part ought to be saved for the sake of future consumption. One can imagine two alternative approaches to the problem: one from the side of consumption and the other from the side of production. Some models (see, Sect. 6.9) determine investment as a result of maximisation of present and future consumption. In Chap. 5, we discuss how investment can be determined from the side of production.

## 2.3 The National Wealth

Every society possesses a huge stock of material and non-material products—the national wealth—which, in the natural form, is a set of things, both tangible (buildings, networks of supply, machinery, transport means, furniture, home appliances and so on) and intangible (principles of organisation of the matter and society, works of the art, literature and others things).

### *2.3.1 Assessments of the Stored Products*

The value of the material and non-material parts of the national wealth can be estimated, if one knows *pure investments*, which are *gross investments* minus the value of the products that disappear for the same unit of time (value of depreciation)

$$\frac{dK_j}{dt} = I_j - \mu K_j, \tag{2.17}$$

$$\frac{dR_j}{dt} = G_j - \mu R_j. \tag{2.18}$$

Here, $I_j$ and $G_j$ are gross production and fundamental investments representing the increase of material and non-material wealth per unit of time. We assume that investments become productive instantaneously. The second terms in relations (2.17) and (2.18) describe the depreciation of the components of national wealth due to wearing and ageing.

Equations (2.17) and (2.18) introduce the stocks of products: $K_j$ is the value of the material assets including basic production equipment (production capital); $R_j$ is the value of the storage of intermediate production materials including the stock of knowledge.[7] It is difficult to give an exact estimate of these quantities because some of these products disappear very quickly, but others keep their value for centuries. Apparently, estimates of the stocks $K_j$ and $R_j$ depend on the choice of the second terms on the right-hand side of Eqs. (2.17) and (2.18). One can assume, for simplicity, that the depreciation is proportional to the amount of national wealth with one and the same coefficients of depreciation $\mu$ for all products in all situations.

The above relations allow one to represent the components of the national wealth in the following form:

---

[7]Set of the quantities $K_j$ can be named as produced capital. The quantities $R_j$ include the so-called *human capital*—an assessment embedded in an individual of capacity to make work, the sum of knowledge, qualifications and skills of the separate worker (Becker [14], P. 8), but do not reduce to it. Among others, intangible products—archive of knowledge, projects, reports of laws and so forth. The quantities $R_j$ also include an assessment of natural resources (natural capital) which possess value by virtue of that efforts and expenses (investments into natural resources) that are made for their purchase and restoration.

$$K_j(t) = \int_0^{\infty} e^{-\mu x} I_j(t - x)\, dx, \tag{2.19}$$

$$R_j(t) = \int_0^{\infty} e^{-\mu x} G_j(t - x)\, dx. \tag{2.20}$$

One can see that the national wealth represents accumulated investments, especially, investments of the recent past, as the earlier produced commodities disappear. The quantity

$$k_j(t, t - x) = e^{-\mu x} I_j(t - x)$$

is a part of the existing fixed production capital, which was introduced during a unit of time at the moment of time $t - x$. This quantity is the smallest part of capital stock, which can be considered in macroeconomic theory.

Relations (2.17)–(2.18) and (2.19)–(2.20) connect with each other two kinds of quantities: the fluxes: $I_j$, $G_j$ and the stocks: $K_j$, $R_j$. Only one set of quantities, namely, fluxes can be estimated directly. The other quantities, stocks, are usually calculated in value units. But this does not mean that stocks are theoretical constructs: they are realities, which can be measured by natural units of products. However, apparently, it is difficult to give a precise direct assessment of value of the stored products, especially non-material products.

The total value of the national wealth is the sum of the quantities, which were defined above

$$W = \sum_{j=1}^{n} (K_j + R_j). \tag{2.21}$$

The national wealth consists of products which were produced at various moments of time and under various conditions of production, which implies the different bygone current prices. The value of national wealth $W$ is a characteristic of the set of the products which depends of the history of bygone prices. In other words, the value of national wealth cannot be a function of amounts of products. However, one can introduce such a function for a set of products or a function of a state—the utility function—which is closely related to value (see Chap. 11, Sect. 11.2). The utility function $U$ replaces the non-existing value function in theoretical considerations.

### 2.3.2 *Structure of Produced Capital*

The national wealth is created by the production system that is a real engine of the economic system, and production capital, which was discussed very thoroughly by many researchers, appears to be a very important part of national wealth. Note that different approaches to the concept of capital stock can be accepted. In a wider sense, capital stock includes all material national wealth; in a narrower sense, the capital stock can be understood as the value of basic production equipment, one can say, the core production capital. To illustrate application of the theory, we shall apply the wider concept of capital stock.

The accumulation of invested products (2.15) determines the production capital (capital stock) via the equations

$$\frac{dK_j^i}{dt} = I_j^i - \mu K_j^i, \quad i, j = 1, 2, \ldots, n, \tag{2.22}$$

where $K_j^i$ stands for value of production equipment of type $j$ in the sector $i$. One can see that the production equipment can be considered as a matrix with components $K_j^i$

$$\mathsf{K} = \begin{Vmatrix} K_1^1 & K_1^2 & \ldots & K_1^n \\ K_2^1 & K_2^2 & \ldots & K_2^n \\ \ldots & \ldots & \ldots & \ldots \\ K_n^1 & K_n^2 & \ldots & K_n^n \end{Vmatrix}. \tag{2.23}$$

The total amount of product of the type $j$ in all sectors is defined as

$$K_j = \sum_{i=1}^{n} K_j^i, \quad j = 1, 2, \ldots, n.$$

Quite similarly, we can calculate the total amount of the production capital in sector $i$ as

$$K^i = \sum_{j=1}^{n} K_j^i, \quad i = 1, 2, \ldots, n.$$

The production capital of the whole economy is now defined as

$$K = \sum_{i=1}^{n} K^i = \sum_{j=1}^{n} K_j = \sum_{i,j=1}^{n} K_j^i. \tag{2.24}$$

One can sum Eq. (2.22) over suffixes $i$ or $j$ to obtain equations for the dynamics of the total amount of equipment labelled $j$ and for dynamics of the fixed capital in sector $i$, correspondingly,

$$\frac{dK_j}{dt} = I_j - \mu K_j, \quad j = 1, 2, \ldots, n, \tag{2.25}$$

$$\frac{dK^i}{dt} = I^i - \mu K^i, \quad i = 1, 2, \ldots, n. \tag{2.26}$$

Remember that all dynamic equations in this section are written for the case, where the depreciation is proportional to the amount of national wealth with one and the same coefficients of depreciation $\mu$ for all products in all situations. Generally speaking, coefficients of depreciation are different for different equipment in various sectors.

### 2.3.3 *Estimates of Produced Capital*

Formulae (2.17) and (2.18) give a basis for approximate formulae, according to which the separate parts of the national wealth can be estimated. In a simple case, when one considers the three-sector model described in Sect. 2.1, Eqs. (2.17) and (2.18) reduce to equations for two components of national wealth. We shall first estimate stock of basic equipment $K$

$$\frac{dK}{dt} = I - \mu K. \tag{2.27}$$

It is easy to see that at the given fluxes $I$, the calculated amounts of stock of basic equipment $K$ must depend on the choice of the value of the depreciation coefficient $\mu$, which is neither a quite arbitrary not a well-known quantity. The time series for capital $K$ and investment $I$ for the U.S. economy is known (see Appendix B) and allows us to calculate values of the rate of capital depreciation $\mu$ by using Eq. (2.27). The results are shown in Fig. 2.6. The website of the U.S. Bureau of Economic Analysis (www.bea.doc.gov) also contains estimates of depreciated capital $\mu K$ which allow us to calculate the rate of capital depreciation $\mu$ in a different way, as the ratio of depreciated amount of capital to the total amount. These results are also depicted in Fig. 2.6. These estimates allow us to consider the depreciation coefficient as an increasing function of time which has value $\mu = 0.026$ in year 1925 and increases linearly from 0.026 to 0.07 over years 1925–2000. However, the results show inconsistency of the primary data: the two estimates from the same source differ from each other, also the depreciation coefficient cannot be negative. We have chosen to consider empirical values of investment and capital, depicted in Fig. 2.3, to be 'correct' values and to exploit the calculated values of the depreciation coefficient, while using a local averaged values (dashed line in Fig. 2.6) instead of negative ones.

The calculated time dependence of capital as well as gross investment for the entire U.S. economy is shown in Fig. 2.3. The time dependence of production capital can be approximated by the exponential function

$$K = 5.49 \times 10^{12} \cdot e^{0.0316\,t} \quad \text{dollar(1996)}, \tag{2.28}$$

where time $t$ is measured in years, and $t = 0$ corresponds to year 1950.

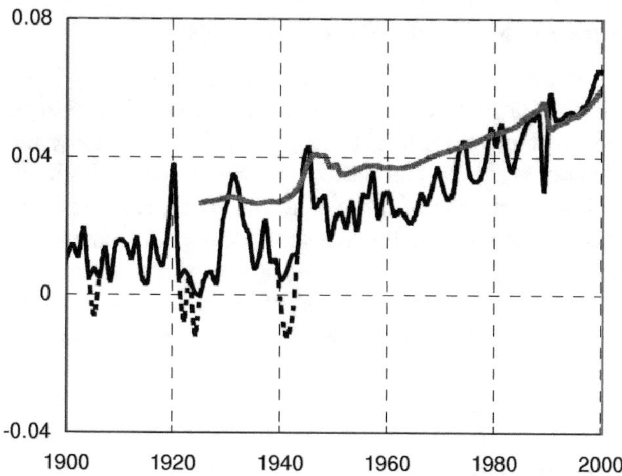


**Fig. 2.6** The depreciation coefficient in the U.S. economy. The direct estimates of the quantity as the ratio of depreciated amount of capital stock to the total amount (the shorter curve) and estimates due to Eq. (2.27) (pulsating curve) with use of values of investment and capital. The dashed lines represent values to be corrected

### 2.3.4 Estimates of Fundamental Wealth

In a simple case, when the three-sector model described in Sect. 2.1 is considered, Eq. (2.18) reduces to the equation for value of the fundamental component of the national wealth $R$ that includes the archive of knowledge and projects

$$\frac{dR}{dt} = G - \mu R, \tag{2.29}$$

where $G$ is a quantitative estimate of all efforts on the creation of long-term projects, including development of the principles of organisation, that is, investments into a science, researches, education and progress. At the assumption that the flux $G$ (which has been described in Sect. 2.2.4) is fixed and also the factor of depreciation $\mu$ of archive of knowledge is known, Eq. (2.29) allows to estimate value of $R$ and to define time structure of this component of national wealth, just as Eq. (2.27) allows to define structure of the production equipment.

Value of this component of the national wealth $R$ for the USA is estimated according to Eq. (2.29), in which the factor of depreciation, on the assumption, has the same values as material products. The results are demonstrated in Fig. 2.5.

## 2.4 Workers in the Production Processes

The human's work is the most important production factor. Its role in production was thoroughly investigated in systems of concepts of political economy and neoclassical economics. 'Labour is, in the first place, a process in which both man and nature participate, and in which man of his own accord starts, regulates, and controls the material reactions between himself and nature. He opposes himself to nature as one of her own forces, setting in motion arms and legs, head and hands, the natural forces of his body, in order to appropriate nature's productions in a form adapted to his own wants. By thus acting on the external world and changing it, he at the same time changes his own nature. He develops his slumbering powers and compels them to act in obedience to his sway. ... At the end of every labour process, we get a result that already existed in the imagination of the labourer at its commencement. ... Besides the exertion of the bodily organs, the process demands that, during the whole operation, the workman's will be steadily in consonance with his purpose.' ([5], Vol. 1, Chap. 7, Sect. 1). '... however varied the useful kinds of labour, or productive activities, may be, it is a physiological fact, that they are functions of the human organism, and that each such function, whatever may be its nature or form, is essentially the expenditure of human brain, nerves, muscles, and c.' ([5], Vol. 1, Chap. 1, Sect. 4).

### *2.4.1 Consumption of Labour*

Modern technology assumes that man is installed into the production process and works inside it. The true measure of the production factor called traditionally *labour* is work (in a physical sense, in energy units) done by a workman, but practically, this production factor is measured by working time, so that it is important to estimate the work, which can be done by a workman per hour. In a sedentary state, the human organism (an adult male) requires about 2500 kcal/day or about $10^6$ kcal/year $\approx 4 \cdot 10^9$ J/year.[8] Extra activity requires an extra supply of energy. The energy needed for a working man can be up to two times more than the energy needed for a resting man ([15], Chap. 26; [16]). Though some types of work require significant energy consumption, we accept the value of the work done by a workman to be approximately 100 kcal/h or $4.18 \times 10^5$ J/h. The possibilities of the human engine were lower in earlier times, as was shown by Fogel and Costa [17] on the basis of historical data for France and Britain for years 1785 and 1790, correspondingly.

Therefore, labour is measured in man-hours, while corrections due to character of labour (heavy or light), intensity of work and other factors are considered to have been taken into account. For the last statement, I rely on Scott [18], who in his turn refers to other researchers. As an example, according to the data compiled in Appendix B, the amount of man-hours per year (labour consumption) in the economy of the U.S. is shown in Fig. 2.7 as a function of time. The dependence can be approximated by

[8] 1 cal = 4.18 J.

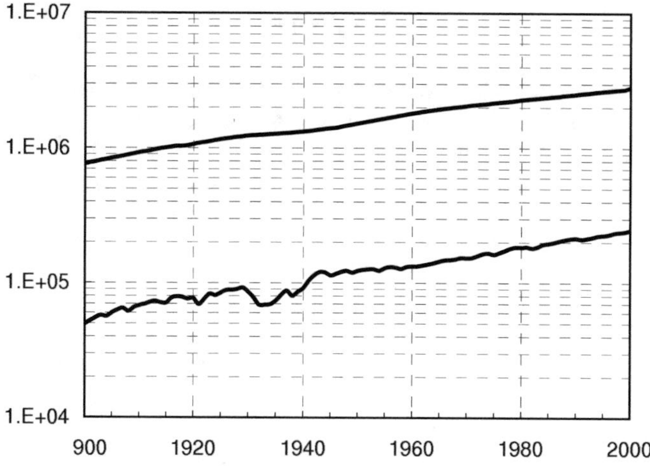


**Fig. 2.7** Population and consumption of labour in the U.S. The upper curve represents population in hundreds of persons. The lower curve represents consumption of labour in million man-hours per year. The latter dependence can be approximated by exponential function (2.30)

a straight line, especially after year 1950, so that for this period

$$L = 1.23 \times 10^{11} \cdot e^{0.0147\,t} \text{ man} \cdot \text{ h/year}, \tag{2.30}$$

where time $t$ is measured in years, and $t = 0$ at 1950.

According to Marx [5], labour is a commodity that produces value. The bulk productivity of labour, that is, the value produced per unit of labour, due to formulae (2.14) and (2.30), can be approximated for the U.S. economy as

$$Y/L = 13.74 \cdot e^{0.0179\,t} \text{ dollar(1996)/man} \cdot \text{ h}. \tag{2.31}$$

One can estimate that productivity of labour in the U.S. economy has grown by six times during the past century. This growth of productivity cannot be explained without taking into account that there is another production factor—work of production equipment—with the property to substitute labour and produce value. We believe that the increase in the labour productivity is connected with the use of newer and newer sources of energy by human beings.

## 2.4.2 *Population and Labour Supply*

The supply of the labour is the potential amount of labour $\tilde{L}$, available at given wage $w$, in other words, at a given price of labour. The labour supply is conventionally considered to be connected with the whole population $N$

$$\tilde{L} = f(w)\,N. \tag{2.32}$$

The population is a reservoir (a pool) from which labour is supplied. The increasing function $f(w)$ changes from zero at $w = 0$ to a certain limiting value, which is usually about 0.5 for developed countries.

The dynamic equation for the change in population can be written as

$$\frac{dN}{dt} = (b - d)N, \tag{2.33}$$

where $b - d$ is the birth rate minus the death rate, that is, the growth rate of population.

To obtain an equation for the labour supply, one ought to differentiate relation (2.32) to get

$$\frac{d\tilde{L}}{dt} = \tilde{\nu}\left(N, b-d, w, \frac{dw}{dt}\right)\tilde{L}. \tag{2.34}$$

The growth rate of the labour supply is determined by the growth of population and changes in the level of wage

$$\tilde{\nu} = (b-d)f(w) + Nf'(w)\,\frac{dw}{dt}.$$

Note that the total amount of wages $wL$ also includes, generally speaking, investments in capital, so that the amount of subsistence $cL$, that is, the amount of expenses that are needed to provide a living and training of workers is less than $wL$ (see also Sect. 2.2.4.2).

## 2.5 Energy in the Production Processes

Energy, as has been discussed repeatedly and for a long time (see, for example, [19, 20]), is vital for the performance of the production system. The socially organised stream of energy begins with identification of primary energy carriers: coal, oil, potential energy of falling water—all that humans find in the nature and that costs nothing, until it is not recognised yet, how to extract work from energy carriers.

### *2.5.1 Work and Quasi-work in a National Economy*

A primary energy carrier is a something that contains potential energy: the chemical energy embodied in fossil fuels (coal, oil and natural gas) or in biomass; the potential energy of a water reservoir; the electromagnetic energy of solar radiation; the energy stored in the nuclei of atoms. The total of the primary energy carriers used by humans and estimated in power units is listed in handbooks as the quantity of used[9] *primary energy*. The *primary energy* consumption is the consumption of energy carriers as they can be taken from nature.

As an illustration, Fig. 2.8 shows the total consumption of primary energy carriers (with a solid top line), as calculated by official statistics of the U.S. Department of Energy (see Appendix B). Apparently, the primary energy carriers (for simplicity, one speaks about consumption of primary energy $E$) in public facilities are used for the most variety of tasks. So, for example, 0.55 quad[10] of oil products from the total amount of about 97 quad of primary energy consumed in the U.S. economy in year 1999 was laid on the roads. It is clear that it is not even the energy content that is important in this case, but the property of oil products as specific materials.

For the most part, primary energy is not used directly but is first transformed and converted into fuels and electricity—*final energy*—which can be transported and distributed to the points of final use. The final energy consumption provides energy services for manufacturing, transportation, space heating, cooking and so on.[11] Extensive investigations of the consumption of primary and final energy in the U.S. economy were conducted by Ayres with collaborators [21, 23].

The total of the primary energy carriers can be broken into two parts according to their role in productions. It is possible to allocate a part which is used for operating various adaptations allowing substitution of labour efforts by work of the production equipment. This quantity can be called *primary substitutive work* $E_P$. True *substitutive work* or *productive energy* $P$, which really replaces workers' efforts, is a small part of the consumed primary productive energy $E_P$, and the coefficient of efficiency $P/E_P$ depends on exploited technology. In the United States to the end of the last century, for example, nearly $3 \cdot 10^{19}$ J, about a third of all consumed energy (nearly $1 \cdot 10^{20}$ J), went to substitution of workmen's work. At an efficiency ratio equal to 0.01, true substitutive work made nearly $3 \cdot 10^{17}$ J.

The other part of the socially organised stream of energy, called quasi-work, is used directly in production and in households for illumination, heating, chemical transformations and other tasks.

---

[9]It is customary to speak about the consumption of energy in a national economy. For precision, the word *consumption* should be replaced by the word *conversion*. Energy cannot be *used up* in the production process; it can only be converted into other forms: chemical energy into heat energy, heat energy into mechanical energy, mechanical energy into heat energy and so on. The measure of converted energy (work) is exergy.

[10]Primary energy is the name for primary energy carriers (oil, coal, running water, wind and so on) measured in energy units. It is convenient to measure huge amounts of energy in a special unit quad (1 quad $= 10^{15}$ Btu $\approx 10^{18}$ J), which is usually used by the U.S. Department of Energy.

[11]The problems arising in the estimation of the amount of energy, which is converted (used up) in production processes to do useful work, are discussed by Patterson [24], Nakićenović et al. [25], Zarnikau et al. [26], Ayres [27] and many others. According to Nakićenović et al. [25], the global average of primary to final efficiency was about 70% in year 1990, while it was higher in developed countries. Data collected by Ayres ([27], Table 2) demonstrates that efficiency of energy conversion increased during the last centuries.

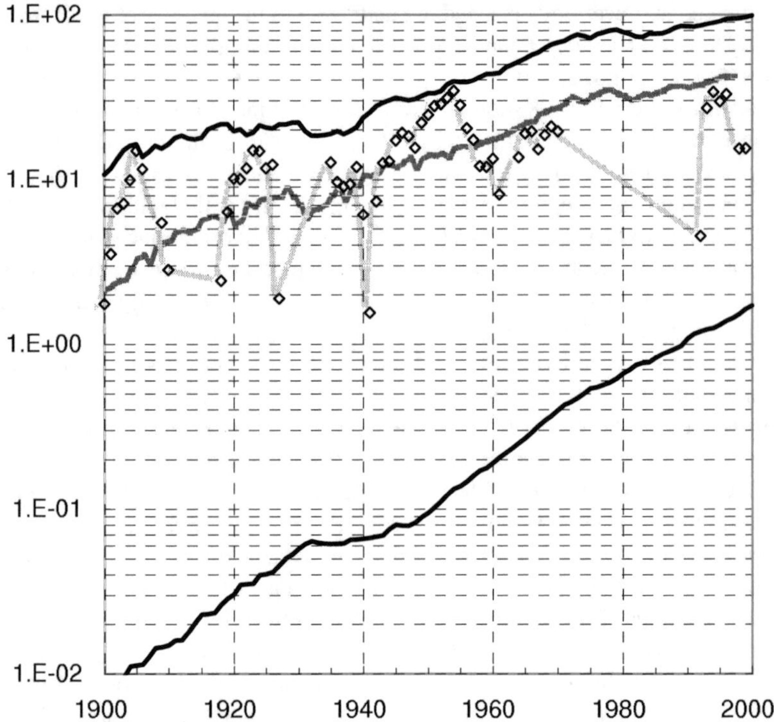


**Fig. 2.8** Consumption of energy in the U.S. economy. The solid lines represent consumption of energy careers (primary energy, top curve) and productive consumption of energy (substitutive work, bottom curve). The dashed line depicts primary energy (exergy) needed for work of production equipment, estimated on the data of Ayres et al. [21] as the sum of non-fuel usage of oil products and consumption of half of the net electricity and energy by other prime movers. Primary substitutive energy is also calculated (and depicted by symbol $\diamond$) as a part of primary energy, which is anti-correlated with labour (see Sect. 7.1.4). All quantities are estimated in quads per year (1 quad $= 10^{15}$ Btu $\approx 10^{18}$ J). The primary energy and substitutive work from year 1950 can be approximated by exponential function (2.36). Adapted from [22]

## 2.5.2 *Direct Estimation of Substitutive Work*

Although one can easily find estimates of the total amount of primary energy carriers, the biggest interest for our aims is caused by possible assessments of the quantity of energy going to the substitution of workers' efforts in production processes. Based on the results of fundamental investigations [21, 23] of the usage of primary and final energy in the U.S. economy, one can estimate the amount of substitutive work in this case.

The substitutive work (productive energy) $P$ could be generally interpreted as capital services. The most important property of this quantity is its ability to substitute labour services, which are variety of efforts of humans in production processes, and the substitutive work itself should be defined as an amount of work which is done by external energy sources with the help of production equipment instead of workers' efforts. To estimate substitutive work, we have to consider humans efforts, which, we assume, can be replaced by the work of production equipment driven by external energy sources. We can divide all efforts into three groups.

#### 2.5.2.1 Efforts on Displacements of Substances and Bodies (Including Human Bodies)

These efforts were substituted by the work of animals, wind and moving steamer engines in the past. Now in the U.S., they are substituted mainly by the work of self-moving machines—automobiles, trucks, airplanes and other mobile equipment—driven by the products of oil. Estimates of energy used for this purposes can be obtained for the U.S. economy as the sum of energy of consumed distillate fuel oil, jet fuel and motor gasoline. According to the U.S. Department of Energy data (www.eia.doc.gov), the amount was 19.46 quad in year 1998. This is the energy content of fuel; the amount is different from the amount of work (service energy) which is needed to move vehicles. The service delivery efficiency for transportation was analysed by Ayres [27], and the ratio of the energy delivered to wheels to the fuel energy was estimated as 0.06. The ratio of the useful work (substitutive work) to fuel energy is much less; it is close, one can suppose, to the Ayres [27] technical efficiency that was 0.015 for transportation (much less for farming and construction) in year 1979. According to Ayres et al. [21], efficiency has been improving beginning with 1975, so that one can estimate the contribution to substitutive work from transportation. The genuine work of transportation vehicles due to energy carriers can be calculated as 0.1 quad in year 1998, though the amount of energy carriers needed to provide this work was about 19.46 quad.

#### 2.5.2.2 Efforts on Transformation and Separation of Substances and Bodies

These are efforts in the production of clothes, tools, different appliances and so on—much, if not all, manufacturing. Animal-driven, wind-driven, water-driven and steam engine-driven power were used to do work instead of humans in previous centuries. Nowadays, the same work is mainly done by machines with electric drives. According to the U.S. Department of Energy (http://www.eia.doe.gov), motor-driven equipment accounts for about half of the electricity in the manufacturing sector. Non-industrial motors, driving pumps, compressors, washing machines, vacuum cleaners and power tools also account for quite a lot of electricity consumption. Part of the electricity consumed by clothes washers and dishwashers provides mechanical movement. So, we can account that more than half of the consumed site electricity in the U.S. economy, that is about 6 quad in 2000, is taken by motors. In the best cases, electricity in a machine drive can be recovered into rotational motion with an efficiency of up to 0.8–0.9 [21]. However, the result of the work of a machine tool, for example, is a component or detail of another machine, and one has to consider the whole procedure of making something: installation, stop–start movements, measurement and so on. It is difficult to get an absolute measure of efficiency in this case, but one can imagine that there is a certain amount of work, which has to be done to obtain the necessary effect. Presumably, it is the work of a human who can obtain this effect on his own. The efficiency of machine drives was estimated by Ayres [21] as about 0.002 in years 1960–1970. At manual operation, the efficiency is low, but automated control and operation allow increases in efficiency. One assumes that the introduction of information processors into the production could affect the efficiency of the processes, which could reach 0.005 in year 2000. This gives an estimate for the contribution to substitutive work from machine drives to be 0.2–0.3 quad per year 2000.

#### 2.5.2.3 Efforts on Sense-Based Supervision and Coordination, Development of Principles of Organisation

While the human efforts listed in the preceding two groups have been successfully substituted by work of other sources of energy from ancient times, attempts to mechanise the functions of the brain were mainly unsuccessful until the advent of computers (information processors) in the twentieth century. Up until recent times, these functions were considered as essentially human functions. Now, the work of the brain is being substituted by information processors driven by electricity. According to the U.S. Department of Energy (http://www.eia.doe.gov), the consumption of electricity by computers and office equipment in the commercial sector of the U.S. economy in year 1999 was 0.4 quad. In the residential sector, electricity was consumed by computers and electronics in the amount 0.35 quad in year 1999. There is no data on consumption of electricity by computers in the industrial sector, though one can hardly have any doubt about the presence of the appliances of information technology in this sector and the sector of transportation. To the sum of the above figures—0.75 quad—one has to add the amount of electricity consumed by other office and communication equipment in all sectors. In total, one can estimate the consumption of electricity by computers, electronics and office equipment to be about 1 quad in year 1999. This figure estimates, at least, a scale of phenomenon. One cannot directly measure the work produced by the devices of information technology to measure the efficiency, but one can see some signs that the useful effect per unit of consumed energy (efficiency) has been increasing. For example, the consumption of electricity by one computer decreased from 299 kWh/year in 1985 to 213 kWh/year in 1999 [28, 29]. This means that consumption of electricity by a computer was decreasing with the average rate 0.025. Simultaneously, the number of computers and consumption of electricity increased with average rate of growth 0.027 between years 1990 and 1999, as can be calculated from the data of Koomey et al. [28] and Kawamoto et al. [29]. All this means that the useful effect from the consumption of electricity by computers has been growing in recent times with a growth rate of more than 0.052, which is the sum of the rate of growth of consumption of electricity, 0.027, and the rate of decreasing of consumption of electricity by one unit, 0.025, plus the estimate of improving the unit performance. Similar considerations can be made for all devices of information technology from the collection of data by Koomey et al. [28] and Kawamoto et al. [29]. The efficiency of computers is certainly less than unity, but they may be more efficient than many other appliances. It is difficult to judge what part of the resulting amount 1 quad per year can be attributed to substitutive work itself, but, perhaps, an estimate of 0.5 quad per year is realistic. This huge amount of energy was spent usefully in year 1999 to produce instructions to humans and apparatuses in the U.S. economy.

#### 2.5.2.4 Final Remarks

Summing up, the total amount of substitutive work in the U.S. economy in 1999 can be estimated as 1 quad per year. It is approximately one hundred times less than the total (primary) consumption of energy, which was about 97 quad in 1999. However, the amount of primary energy (energy carriers) needed to provide this amount of substitutive work is about 25 quad, which is about 26% of the total primary consumption of energy. This number corresponds to the estimates by Ayres ([27], Table 1), who found that the part of energy which can be considered as the primary production factor (machine drive, transport drive, farming and construction) in the U.S. economy was 9% in year 1800, 23% in 1900 and about 32% in 1991.

### *2.5.3 Energy Carriers as Intermediate Products and Energy as a Production Factor*

Energy carriers are consumed now in great amounts in production processes and are considered to be products that are circulating in the production system and, thus, must be included in the balance table (Table 2.1). From the conventional economic point of view, all consumed energy carriers can be considered as intermediate or, sometimes, final products.

Electricity as an energy carrier, for example, is the most important intermediate product in the production of aluminium, metallurgical operations, and some chemical processes, among others. Electricity consumed for lighting, comfort and process heating must be considered either as final product (in residential sector), or as intermediate products (in commercial and other sectors). In all cases of production consumption, the cost of energy is included in the cost of the final products, and energy contributes to the value of produced commodities no more than other intermediate products participating in the production process.

However, it has long been argued [19, 20] that aside from regarding the energy carriers as intermediate or final products, the delivered energy is universally vital to the performance of the economy and must be included in the theory of production as an important production factor. Apart from being a commodity, in some cases, energy from external sources plays a special role, substituting for efforts of workers in the technological processes. Energy-driven equipment works in the place of workers, and energy can be ascribed all the properties of labour, including the property to produce surplus value. In these cases, the substitutive work or productive energy that apparently is only a part of the total (primary) consumption of energy, has to be specified in the conventional economic terms as a value-creating production factor.

Thus, one can define the various roles of the consumed energy carriers in the production processes. In any case, energy carriers participate in the production processes as usual commodities. However, simultaneously part of the consumed energy $P$—it is called *productive energy* or *substitutive work*—has to be considered not only as an ordinary intermediate or final product, but also as a value-creating factor, which has to be introduced in the list of production factors equally with the production factors of conventional neoclassical economics: capital $K$ and labour $L$. This production factor, substitutive work $P$, is not primary energy and, moreover, not even energy delivered to production equipment. It has to be considered as genuine work done by production equipment with the help of external sources of energy instead of workers. This quantity can also be considered as capital service provided by capital stock.

The substitutive work $P$ defined in this way has a special price, different from the prices of energy carriers as a usual intermediate or final products. It is clear that the amount of consumed products which are needed to support substitutive work $P$ is valued as $\mu K$, so that the price of substitutive work, as a production factor, is

$$p = \frac{\mu K}{P} \tag{2.35}$$

### 2.5.4 Estimates of Primary Energy and Substitutive Work

There are plenty of data on the total consumption of primary energy $E$ in different countries (in the Energy Statistics Yearbook, for example), but little is known about the productive part of consumption $P$ that is a true value-creating production factor. Methods of evaluation of substitutive work should be developed, however, at the moment there is a method of estimation of past values of substitutive work $P$. This method, which is based on a relation between the rates of growth of production factors (Eq. 5.20) and described in detail in Sect. 7.1.2, allows one to calculate the growth rate $\eta$ of substitutive work, if one knows the rates of growth of output, capital and labour consumption. Then, one can restore the time dependence of substitutive work, if the absolute value of the quantity itself is known in one of the moments of time.

As an illustration, Fig. 2.8 shows the total consumption of primary energy carriers—as shown by official statistics of the U.S. Department of Energy (see Appendix B), and the calculated usage of substitutive work [30] in the U.S. economy according to official estimation of the empirical situation. The method does not allow one to calculate absolute values of substitutive work, it was taken to be about 1 quad at the end of the century, as was estimated in Sect. 2.5.2. The extra growth rate of substitutive work in the U.S. economy in years 1950–2000 in comparison with the primary consumption of energy was about 0.04 per year in the second half of the century. The dependence of the total and productive consumption of energy from year 1950 can be approximated by the functions

$$E = 33.3 \cdot e^{0.0205\,t} \text{ quad/year,} \tag{2.36}$$

$$P = 1.96 \cdot e^{0.0585\,t} \text{ quad/year,} \tag{2.37}$$

where, as in previous examples, time $t$ is measured in years, starting from year 1950. It is possible to estimate the productive consumption of energy for a unit of labour. For the U.S. economy, since 1950,

$$P/L = 6.42 \times 10^5 \cdot e^{0.0441\,t} \text{ J/man} \cdot \text{h.} \tag{2.38}$$

Figure 2.9 shows the ratio of the work executed by the production equipment to energy estimates of the efforts of the workers, taking into account an estimate of an hour of work, obtained in Sect. 2.4.1. One can find that by the present time, the efforts of every worker in the economy of the U.S. are amplified more than 10 times. This is a rule: consumption of energy from external sources exceeds the work done by man by a few times in all developed countries. The average productivity of substitutive work for the U.S. economy can be approximated by the function

$$Y/P = 2.14 \times 10^{-5} \cdot e^{-0.0259\,t} \text{ dollar(1996)/J.} \tag{2.39}$$

The better characteristics of labour and energy productivity are marginal productivities, which will be introduced and estimated in Chap. 7.

### 2.5.5 *Stock of Knowledge and Supply of Substitutive Work*

While the labour supply $\tilde{L}$ can be related to the population, which can be considered to be a pool from which the labour force emerges (see Sect. 2.4.2), the productive energy supply $\tilde{P}$ can be related to the stock of knowledge which is playing a role of a reservoir (pool) from which applications of energy emerge. Indeed, one ought to have available sources of energy and appliances, which allow the use of energy in production aims. Some devices ought to be invented, made and installed for work. Human imagination provides methods of using energy in the production tasks. Therefore, the base for the energy supply lies in a deposit of knowledge which is fallow, unless it is used in a routine production process. This deposit determines the possibility of

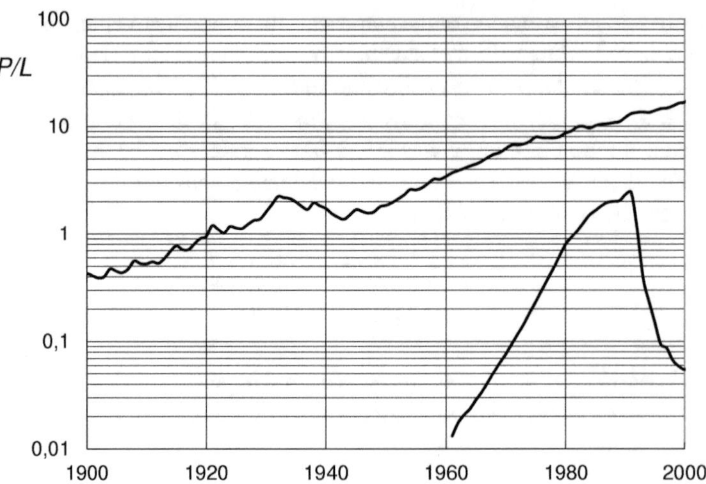


**Fig. 2.9** The ratio of substitutive work to workers' efforts. The ratio of substitutive work to estimates of workers' efforts for the U.S. economy (the upper curve) and for the Russian economy (the lower shorter curve). Adapted from [31] with corrections of values for Russia

the society attracting the extra energy to production. The stock of knowledge should be considered as a resource.

To describe the process of development of the energy supply, one can refer to the simple three-sector model of the production system introduced in Sect. 2.1. Discovering the principle of organisation and developing projects of technological processes is included in the content of activity of the second sector. One can consider that this stock of knowledge, that is, fundamental results of science, results of research, project works and so on (stock of principles of organisation) are measured by their total value $R$, which is governed by Eq. (2.28). Alternatively, the stock of knowledge can be measured directly in terms of natural units, that is, by numbers of patents issued, numbers of technical journals, number of books in print and so on. The knowledge is embodied in organisations and cultures more than in individuals, although individual skills are also part of this category. Can the value of stock of knowledge $R$ be a measure of the information contained in all this?

Then, the first sector materialises the projects. One can find plenty of brilliant examples of 'transformation' of knowledge into useful work in the history of technology and one can try to formalise this process, considering the stock of knowledge as a resource or as a reservoir (pool) from which applications of energy emerge. One can assume, noting an analogy of Eq. (2.28) with (2.33), that an equation for energy supply $\tilde{P}$, that is, the amount of energy which can be used in production processes as substitutive work, can be written similarly to Eq. (2.34) in the form

$$\frac{d\tilde{P}}{dt} = \tilde{\eta}(\varepsilon, R)\tilde{P}. \tag{2.40}$$

One can assume that the rate of potential growth of substitutive work $\tilde{\eta}$ depends on the stock of knowledge $R$ and on the price of introducing of substitutive work into production $1/\varepsilon$ (see Sect. 5.2, Eq. 5.16). The price of transformation and materialisation of deposited massages, that is the price of attracting the energy, has been appearing on the stage of materialisation of principles of organisation. The function $\tilde{\eta} = \tilde{\eta}(\varepsilon, R)$ remains unknown; one can assume a simple dependence

$$\tilde{\eta} = g(\varepsilon)R. \tag{2.41}$$

However, in a situation of uncertainty, the growth rate of potential energy, or the energy supply itself, ought to be given.

Though it is indisputable that knowledge makes energy available for humans, the question remains of how to describe it in quantitative terms. Does function (2.41) really exist and, if it exists, which is its asymptotic behaviour? One may think that the current attention to the stock of knowledge, as to the genuine source of economic growth [32–34] (see also the textbook [35]) can help to solve the problem. However, we do not know whether the available energy is limited or not. One can imagine and consider two scenarios of development: the energy supply $\tilde{P}$ as a function of time has or does not have a limit value. There is apparently no question of lack of energy. It is a question of ways of utilisation of energy to get the desired effect. This question is clearly connected with the other question: can the stock of knowledge be limited?

## 2.6 Natural Processes in the Human-Designed Production System

The production system is embedded in the natural environment. In the beginning of the production cycle, raw materials are extracted from the natural environment, while at the end of the production cycle, the wastes and useless by-products are thrown out into nature. The flow of substances starts and finishes in the natural environment (see Fig. 1.1), thus one has to consider the interaction of the production system with the environment.

Some industries (agriculture and forestry, for example) use natural processes to provide the production of commodities. Some natural things are even used as production equipment. Soil (land) is used to produce corn, cows are used to produce milk and so on. The natural things are considered as production capital, and their value is estimated in the same way as value of all other capital products.

The sector theory of production, considered in Sect. 2.2, assumes that some natural processes are included in the production system. To consider the interaction between the environment and the production system in more detail, one has to admit that some of the variables $X_j$ represent amounts of natural products. It is convenient to assume that, in consistency with the definitions of Sect. 2.2, $X_j$, $j \leq r$ is the gross output of artificial products in money units and $N_j$, $j > r$ is gross output of natural products measured in natural units. The gross output $X_j$ both of artificial and natural products can be distributed (similar to Eq. 2.3) as

$$X_i = \sum_{j=1}^{r} X_i^j + \sum_{j=r+1}^{n} X_i^j + Y_i, \quad i = 1, 2, \ldots, r \tag{2.42}$$

$$N_i = \sum_{j=1}^{r} N_i^j + \sum_{j=r+1}^{n} N_i^j + \frac{Y_i}{p_i}, \quad i = r+1, r+2, \ldots, n \tag{2.43}$$

where $X_i^j$ is an amount of artificial product labelled $i$ used for the production of product labelled $j$ and, similarly, $N_i^j$ is an amount of natural product labelled $i$ used for the production of product labelled $j$, while there is a residue $Y_i$ called final output. We assume that the price $p_i$ and money measure might be introduced for those of the natural products which are supported by human activity.

Now, one can write the second set of balance equations, which, as in Eq. (2.4), represent the balance of production of value in sectors of production of both artificial and natural products

$$X^j = \sum_{l=1}^{r} X_l^j + \sum_{l=r+1}^{n} p_l N_l^j + Z^j, \quad j = 1, 2, \ldots, r, \tag{2.44}$$

$$p_j N^j = \sum_{l=1}^{r} X_l^j + \sum_{l=r+1}^{n} p_l N_l^j, \quad j = r+1, r+2, \ldots, n, \tag{2.45}$$

where $Z^j$ is production of value in sector $j$, and we admit that there is no production of value in the sectors of natural production.

It is convenient to define the amounts of value of gross product and intermediate consumption for products of the natural processes

$$X_j = p_j N_j, \quad X_l^j = p_l N_l^j$$

to include all considered quantities in the more detailed (in comparison with Table 2.1) balance table, that is Table 2.3. However, the majority of natural products are traditionally regarded as zero-price products and are not included in the balance scheme.

To determine the production of value $Z$ and components of the gross output $Y$ in this case, we sum Eq. (2.42) over index $i$ from 1 to $r$ and also Eq. (2.44) over index $j$ from 1 to $r$. After comparing the results, one obtains

$$Z = \sum_{j=1}^{r} Y_j + \sum_{j=r+1}^{n} \left( \sum_{l=1}^{r} X_l^j - p_j \sum_{l=1}^{r} N_j^l \right). \tag{2.46}$$

**Table 2.3** Balance of artificial and natural products

| Gross output | $X^1\ X^2\ \cdots\ X^r$ | $X^{r+1}\ X^{r+2}\ \ldots\ X^n$ | Final output |
|---|---|---|---|
| $X_1$ | | $X_1^{r+1}\ X_1^{r+2}\ \ldots\ X_1^n$ | $Y_1$ |
| $X_2$ | MAN-CREATED | $X_2^{r+1}\ X_2^{r+2}\ \ldots\ X_2^n$ | $Y_2$ |
| $\ldots$ | PROCESSES | $\ldots\ \ldots\ \ldots\ \ldots$ | $\ldots$ |
| $X_r$ | | $X_r^{r+1}\ X_r^{r+2}\ \ldots\ X_r^n$ | $Y_r$ |
| $X_{r+1}$ | $X_{r+1}^1\ X_{r+1}^2\ \cdots\ X_{r+1}^r$ | | $Y_{r+1}$ |
| $X_{r+2}$ | $X_{r+2}^1\ X_{r+2}^2\ \cdots\ X_{r+2}^r$ | NATURAL | $Y_{r+2}$ |
| $\ldots$ | $\ldots\ \ldots\ \ldots\ \ldots$ | PROCESSES | $\ldots$ |
| $X_n$ | $X_n^1\ \ X_n^2\ \cdots\ X_n^r$ | | $Y_n$ |
| Production of value | $Z^1\ \ Z^2\ \cdots\ Z^r$ | $0\ \ 0\ \cdots\ 0$ | $Y$ |

The conventional characteristic of efficiency of the production system, the final output $Y = \sum_{j=1}^{n} Y_j$ is considered to be equal to the production of value $Z = \sum_{j=1}^{n} Z^j$. Thus, the right-hand side of Eq. (2.46) can be considered as the sum of components of the vector $\mathsf{Y}$, which can be determined as

$$Y_j = \begin{cases} Y_j, & j = 1, 2, \ldots, r \\ \sum_{l=1}^{r} X_l^j - p_j \sum_{l=1}^{r} N_j^l, & j = r+1, r+2, \ldots, n. \end{cases} \tag{2.47}$$

The quantity $X_l^j$, at $l \le r$, $j > r$ is the amount of artificial product labelled $l$ supporting the production of natural product $j$, so that the sum $\sum_{l=1}^{r} X_l^j$ is a total amount of the artificial products supporting the production of the natural product $j$. On the other hand, $N_j^l$, at $j > r$, $l \le r$ is an amount of the natural product labelled $j$ needed for production of the artificial product $l$, so that the sum $\sum_{l=1}^{r} N_j^l$ is the total amount of natural product $j$ used in production in all sectors. Therefore, one can see that the components of final output (2.47), at $j > r$, are characteristics of our interactions with nature. Values of the characteristics depend on prices of the natural products, so that it is very important to use the right prices for estimation of the interaction characteristics. Because nature does not have a representative agent on the market, there is no market evaluation of the prices of natural products and one can choose the prices arbitrarily. It is natural to choose the right prices in such a way that in the situation of balance, all the components (2.46) at $j > r$ vanish. This requirement is followed by a definition of the balancing price of natural product $j$ as

$$p_j = \frac{\sum_{l=1}^{r} X_l^j}{\sum_{l=1}^{r} N_j^l}, \quad j = r+1, r+2, \ldots, n. \tag{2.48}$$

At these prices, the interaction characteristics can be positive or negative: the former case means that humans invest in the environment, whereas the latter case means that there is a damage to the environment, or this can be interpreted as our debt to the nature.

However, whatever the prices of natural products, one always assumes that $Y_j = 0$ at $j > r$. The production of natural sectors is not usually accounted for at all, so the national statistics can show a truncated produced value $Y = \sum_{j=1}^{r} Y_j$ instead of the real amount $Y = \sum_{j=1}^{n} Y_j$. The underestimation of prices of natural products in comparison with the balancing price (2.48) result in a deficiency of gross investment in nature. In both our century and the previous ones, the production system was contained in the environment but in the previous years, the interaction with the environment was not as large in scale as it is in our times, and it was mainly local. Nowadays, ecological problems have appeared, which seem to stem from underestimation of the prices of the natural products. A proper social mechanism of regulation of our interaction with the environment does not exist at the moment; it has to be invented and implemented in reality.

## References

1. Leontief, W.W.: Quantitative input and output relations in the economic system of the United States. Rev. Econ. Stat. **18**, 105–125 (1936)
2. Leontief, W.W.: The Structure of the American Economy 1919–1939. Harvard University Press, Cambridge (1941)
3. Leontief, W.W.: Input-Output Economics, 2nd edn. Oxford University Press, New York (1986)
4. Sraffa, P.: Production of Commodities by Means of Commodities: Prelude to a Critique of Economic Theory. Cambridge University Press, Cambridge (1975)
5. Marx, K.: Capital. Encyclopaedia Britannica, Chicago (1952). English translation of Karl Marx, Das Kapital. Kritik der politischen Oekonomie, Otto Meissner, Hamburg (1867)
6. Tougan-Baranovsky, M.: Theoretische Grundlagen des Marxismus. Dunker and Humbolt, Leipzig (1905)

7. Bortkiewicz, L.: Value and price in the Marxian system. Int. Econ. Pap. **2**, 5–60 (1952). English translation of von Bortkiewicz, L.: Wertrechnung und Preisrechnung im Marxschen System. Archiv fur Sozialwissenschaft und Sozialpolitik **3**, XXIII and XXV (1906)
8. Lietaer, B.: The Future of Money: A New Way to Create Wealth, Work, and a Wiser World. Century, London (2001)
9. Samuelson, P.: Understanding the Marxian notion of exploitation: a summary of the so-called transformation problem between Marxian values and competitive prices. J. Econ. Lit. **9**(2), 399–431 (1971)
10. Studenski, P.: The Income of Nations: Theory, Measurement and Analysis: Past and Present. New York University Press, Washington (1961)
11. Afriat, S.N.: The Price Index. Cambridge University Press, Cambridge (1977)
12. Smith, A.: An Inquiry into the Nature and Causes of the Wealth of Nations. Clarendon Press, Oxford (1976). In two volumes
13. Machlup, F.: Knowledge and Knowledge Production. Knowledge, its Creation, Distribution, and Economic Significance, vol. 1. Princeton University Press, Princeton (1980)
14. Becker, G.: Human Capital. Columbia University Press, New York (1964)
15. Harrison, G.A., Weiner, J.S., Tanner, I.M., Barnicot, N.A., Reynolds, V.: Human Biology: An introduction to Human Evolution, Variation, Growth and Ecology. Oxford University Press, Oxford (1977)
16. Rivers, J.P.W., Payne, P.R.: The comparison of energy supply and energy needs: a critique of energy requirements. In: Harrison, G.A. (ed.) Energy and Effort, pp. 85–105. Taylor and Francis, London (1982)
17. Fogel, R.W., Costa, D.L.: A theory of technophysio evolution, with some implications for forecasting population, health care costs, and pension costs. Demography **34**, 49–66 (1997)
18. Scott, M.F.G.: A New View of Economic Growth. Clarendon Press, Oxford (1989)
19. Allen, E.L.: Energy and Tconomic Growth in the United States. MIT Press, Cambridge (1979)
20. Cottrell, W.F.: Energy and Society: The Relation between Energy, Social Change and Economic Development. McGraw-Hill, New York (1955)
21. Ayres, R.U., Ayres, L.W., Warr, B.: Exergy, power and work in the U.S. economy, 1900–1998. Energy **28**, 219–273 (2003)
22. Pokrovski, V.N.: Productive energy in the US economy. Energy **32**, 816–822 (2007)
23. Ayres, R.U., Ayres, L.W., Pokrovski, V.N.: On the efficiency of electricity usage since 1900. Energy **30**, 1092–1145 (2005)
24. Patterson, M.G.: What is energy efficiency? Concepts, indicators and methodological issues. Energy Policy **24**, 377–390 (1996)
25. Nakićenović, N., Gilli, P.V., Kurz, R.: Regional and global exergy and energy efficiencies. Energy **21**, 223–237 (1996)
26. Zarnikau, J., Guermouche, S., Schmidt, P.: Can different energy resources be added or compared? Energy **21**, 483–491 (1996)
27. Ayres, R.U.: Technological progress: a proposed measure. Technol. Forecast. Soc. Chang. **59**, 213–233 (1998)
28. Koomey, J.G., Cramer, M., Piette, M.A., Eto, J.H.: Efficiency improvements in U.S. office equipment: expected policy impacts and uncertainties. Lawrence Berkeley National Laboratory 37383, Berkeley (1995). http://enduse.lbl.gov/projects/offeqpt.html. Accessed 17 March 2011
29. Kawamoto, K., Koomey, J.G., Nordman, B., Brown, R.E., Piette, M.A., Ting, M., Meier, A.K.: Electricity used by office equipment and network equipment in the U.S. Energy **27**, 255–269 (2002)
30. Pokrovski, V.N.: Energy in the theory of production. Energy **28**, 769–788 (2003)
31. Beaudreau, B.C., Pokrovskii, V.N.: On the energy content of a money unit. Physica A **389**, 2597–2606 (2010)
32. Romer, P.M.: Increasing returns and long-run growth. J. Polit. Econ. **94**, 1002–1037 (1986)
33. Romer, P.M.: Endogenous technological change. J. Polit. Econ. **98**, 71–102 (1990)
34. Lucas, R.E.: On the mechanics of economic development. J. Monet. Econ. **22**, 3–42 (1988)
35. Aghion, P., Howitt, P.: The Economics of Growth. MIT Press, Cambridge (2009)

# Chapter 3
# Monetary Side of Social Production

**Abstract** The input–output model of a production system assumes that a specific motion takes place: natural substances transform into finished and semi-finished things, the latter transform into other things and so on, until all this is finally consumed, and the substances return into the environment as waste. Simultaneously with the motion of products, one discovers the motion of money, which has to be considered as a separate, special artifact. The money is circulating in the economy, providing the exchange of products. To describe the phenomenon of money circulation, in this chapter, we are considering fluxes of money in a simplified system, consisting of the government and many production firms, while the subsystem, which produces money, consists of a central bank and many commercial banks. A set of dynamic equations is formulated and investigated for both steady-state and unsteady situations. The description of money circulation is impossible beyond the description of real production fluxes, though the basic features of the real production can be described on their own.

## 3.1 Architecture of the Money System

The production system, which is needed to maintain the existence of the human society, used to be described, due to the works of Leontief [1] (see also Chap. 2), as a system of interacting sectors, each of them creating its own product. In the simplest case, one can consider a system consisting of three sectors, as was described in Sect. 2.1. The first sector creates basic production equipment ($K$, production funds), the second one creates fundamental and intermediate products ($S$), consumed by the other two sectors and stored in warehouses and depositories for future production and non-production consumption, and the third sector creates products for the direct consumption by humans ($C$). In accordance with the speculations in Chap. 2, one can assume that the output of each sector is needed to maintain the production of, generally speaking, all other sectors, so that the gross products $X_K$, $X_S$ or $X_C$ are generally distributed among three sectors, and the balance relation for the products can be written as



V. N. Pokrovskii, *Econodynamics*, New Economic Windows,
https://doi.org/10.1007/978-3-319-72074-6_3

$$\begin{aligned} X_{\mathrm{K}} &= X_{\mathrm{KK}} + X_{\mathrm{KS}} + X_{\mathrm{KC}} + I \\ X_{\mathrm{S}} &= X_{\mathrm{SK}} + X_{\mathrm{SS}} + X_{\mathrm{SC}} + G \\ X_{\mathrm{C}} &= C \end{aligned} \tag{3.1}$$

where $I$, $G$ and $C$ are components of final products, which are planned for sale beyond the intermediate production usage; $I = I_{\mathrm{K}} + I_{\mathrm{S}} + I_{\mathrm{C}}$ is the value of the investment products, distributed over the three sectors. For simplicity, it is assumed that the product of the third sector in the amount $C$ is completely consumed.

Equation (3.1) describe motion of products between sectors, which, as we know, is accompanied with motion of money that is moving in the opposite direction. We consider here the production system immersed in the money system of the society, as shown schematically in Fig. 3.1. There is a correspondence between fluxes of money and fluxes of products, and also as production subsystem creates real value, the bank subsystem generates corresponding amount of money. But there is no sign of the activity of bank subsystem in balance equation (3.1), which are written on the assumption that the money is moving without any expenses, and the banks, if present, acts free. There is no sign of money also in the expression for Gross Domestic Product, defined in Sect. 2.2.4. Our first task to fulfill is to make fluxes of money explicit, that are in line with the product circulation, write equations for money circulation. It allows to generalise the expression for the Gross Domestic Product and gives a solid base to understand interaction between the real and the financial sides of the economy.

The main organisers and managers of money circulation are a central bank and commercial banks. The central bank issues the banknotes and coins—primary money. These banknotes and coins are distributed to the commercial banks, which supply many customers with cash money. The central bank also provides the commercial banks with credits, so that the commercial banks can provide the customers with credit money. The records on the accounts of the customers are non-paper money, which are created by the commercial banks. So, the central bank and commercial banks introduce an uncertain arbitrary amount of the circulating money in coins, banknotes and cashing deposits into the system consisting of the government and many customers of the commercial banks.

Although the money system contains many commercial banks, each with many customers, for simplicity, we consider all commercial banks together as the only commercial bank; further, instead of many customers, we consider four groups of customers. One can separate all accounts in the commercial banks into groups: a group of producers of main production equipment ($K$), a group of producers of fundamental products ($S$), a group of producers of products for immediate consumption ($C$) and a group of final consumers ($L$).

Economic subjects interact with each other using money as a tool for the purchase of resources, both for consumption and for production. The major function of the bank system resides in the redistribution of money, in particular, in directing money from investors to firms that require finances for forthcoming projects. One can assume an elementary diagram of monetary streams, in which only banks are accumulators

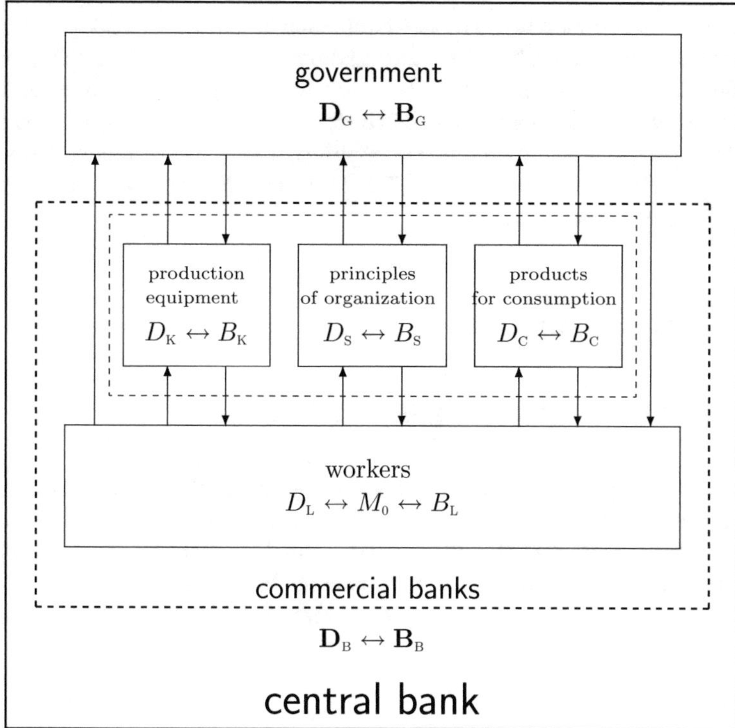


**Fig. 3.1** The scheme of the money system. The central bank and the commercial banks create a money medium for activity of the economic agents. The three sectors of the production system create all products and originate the fluxes of money, which are not shown here, between the sectors and to workers in the form of wages $W_K$, $W_S$, $W_C$. The workers are buying products, and money is returning to the producers. The government receives its part of created value in the form of taxes $T = \theta_K Y_K + \theta_S Y_S + \theta_C Y_C + \theta_L W$, which in different amounts are returning to the economic agents. To each arrow representing a flux of money, there corresponds an arrow of opposite direction, presenting fluxes of labour force and products. There is a bargain every time when money is exchanged for products and labour force

of expenses and incomes. The system of monetary circulation is described using the assumptions that the central bank has no aim to receive any profit, and the commercial banks are limited to obtaining a reasonable modest percent. Actually, the situation is more complex: the commercial banks are aspiring to increase in profit by increasing their own capital, due to a share issue, and are engaging in speculative operations. All of this can lead to an essential discrepancy of the monetary circulation with production needs, which reveals itself as a crisis.

Due to its huge practical importance, the phenomenon of money circulation has been studied thoroughly throughout the centuries [2]. In addition to some seminal monographs [3–6], there is a huge amount of books and articles devoted to different aspects of the problem. A paramount understanding of the problem provokes a closed mathematical description of the money circulation or a mathematical monetary theory of production, which has been successfully developed [7–9]. In this chapter, following the previous edition of this book and work [10], a system of equations for the simplest system, described above, will be formulated and investigated. The results of the theory depend on the specific assumptions of the system architecture and the preferences of the process participants. The derived parities present a schematization of the processes of money circulation.

## 3.2 Participants of the Money System

### *3.2.1 Customers of the Commercial Banks*

For expansion of production and consumption, the clients of commercial banks need money; thus, at each given point in time, clients should determine whether a financial source of possible expenses should be money from their own account, or a loan from a commercial bank. The customers of commercial banks create a demand for credit money, and they appear to be the basic movers of the progress of the economic system.

#### 3.2.1.1 Producers

The product fluxes are accompanied by money fluxes, which are moving in the opposite direction. Each production sector receives money from the sales of its product from the government, workers and all production sectors, including payments from its own sector,

$$\begin{aligned}
&M_{K\to K} + M_{S\to K} + M_{C\to K} + Y_K,\\
&M_{K\to S} + M_{S\to S} + M_{C\to S} + Y_S,\\
&Y_C
\end{aligned} \tag{3.2}$$

The quantities $Y_K$, $Y_S$, $Y_C$ can be considered to be components of the final output—Gross Domestic Product (GDP), which, in addition to assessments of production investments $I$, the governmental expenses $G$ and the consumed products $C$, includes also an assessment of services of moving money—the end product of financial activity.

Simultaneously, each production sector pays (symbolised by minus sign) wages to workers and money for the products of all the sectors,

$$
\begin{aligned}
&-M_{K\to K} - M_{K\to S} - I_K - W_K - \theta_K Y_K \\
&-M_{S\to K} - M_{S\to S} - I_S - W_S - \theta_S Y_S \\
&-M_{C\to K} - M_{C\to S} - I_C - W_C - \theta_C Y_C
\end{aligned} \tag{3.3}
$$

Here, we take into account that the producers have to pay taxes to the government in the amounts $\theta_K Y_K$, $\theta_S Y_S$ and $\theta_C Y_C$.

Before writing the payment balance for the sectors, note that, though the receiving and payments of the money occur at one and the same time, due to the time involved for production, marketing, transportation, investment, consumption and so on, the symbols in Eqs. (3.2) and (3.3) present payments for amounts of products produced at different times. For simplicity, one can consider the symbols for intermediate products within the sectors to have identical meaning, so that the payment balance for every production sector can be written as

$$
\begin{aligned}
0 &= Y_K + M_{S\to K} + M_{C\to K} - M_{K\to S} - I_K - W_K - \theta_K Y_K \\
0 &= Y_S + M_{K\to S} + M_{C\to S} - M_{S\to K} - I_S - W_S - \theta_S Y_S \\
0 &= Y_C - M_{C\to K} - M_{C\to S} - I_C - W_C - \theta_C Y_C
\end{aligned} \tag{3.4}
$$

Regarding the interaction of the production units with banks, for simplicity we assume that the commercial bank is the only source of financing of production activity, not considering the issuing of shares and bonds.[1] Consequently, we consider that the financial state of the producers is defined by the amounts of deposits $D$ and debts $B$ (with corresponding superscripts) in the commercial banks. These quantities are connected with the balance equations

$$
\begin{aligned}
\frac{dD_K}{dt} &= r_K D_K + Y_K + M_{S\to K} + M_{C\to K} - M_{K\to S} \\
&\qquad - I_K - W_K - \theta_K Y_K - q_K B_K + \frac{dB_K}{dt} \\
\frac{dD_S}{dt} &= r_S D_S + Y_S + M_{K\to S} + M_{C\to S} - M_{S\to K} \\
&\qquad - I_S - W_S - \theta_S Y_S - q_S B_S + \frac{dB_S}{dt} \\
\frac{dD_C}{dt} &= r_C D_C + Y_C - M_{C\to K} - M_{C\to S} \\
&\qquad - I_C - W_C - \theta_C Y_C - q_C B_C + \frac{dB_C}{dt}
\end{aligned} \tag{3.5}
$$

[1] Production units distributing *shares* can receive money to cover of expenses directly from consumers. These primary securities are promissory notes on which emitters undertake to pay the cost of the securities and percentage on them through a certain time and in a certain way. Money from securities is directed by the emitters to cover of investment expenses, which after a while results in an additional product.

Here, we use the designations $q_K$, $q_S$, $q_C$ for the norms of payments to banks for credit, and $r_K$, $r_S$, $r_C$ for the norms of payments of banks for customer deposits. These quantities are established by the commercial banks to adjust the quantities of deposits and debts.

To exclude intermediate products from discussion, we introduce notation for the total amount of deposits and debts of the production customers in the commercial banks

$$D_P = D_K + D_S + D_C, \quad r_P D_P = r_K D_K + r_S D_S + r_C D_C,$$
$$B_P = B_K + B_S + B_C, \quad q_P B_P = q_K B_K + q_S B_S + q_C B_C.$$

Summing up Eq. (3.5), we get

$$\frac{dD_P}{dt} = r_P D_P + Y - I - W_P - \theta_P Y - q_P B_P + \frac{dB_P}{dt}, \tag{3.6}$$

Notice, that, the specified in the Eq. (3.4) as identical, the paid and received payments (for example, $M_{S\to K}$ in lines 2 and 1) are different with the amount of commission fee (when bank's intermediary is taking into the account), and this came to existence additional negative term, that is not designated here, in the right part of the Eq. (3.6). This term is a contribution to the quantity $\kappa(D + M_0)$ in Eq. (3.8).

The loans allow the producers to avoid a disruption between receiving and payment. They are needed to provide the expenditures $I$ and $W_P$, so one can consider that the amount of loan is connected with the amount of the expenditures.

#### 3.2.1.2 Consumers

The financial state of the consumers is determined by the amounts of deposits and debts, $D_L$ and $B_L$, in the commercial banks. In addition, they are holders of paper money in the amount $M_0$—this is cash money, which is in circulation at the moment. The consumers use their money and possible loans to acquire products, so that the balance equations for the consumers can be written as

$$\frac{\mathrm{d}D_L}{\mathrm{d}t} + \frac{\mathrm{d}M_0}{\mathrm{d}t} = r_L D_L + W - C - \theta_L W - q_L B_L + E_0 + \frac{\mathrm{d}B_L}{\mathrm{d}t}, \tag{3.7}$$

where $W = W_K + W_S + W_C + W_G$ is a flux of money to workers in the form of wages, which are received from the production sectors ($W_K$, $W_S$, $W_C$) and the government ($W_G$). The bank emission of paper money $E_0$, generally speaking, is not equal to a change of the amount of circulating paper money, but if we do not consider processes of transformation of paper money in non-cash and on the contrary, we have to suppose that $E_0 = \frac{\mathrm{d}M_0}{\mathrm{d}t}$, what we accept in the next. The consumers pay money in the amount $C$ to the third sector for the consumption products, which were created some time ago, and to the government in the form of taxes, $\theta_L W$, so that, in the balance situation, one has

$$C = (1 - \theta_L)W.$$

In a reality, a situation can be somewhat more complex; part of wages can be used for purchase shares of the enterprises, that is, for investment in various sectors, which we do not consider here.

The right-hand sides of Eqs. (3.5) and (3.7) contain payments to and by the commercial banks. The banks ask the interest rates $q_P$, $q_L$ for debts and gives the interest rates $r_P$, $r_L$ to customers for their deposits. So, as there is a payment for debts, customers try to reduce quantity of debts as far as possible and to keep some money on the depositary accounts in commercial banks.

#### 3.2.1.3 The United Customer of Commercial Bank

For simplicity of consideration, we unite deposits and credits of producers and consumers in commercial bank in united quantities and introduce new variables according to the rules

$$D = D_P + D_L, \quad r = (r_P D_P + r_L D_L)/(D_P + D_L),$$
$$B = B_P + B_L, \quad q = (q_P B_P + q_L B_L)/(B_P + B_L).$$

The balance equation for these variables is recorded as the sum of the Eqs. (3.6) and (3.7), so that

$$\frac{\mathrm{d}D}{\mathrm{d}t} = rD - qB + Y - I - C + W_G - T + A_0 - \kappa(D + M_0) + \frac{\mathrm{d}B}{\mathrm{d}t}. \tag{3.8}$$

Besides the sum, the equation contains two additional terms. To take into account the commission fees for the transactions between subjects (which have been omitted at aggregation of variables), and various gathering for carrying out of operations, the quantity proportional to the amount of money $D + M_0$ is put in into Eq. (3.8). The factor of proportionality $\kappa$ represents an assessment of efforts on maintenance of one monetary unit under circulation and is a characteristic of the system. This equation is also added by the quantity $A_0$, which represents a possible flux of money into the accounts of clients from external sources. The quantities of such type should be present also at the Eqs. (3.6) and (3.7), but, at summation, the additional quantities reduce to the additives, specified in Eq. (3.8).

Similar to Eqs. (3.6) and (3.7), the Eq. (3.8) contains payments for services of commercial banks. Banks establish norms of payments for credits and deposits of clients $q$ and $r$ which are, generally speaking, functions of the amounts of deposits and credits; they are established by commercial banks from the requirement to receive some profit on bank operations (see the following section).

### 3.2.2 Commercial Bank as a Customer of the Central Bank and a Producer of Credit Money

We consider the commercial banks are intermediaries in the interaction among the production sectors and consumers within the economic system. The money deposits with commercial banks and loans from the banks are the means to organise and facilitate the interaction of economic actors. The commercial banks are supported by central bank and are motivated by desire to get profit from the operations with the customers.

#### 3.2.2.1 Balance of Commercial Bank

Considering all commercial banks as the only bank, we assume that the bank has customers' deposits and debts in amounts $D$ and $B$, correspondingly. One can assume that the commercial bank has the only account with the central bank $D_{\mathrm{B}}$, on which it holds all its reserve, including the amount of mandatory deposit of the commercial bank $\xi D$, where $\xi$ is a norm of the mandatory reserve deposit that is set up by the central bank. The commercial banks have also a debt $B_{\mathrm{B}}$ to the central bank.

The state of the commercial bank is determined by its actives: $K_{\mathrm{KB}}$, $D_{\mathrm{B}} - \xi D$, $B$ and passives: $B_{\mathrm{B}}$, $D$. The income of the bank, neglecting the income from the bank's capital $K_{\mathrm{KB}}$ and any other operations, can be written as

$$r_{\mathrm{B}}(D_{\mathrm{B}} - \xi D) - q_{\mathrm{B}} B_{\mathrm{B}} + qB - rD. \tag{3.9}$$

The central bank fixes the interest rate given to the commercial banks for their deposits $r_{\mathrm{B}}$ and the interest rate $q_{\mathrm{B}}$ asked by the central bank for debts of the commercial bank. In its turn, the commercial bank sets the interest rate given by banks (to customers) for their deposits $r$ and the interest rates $q$ asked by banks for debts of its clients. In any case, it is expected, that value of $q$ with any index will appear greater, than value of $r$ with an appropriating index. Usually, the central bank does not pay for mandatory deposits of commercial banks and sets up a high level of the refinancing rate $q_{\mathrm{B}}$. The norm of the mandatory reserve deposit $\xi$ and the refinancing rate $q_{\mathrm{B}}$ are considered as main regulators of the amount of non-paper money.

The deposit $D_{\mathrm{B}}$ changes due to its income (3.9), the changes of the debt to the central bank $B_{\mathrm{B}}$ and operations with the customers of commercial banks, so that the balance equation can be written as

$$\frac{\mathrm{d}D_{\mathrm{B}}}{\mathrm{d}t} = r_{\mathrm{B}}(D_{\mathrm{B}} - \xi D) - q_{\mathrm{B}} B_{\mathrm{B}} + qB - rD + \frac{\mathrm{d}B_{\mathrm{B}}}{\mathrm{d}t} + \frac{\mathrm{d}(D - B)}{\mathrm{d}t} + A_{\mathrm{C}}. \tag{3.10}$$

Here we assume, that the stream of money into the account of commercial bank from external sources $A_{\mathrm{C}}$ is possible. The expression (3.10) ought to contain commission payments and other gathering, which ought to be included in the sum $\kappa(D + M_0)$, but we assume that this income, also as the income from the bank's capital $K_{KB}$, completely goes on employee wages and current maintenance of the bank.

By virtue of the Eq. (3.8), the Eq. (3.10) can be rewritten in the form

$$\frac{dD_B}{dt} = r_B(D_B - \xi D) - q_B B_B + Y - I - C + W_G - T - \kappa(D + M_0) + A_0 + E_C. \quad (3.11)$$

Here we introduce a symbol for emissions of credit money, taking into account the external flux of money,

$$E_C = A_C + \frac{dB_B}{dt}. \quad (3.12)$$

#### 3.2.2.2 Mechanism of Creation of Credit Money

The primary activity of commercial banks is connected with crediting the clients. Usually the aggregate amounts of loans $B$ and credits $D$ appear to be greater than available banks' reserves $D_B - \xi D$; commercial banks create credit money out of nothing—the evidence that ought to be taken into account. According to Werner [11], there are three main theories of banking activity. A first theory states that banks are merely intermediaries like other non-bank financial institutions, collecting deposits that are lent out then. A second, the fractional reserve theory of banking is a generalisation of the first theory and asserts that banks collectively end up creating money through systemic interaction. A third theory, which is proponed by Werner and supported by researches of Bank of England [12], maintains that each individual bank has the power to create money 'out of nothing' and does so, when it extends credit (the credit creation theory of banking).

According to the modern representation of banking activity [11, 12], the individual commercial banks, while providing loans, do not feel any direct constraints from the central bank; the reserve account in the central bank remains untouched [11]. To increase the profit, the commercial banks are motivated to produce more credits to their customers, but increase in credits $B$ to the customers apparently meets some restrictions: the operation ought to be acceptable both for the commercial bank and the customers. But ultimate constraint for money creation, as the researchers [12] assert on p. 4, is monetary policy of the central bank. However, it remains unclear what instruments that the central bank does use to influence on the money creation, if each individual commercial bank does not pay any attention on the amount of its reserve. This is a problem in the credit creation theory of banking that ought to be cleared by its proponents.

In contrast to it, the fractional reserve theory, in which the restrictions are connected with available reserves, is developed in all details. According to the well-known mechanism of multiplication (see, for example, Samuelson and Nordhaus [13], p. 240), enlargement of amount of available money $\Delta A = \Delta\,(D_B - \xi D)$ from the central bank allows the commercial bank to lend $(1 - \xi)\Delta A$ to clients and other banks, whereas the part $\xi \Delta A$ of the total amount must be reserved in the central bank.

The banks use the amount $(1-\xi)\Delta A$ for further lending, leaving the part $\xi(1-\xi)\Delta A$ of the amount in the central bank as a reserve. The process is continuing, so that the banks are creating money on the customers' deposits in the total amount

$$\Delta A + (1-\xi)\Delta A + (1-\xi)^2\Delta A + \ldots = \frac{1}{\xi}\Delta A,$$

and one can write the relation

$$\frac{\mathrm{d}B}{\mathrm{d}t} \leq \frac{1}{\xi}\frac{\mathrm{d}\,(D_\mathrm{B} - \xi D)}{\mathrm{d}t}. \tag{3.13}$$

This equation defines restriction on release of credit money. The quantity $1/\xi$ appears the multiplicator showing a possible increase of credit money. Apparently, the mechanism of multiplication works at large number of commercial banks and Eq. (3.13) is valid for the entity of banks, not for individual bank.

There are apparently some discrepancies in the explanation of credit money creation by the existing theories. One can get such an impression that the fractional reserve theory and the credit creation theory describe the money creation from different points of view: the first one operates with aggregate quantities (macroeconomic approach), while the second considers variables and concepts describing individual banks (microeconomic approach). It means that each individual commercial bank acts, as it is described by Werner [11], freely and its credits can even overpass its reserves, but loans issued by all commercial banks collectively cannot be greater than their aggregate reserve. One can think that, after considering the proper role of the central bank and aggregation, the constraint in the credit creation theory could be formulated, along with some relation among the aggregate quantities that could be similar to inequality (3.13).

If one considers aspiration of commercial banks to expand the credit and some rationality of their behaviour, it is possible to expect that the inequality (3.13) trends to become equality. Anyway, it is possible to introduce an effective quantity $\xi^* > \xi$, at which the parity (3.13) is read as equality. The ratio $\xi/\xi^*$ shows the breadth (depth) of propagation of credit money after some permutation. Then, it follows, from a relation (3.13) (at constant $\xi$), a parity between derivatives of the quantities

$$\frac{\mathrm{d}B}{\mathrm{d}t} = \frac{1}{\xi^*}\frac{\mathrm{d}D_\mathrm{B}}{\mathrm{d}t} - \frac{\mathrm{d}D}{\mathrm{d}t}. \tag{3.14}$$

At the fixed value $\xi^*$, the Eq. (3.14) defines a restrictive condition on a possibility of banks to increase credits. There are, apparently, some other restrictive parities on the quantity of loans to clients of commercial bank $B$, imposed by Eqs. (3.8) and (3.11). Within the limits of these restrictions, commercial banks define amount of the loans $B_\mathrm{B}$ from the central bank, and the clients of commercial banks define quantities of the deposits $D$ and debts $B$. At this, the bank requires a quantity of a seed capital $K_\mathrm{KB}$ to start the operations.

### 3.2.3 The Government as a Customer of the Central Bank and the Central Bank as a Producer of Paper Money

The institution that is crucial in the organisation of money circulation in a society is the central bank, which is a bank of the commercial banks and the bank of the government. The activity of the central bank is closely connected with the activity of the government and is based on the credit to the government and the central bank's assets. To organise the money circulation, the central bank issues money in the form of paper notes (coins) and credits to commercial banks. The central bank creates fiat money that sets up a scale of value.

#### 3.2.3.1 The Central Bank

It is supposed that the central bank is established for the aim to organise, together with the government, circulation of money in the system. Besides that, the central bank accounts the incomes and expenses of the government, financial state of which is fixed, in an elementary case, by two quantities: amount of available money $D_G$ and debts $B_G$. The state of the central bank is fixed by its actives: $K_{CB}$, $B_G$, $B_B$ and passives: $D_G, D_B, M_0$. On the disposal of the central bank and the government, there is a profit from the basic activity

$$q_G B_G + q_B B_B - r_G D_G - r_B (D_B - \xi D), \tag{3.15}$$

where $q_B B_B$ is the payment of the commercial banks for use of credits of the central bank, $q_B$ is a refinancing rate. The rate of interest for debts and deposits of the government, $q_G$ and $r_G$, in Eq. (3.15) are specified by agreement of the central bank with the government. Due to its close relationship with the government, the central bank does not intend to get any profit from the service to the government.

#### 3.2.3.2 Balance of the Government

The government bothers to have enough money at its disposal for financing national projects $G$ and salary payments to the civil servants $W_G$. The main account of the government with the central bank presents the governmental budget and reflects motion of money to and from the government. The incoming fluxes of money include taxes (and other incomes) $T$ into the budget, which are the payments from the producers and consumers[2]

$$T = \theta_P Y + \theta_L W,$$

[2]The different system of taxation according to consumption of social resources is discussed in Sect. 13.2.

where $Y = Y_K + Y_S + Y_C$ is the gross domestic product (GDP) with contribution from the three production sectors and $W = W_K + W_S + W_C + W_G$ is the total amount of wages paid to workers and civil servants. The government supervises norms of the taxation $\theta_P$ and $\theta_L$ to provide the government's expenses $G$ that represent investments in various national projects and wage payments to the civil servants $W_G$. Note that the government pays money at moment $t$, but it receives taxes from the earlier activity. It ought to be taken into account under a detailed analysis.

The profit of the central bank (3.15), minus expenses on the organisation of the circulation of money, comprises the amount of money on the government's disposal $D_G$ that obeys the balance equation

$$\frac{dD_G}{dt} = q_B B_B - r_B(D_B - \xi D) + T - G - W_G + \frac{dB_G}{dt} + A_G + E_0, \tag{3.16}$$

For financing the activity, the government can let out paper money $E_0$ and (or) address to creditors $\frac{dB_G}{dt}$. It is supposed also, that external loans—a flux of money $A_G$—from external sources are possible, so that, alongside with the symbol for emission of paper money, a new symbol is introduced

$$E_0 = \frac{dM_0}{dt}, \quad E_G = A_C + \frac{dB_G}{dt}. \tag{3.17}$$

The loan is needed to provide the governmental expenditures $G$ and $W_G$, so that one can assume that the amount of the loan is connected with the amount of the expenditures. The government can stimulate the production sectors by some money interventions. For simplicity, it is assumed further that all money is coming to the second sector producing non-material products.

## 3.3 Money Circulation and Production

The assumptions about the composition and architecture of the closed system, consisting of the government, the central bank, commercial banks and many production and consumption units, allow us to start describing money circulation within the system. The economic subjects are interacting with each other by means of fluxes of money. To create money, the central bank issues coins and paper money in the amount of $M_0$. Besides, it credits the commercial banks in the amount of $B_B$, producing the means of non-paper exchange $D_B$. The sum of the issued paper and non-paper money $M_0 + D_B$ is called the *monetary base*. The credit of the central bank $B_B$ to the commercial banks provides an opportunity to credit the producers and consumers, thus creating deposits $D_P + D_L$, which can be called non-paper money. The non-paper money can be converted into paper money and, likewise, the paper money can be converted into non-paper money, so that the characteristic quantity is the sum of all deposits in commercial banks $D_P + D_L$ and paper money $M_0$. The total is called the *monetary mass*, for which the conventional symbol $M_2 = M_0 + D_P + D_L$ is used. The process of introducing and circulating money is described by the equations formulated in Sect. 3.2, and our task now is to estimate the amounts of both paper and non-paper money needed for the proper functioning of the production system.

### *3.3.1 Program of Progress of Production and Consumption*

In the 'basement' of the program of economic activity, one can find apparently the real consumption and production. John Maynard Keynes wrote in his *Treatise on Money* that '[h]uman effort and human consumption are the ultimate matters from which alone economic transactions are capable of deriving any significance; and all other forms of expenditure only acquire importance from their having some relationship, sooner or later, to the effort of producers or to the expenditure of consumers' ([14], pp. 120–1). In the basis of any program of economic development, one could find, apparently, the program of consumption and production. It is impossible to exclude, certainly, influence of monetary circulation on the production subsector of a national economy, but, nevertheless, in considered approximation we consider, that the production sector develops under its own laws.

It is natural to believe, that, investigating the actual situation, the producers, the government and the consumers can adapt their programs of development and expenditure, which can be described by means of the time-dependent rates of growth as

$$\frac{\mathrm{d}\underline{I}}{\mathrm{d}t} = \sigma_I \underline{I}, \quad \frac{\mathrm{d}\underline{G}}{\mathrm{d}t} = \sigma_G \underline{G}, \quad \frac{\mathrm{d}\underline{C}}{\mathrm{d}t} = \sigma_C \underline{C},$$

$$\frac{\mathrm{d}\underline{W_P}}{\mathrm{d}t} = \psi_P \underline{W_P}, \quad \frac{\mathrm{d}\underline{W_G}}{\mathrm{d}t} = \psi_G \underline{W_G}. \tag{3.18}$$

Let's notice, that at planning they are interested, as a rule, in the growth of actual output of production and the growth of the actual wages, so that all quantities in the Eq. (3.18), signified by underlining, are estimated by monetary unit of constant purchasing capacity. In a reality, purchasing capacity of monetary unit can change, that usually describe introduction of price indexes. Certainly, the money supply is distributed non-uniformly on the sectors of production, one sector may have a lot of money, while the others - less, so that the usage of several price indexes is necessary, but for simplicity, we use the only price index $\rho$, which is introduced by a relation

$$I + G + C = \rho\,(\underline{I} + \underline{G} + \underline{C}). \tag{3.19}$$

Symbols without the underlining represent assessments of the quantities with the current monetary units.

The rates of growth $\sigma_I$, $\sigma_G$, $\sigma_C$, $\psi_W$, $\psi_G$ in Eq. (3.18) are, generally speaking, functions of time which are estimated or appointed by operating subjects. Average value of the rates of growth can be expressed through the quantities measured by arbitrary monetary unit,

$$\sigma = \frac{1}{I+G+C}(\sigma_I I + \sigma_G G + \sigma_C C),$$
$$\psi = \frac{1}{W_P + W_G}(\psi_P W_P + \psi_G W_G). \tag{3.20}$$

The government, alongside with the expenses connected with maintenance of the general projects $G$ and wages paid to the civil servants $W_G$, plans receiving the income in the form of taxes

$$T = \theta_P Y + \theta_L (W_P + W_G). \tag{3.21}$$

With a view of balancing the state budget, the government establishes norms of taxes $\theta_P$ and $\theta_L$.

In this approach, we focus only on the monetary circulation meaning that $I$, $G$ and $C$ are assumed to be given as a function of time to allow considering monetary fluxes in the system. However, a more detailed definition of these functions would imply a specific theory of production for the three involved sectors, which is considered in Sect. 9.5.

### 3.3.2 Gross Domestic Product

The important characteristic of the system appears to be the Gross Domestic Product $Y$, which is included in Eq. (3.2) and following equations. This quantity has to be presented as pure output $I+G+C$ of production sectors plus contribution of bank system. The form of the last contribution can be found after aggregation of Eqs. (3.8), (3.10) and (3.16), which gives

$$Y = I+G+C-(A_0+A_C+A_G)+\kappa(D+M_0)-\frac{dM_0}{dt}+\frac{d(D_B-B_B)}{dt}+\frac{d(D_G-B_G)}{dt}. \tag{3.22}$$

This formula is, apparently, generalisation of conventional expression of the Gross Domestic Product as the sums of assessments of investments $I$, the governmental expenses $G$ and immediate consumption $C$ (see Eq. 2.11). In addition to these quantities, the expression (3.22) contains export of money by clients, commercial banks and the central bank, $-(A_0+A_C+A_G)$, accordingly. The quantity $\kappa(D+M_0)$ represents an assessment of efforts on maintenance of the circulation of money in amount

$M_0 + D$ (the coefficient $\kappa$ represents an assessment of efforts on maintenance of the circulation of one monetary unit). The last two terms in expression (3.22) show, that the part of the added value is conserved (with a sign plus) on accounts of the central bank.

The first terms of expression (3.22) present average quantities, while the last three terms represent pulsating quantities, which should be anyhow averaged to exclude the pulsating parts from consideration. It is possible to believe that after averaging the Gross Domestic Product can be presented as

$$Y = I + G + C - (A_0 + A_C + A_G) + \kappa(D + M_0). \tag{3.23}$$

The last term in this expression includes all expenses on emission of banknotes and the organisations of monetary circulation. It is assumed, that no residual debts appear after averaging of the last three terms in relation (3.22).

The relation (3.23) can be presented in another form, if one introduces a symbol $R$ for the ratio of an assessment of services of bank system to the pure output of production system.

$$Y = (1 + R)(I + G + C) - (A_0 + A_C + A_G). \tag{3.24}$$

It is possible to believe, that the quantity $R$ changes slowly and can be considered as a characteristic of the system. Apparently, this quantity determines efficiency of a social production: the greater is the quantity $R$, the more expensive is the maintenance of monetary circulation and the lower the efficiency of the system.

In this way, we have integrated the bank's profit into the national income and, therefore, into the money credit process.

### 3.3.3 Price Index and the Quantity Theory of Money

Parities (3.23) and (3.24) allow to establish a relation between total amount of money $M_2 = M_0 + D$, and production output. Being limited to consideration of the average quantities, on the basis of the specified equations, it is possible to record the relation

$$M_0 + D = \frac{R}{\kappa}(I + G + C) = \rho\frac{R}{\kappa}(\underline{I} + \underline{G} + \underline{C}). \tag{3.25}$$

Here, the definition of the price index $\rho$ by relation (3.20) is also used.

Equation (3.25) defines a total amount of money $M_2 = M_0 + D$, circulating in the system at a given value of the real production output estimated in arbitrary, current or constants, monetary units. This equation establishes 'physical' content of monetary unit, which is not known beforehand. The factor of proportionality $R/\kappa$, also as the quantities $R$ and $\kappa$ separately, is the characteristic of the production subsystem, which, it is assumed, changes slowly.[3]

The price index $\rho$ is one of the main quantities describing monetary circulation, its constancy (stability) is a condition and a certificate of normal functioning of a national economy; values of the index exceeding unity provide some temporary income to the government, however values $\rho \gg 1$ destroy economic interactions. For the analysis, it is convenient to use expression for the rate of growth of the price index. Differentiating relation (3.25), we find

$$\frac{1}{\rho}\frac{\mathrm{d}\rho}{\mathrm{d}t} = \frac{1}{M_0 + D}\left(\frac{\mathrm{d}M_0}{\mathrm{d}t} + \frac{\mathrm{d}D}{\mathrm{d}t}\right) - \frac{\kappa}{R}\frac{\mathrm{d}(R/\kappa)}{\mathrm{d}t} - \frac{\sigma_{\mathrm{I}} I + \sigma_{\mathrm{G}} G + \sigma_{\mathrm{C}} C}{I + G + C}. \quad (3.26)$$

This equation determines that the rate of variation of the price index depends subsequently on the fluxes of money emerging within the system, evolution of a state of the money system and the rates of growth of production outputs.

## 3.4 Dynamics of the System

### *3.4.1 System of Evolutionary Equations*

The basis of the system of evolutionary equations comprises the balance equations discussed in the second section. Equations (3.8), (3.11), (3.14) and (3.16) are connecting the state variable of the five interacting economic subjects, every one of which possesses the certain financial actives and has its own tactics of behaviour. We are considering the closed system, taking into account influence of external factors as it is fixed at the formulation of balance equations. By consideration of a financial system in this (quasi-closed) approximation, the quantity $E_0$, $E_{\mathrm{C}}$ and $E_{\mathrm{G}}$ represent now the sources of money, in which the external fluxes $A_0$, $A_{\mathrm{C}}$, $A_{\mathrm{G}}$, that should be set independently, are included. Addressing to the specified relations (3.8), (3.11), (3.14), (3.16) and uniting the Eqs. (3.8) and (3.14), we record system of the evolutionary equations in the form of convenient for the analysis

[3]The relation (3.25) is known as expression so-called *the quantitative theory of money* [3], in which the quantity $R/\kappa$ has been interpreted as average time of the rotation of money (time from manufacture before consumption of an end-product). This relation is also known as Fisher's relation, though, according to Harrod ([15], p. 26) this law was classically exposed in the report of the British Bullion Committee in the year 1810. Moreover, Harrod notes: 'Of course, the Bullion Committee did not invent the quantity theory. Traces of it may be found in writers dating back for centuries before that.'

$$\frac{dD_B}{dt} = r_B(D_B - \xi D) - q_B B_B - \Delta - A_C - A_G + E_C,$$
$$\frac{dD}{dt} = \frac{1}{2}(rD - qB) + \frac{1}{2\xi^*}[r_B(D_B - \xi D) - q_B B_B] - $$
$$- \frac{1+\xi^*}{2\xi^*}[\Delta + A_C + A_G] + \frac{1}{2\xi^*}E_C,$$
$$\frac{dD_G}{dt} = q_B B_B - r_B(D_B - \xi D) + \Delta + E_G + E_0.$$
$$\frac{dB}{dt} = -\frac{1}{2}(rD - qB) + \frac{1}{2\xi^*}[r_B(D_B - \xi D) - q_B B_B] - $$
$$- \frac{1-\xi^*}{2\xi^*}[\Delta + A_C + A_G] + \frac{1}{2\xi^*}E_C,$$
$$\frac{dM_0}{dt} = E_0, \quad \frac{dB_B}{dt} = E_C - A_C, \quad \frac{dB_G}{dt} = E_G - A_G. \tag{3.27}$$

While recording the equations, definition (3.24) for output $Y$ is used. A symbol for excess of incomes of the government over expenses—proficit of the budget—is introduced

$$\Delta = T - G - W_G.$$

The quantities in the right part of this relation are connected with the production program with Eqs. (3.18) and (3.21).

The global characteristics of the production–distribution system $\kappa$ and $R$ ought to be specified. The central bank establishes norm of mandatory deposit $\xi$ that does not coincide with its effective value $\xi^*$, determining the expansion of credit money. These quantities should be set. The central bank establishes also norms of payments $r_B$, $q_B$ for deposit and credit of the commercial bank, which, in its turn, defines norms of payments for deposits and credits, $r$ and $q$, of the clients. The parameters, established by the government, the central bank and commercial bank, are not constant, but depend on the situation and ways of behaviour of economic subjects. Apparently, various models of behaviour of agents are possible, and it is necessary to analyse the actual situation to formulate appropriating dependences, but we shall consider further the parameters constant.

The system of the Eq. (3.27) describes behaviour of seven variables $D$, $D_B$, $D_G$, $B$, $B_B$, $B_G$ and $M_0$ at preset values of the listed parameters, and, eventually, evolution of financial variables is determined by the sources of money $E_0$, $E_C$, $E_G$, which includes the external fluxes $A_C$ and $A_G$. It is remarkable, that the external stream $A_0$, also as the real production defined by the quantities $I$, $G$, $C$, $W_P$ and $W_G$, do not enter into the system of the Eq. (3.27). The financial subsystem is developing autonomously, the linkage with the real production is being established through the sources (emission) of money $E_0$, $E_C$, $E_G$ (see further Sect. 3.4.3).

## 3.4.2 Steady-State Situation

The equations of evolution determine a trajectory of the system at the characteristics of system defined by the activity of the ensemble of participants of actual processes of consumption-production. It is useful to consider, first of all, a stationary case, when all characteristics, as well as all system's variables, have constant values. Economists call such situations equilibrium, but, from the thermodynamic point of view, the considered system is in a stationary, non-equilibrium state. The state of the production-monetary system is defined in this case only by balance equations, without any additional assumptions.

### 3.4.2.1 Steady-State Condition

We consider the established state of a financial system when all variables defined by the Eq. (3.27) are constant, however, the rates of growth of the production subsystem can be any. Provided that $E_0 = 0$, but the external fluxes $A_0$, $A_C = E_C$, $A_G = E_G$ are possible, and the equations of evolution (3.27) reduce to the algebraic equations

$$\begin{aligned} 0 &= r_B(D_B - \xi D) - q_B B_B - \Delta - A_G, \\ 0 &= rD - qB - \Delta - A_C - A_G. \end{aligned} \tag{3.28}$$

The system of equations does not determine a unique point in phase space: the number of variables is more than number of the equations. Equation (3.28) shows that, in the established situation, two variables only from a set $D$, $D_B$, $B$ and $B_B$ ought to be considered independent. When proficit of the budget is excluded from Eq. (3.28), one has a relation

$$A_C = rD - qB + q_B B_B - r_B(D_B - \xi D). \tag{3.29}$$

Equation (3.28) can be presented in the form of

$$\begin{aligned} D_B &= \xi D + \frac{q_B}{r_B}(B_B - B_B^0), \quad B_B^0 = \frac{1}{q_B}(-\Delta - A_G), \\ D &= \frac{q}{r}(B - B^0), \quad B^0 = \frac{1}{q}(-\Delta - A_C - A_G). \end{aligned} \tag{3.30}$$

These equations define quantities of deposits in steady state $D$ and $D_B$, as functions of loans $B$ and $B_B$. The amounts of deposits should be considered non-negative, so that Eq. (3.30) define amounts of loans, which are necessary that the commercial bank could start to function. The amounts of loans with zeroes $B^0$ and $B_B^0$ can be interpreted, accordingly, as a minimum quantity of the capital of commercial bank and the loan from the central bank, which allow bank system to start action.

#### 3.4.2.2 Measure of Propagation of Credit Money

Equation (3.14) allows to find expression for effective quantity $\xi^*$ in a stationary case. Really, this equation is followed (at constant value $\xi^*$) by an expression for the quantity of money, which is on the disposal of commercial bank

$$D_B = \xi^*(B + D) + const. \tag{3.31}$$

Considering this equation at zero deposits, one can define the constant in the equation by means of the first line from system (3.30), so that the equation now can be expressed in the form of

$$D_B = \xi^*(D + B - B^0).$$

This equation defines stationary value of the quantity $\xi^*$

$$\xi^* = \frac{D_B}{rD + B - B^0}. \tag{3.32}$$

Let us remind, that the relations recorded here are valid for the steady-state situation or, as economists speak, for a situation of equilibrium.

### 3.4.3 *The Rules of Emission*

The trajectory of evolution of the monetary system, described by the system of Eq. (3.27), is essentially determined by values of emission of paper and credit money $E_C$, $E_0$, $E_G$, including the fluxes of money in system $A_0$, $A_C$ and $A_G$. The listed quantities are being set by the economic subjects (clients of banks, commercial and central banks, as well as the government) after estimation of the situation, in which the subjects appear to be. In particular, the external fluxes depend on conditions of managing inside and outside of the system; in our approximation we consider values $A_0, A_C$ and $A_G$ to be given. However, the internal emission of paper and credit money are being done by economic subjects, and it is possible to formulate some rules for this emission.

It is possible to believe, that the central bank monitors progress of real production and supervises its accounts, aiming that the total amount of money in the system $D + M_0$ would correspond to the real production output in the sense defined by relation (3.25). It means, that the central bank watches the rate of growth of the price index (inflation), defined by the Eq. (3.26), would not be too great or even was equal to zero. To reach the desirable correspondence, the central bank monitors that the deposits of the government and commercial bank, $D_G$ and $D_B$ would increase according to progress of production subsector, in the simplest case, with the rate of growth of production output $\sigma$. Then, the first and the third equations from system (3.27) define emission of paper and credit money

$$
\begin{aligned}
E_0 + E_\mathrm{G} &= \sigma D_\mathrm{G} - q_\mathrm{B} B_\mathrm{B} + r_\mathrm{B}(D_\mathrm{B} - \xi D) - \Delta, \\
E_\mathrm{C} &= \sigma D_\mathrm{B} - r_\mathrm{B}(D_\mathrm{B} - \xi D) + q_\mathrm{B} B_\mathrm{B} + \Delta + A_\mathrm{C} + A_\mathrm{G}.
\end{aligned} \tag{3.33}
$$

The quantity $E_0$ and $E_\mathrm{G}$ are defined by the government and the central bank, proceeding from the necessity to provide additional financing of the budget area and government projects. The emission of credit money is defined by joint actions of the central and commercial banks.

The lack of money in the budget can be immediately compensated by release of paper money $E_0$ and obligation $E_\mathrm{G}$. However, the emission cannot proceed long time, therefore the government could undertake measures on increase in tax revenues, increasing rates of taxes $\theta_\mathrm{P}$ and $\theta_\mathrm{L}$ to receive the desirable income

$$
T = \theta_\mathrm{P} Y + \theta_\mathrm{L}(W_\mathrm{P} + W_\mathrm{G}),
$$

for example, by rules

$$
\frac{\mathrm{d}\theta_\mathrm{P}}{\mathrm{d}t} = -n_\mathrm{P}, \quad \frac{\mathrm{d}\theta_\mathrm{L}}{\mathrm{d}t} = -n_\mathrm{L}\Delta, \tag{3.34}
$$

where factors $n_\mathrm{P}$ and $n_\mathrm{L}$ define the rate of change of the tax rates. These quantities, as well as the norms of taxes $\theta_\mathrm{P}$ and $\theta_\mathrm{L}$, are fixed by the central bodies: the government and the central bank, which start with their intentions and assessment of the situation in the production subsector.

There is no solid causal relationship between development of production and evolution of monetary system; such linkages are being established by hands of workers of the central bank after some analysis. Apparently, rules of emission can be formulated on the basis of various requirements, for example, specifying small inflation that is a preferential version for the existing governments, so as it provides the additional income into the budget. To support the production development, the central organisations should define the program of issue of credit and paper money. The simple reasons stated above are based on the assumptions, that the government and the central bank in their activity are guided only by the interests of creation of favourable environment for development of national production. Unfortunately, it not always true: if a national monetary system is strongly integrated into the World environment, when economic subjects can interact directly with the external agents, there appears additional reasons connected with the necessity to provide favorable (or adverse) interoperability with external agents. Additional conditions can lead to contradictions with national interests and the big art is required from the bank managers to manoeuver among controversial requirements.

## 3.5 Money System of Russia

The considered scheme of money circulation represents the most general features of functioning of any national economic system. National central banks provide the tremendous statistical information, allowing to estimate adequacy of model. Further,

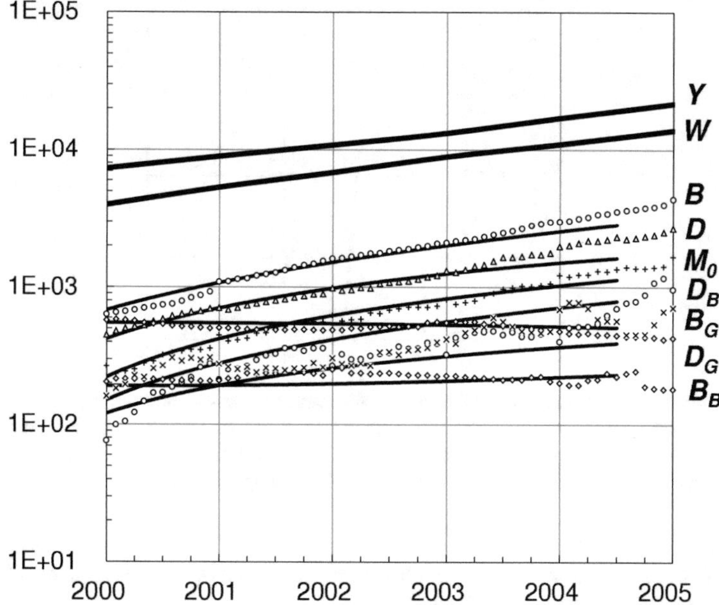


**Fig. 3.2** The monetary situation of Russia. The top curves represent values of the Gross Domestic Product Y and income of working people W. Below, the continuous lines, noted by symbols of variables, show the calculated according to Eq. (3.27) trajectories of evolution of variables. By points, empirical values of variables—cash money, credits and deposits of the central and commercial banks—are presented. All quantities are estimated in $10^9$ of circulating in corresponding year rubles

as an illustrative sample, discussed also in the publication [16], we consider the dynamics of real money system of Russia for a few years since 2000, using the data of the Central bank of Russia [17].

### 3.5.1 *Identification of Variables*

According to the system of Eq. (3.27), we are considering seven variables: $D$, $D_B$, $D_G$, $B$, $B_B$, $B_G$ and $M_0$, the empirical values of which are depicted in Fig. 3.2 with points.

The numbers for cash money $M_0$, monetary base $M_B = M_0 + D_B$ and a money supply $M_2 = M_0 + D$, estimated by the Central Bank of the Russia, are put in basis of the assessment of variables of the model. Values of deposit $D$ in commercial banks and the deposit of commercial bank in the central bank $D_B$ are found as $D = M_2 - M_0$ and $D_B = M_B - M_0$. The calculated in this way values conform to direct assessments of these quantities by the Central Bank of Russia [17]. Values of credits in commercial bank $B$ and credit of commercial bank in the central bank $B_B$, as well as credit and

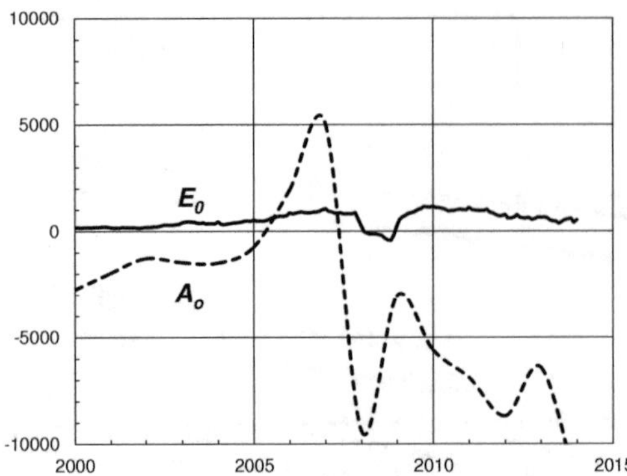


**Fig. 3.3** Fluxes of money into accounts of clients of commercial bank. The continuous line presents quantity of emission of paper money $E_0$, the dashed line represents the external flux of money $A_0$ into accounts of clients of commercial banks, according to the Central Bank of Russia [17]. Unit of estimation of quantities—$10^9$ of circulating in corresponding year rubles in a year

deposit of the government in the central bank, $B_G$ and $D_G$, are estimated directly according to the Central bank of Russia.

It is clear, that under neglecting some quantities, which bring, on the assumption and assessments, smaller contribution, and significant aggregation, the description inevitably is rough, but acceptable for the analysis. Let's notice, that an analysis of the Russian bank system for years 2004–2007 [18] confirms importance of the variables used in the described model, though actual lists of the aggregated variables used by us and in the Pospelov's with co-authors works [18] coincide only partially.

### 3.5.2 *External Fluxes and Their Approximation*

The quantities $A_0$, $A_C$ and $A_G$ in balance equations (3.8), (3.10) and (3.16) represent fluxes of money (taking outflow in view) into accounts of clients of commercial bank, the account of commercial bank in the central bank of Russia and the account of the government in the Central bank, accordingly. The quantities of fluxes estimated directly by the Central bank of Russia are shown in Figs. 3.3, 3.4 and 3.5. Existence of exogenous fluxes characterises 'international' activity of considered economic subjects. Negative values of the quantities mean, that clients and commercial banks prefer to send the reserves on storage not in the central bank, but to other places. The fluxes include accumulation of money in securities or on accounts outside of the system; it is possible both that and another. The reader can pay attention to an essential increase of a negative stream $A_C$ in 2006 when the rule about mandatory sales of currency proceeds to the central bank has been cancelled.

The balance equations (3.10) and (3.16) at known time series for seven variables $D, D_B, D_G, B, B_B, B_G$ and $M_0$ and at known norms of payment of credits and deposits

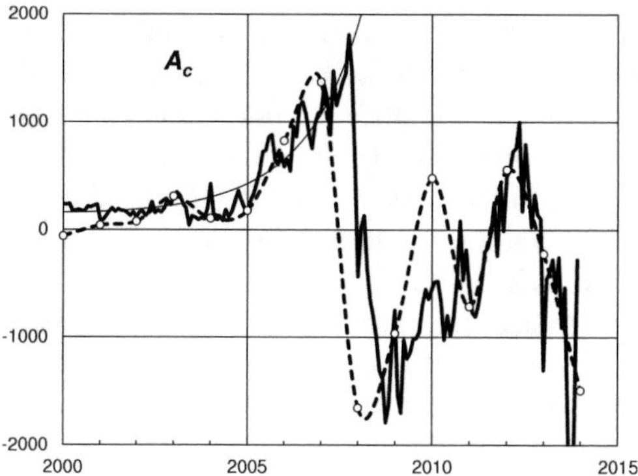


**Fig. 3.4** The flux of external money into accounts of commercial bank. The solid line presents average values of the external flux of money $A_C$, calculated according to Eq. (3.10). By the dashed line with points, the empirical values estimated by the Central Bank of Russia [17] are shown. The quantities are estimated in $10^9$ circulating in corresponding year rubles in a year. A thin solid line shows an approximation of the flux with the exponential function

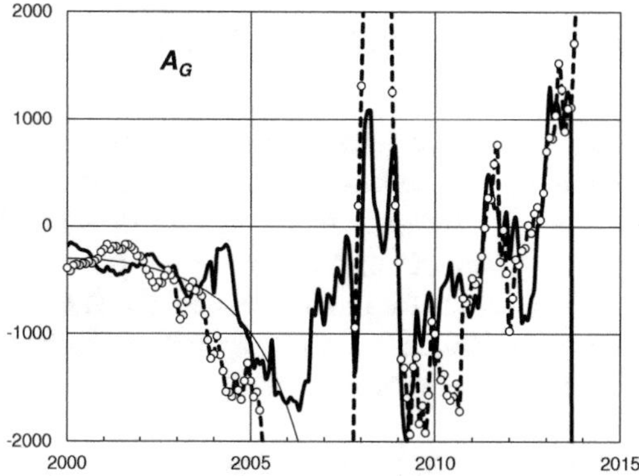


**Fig. 3.5** The flux of external money into the account of the government of Russia. The solid line shows the values of an external flux (in view of outflow) of money $A_G$, calculated according to Eq. (3.16). The dashed line with points represents variation (with a sign a minus) the international reserves of the government. All quantities are estimated in $10^9$ circulating in corresponding year rubles for a year. A thin line shows an optional approximation of the flux by exponential function

allow us to estimate independently the fluxes of money from external sources $A_C$ and $A_G$. The calculated and averaged (the period of averaging is year) values of the external fluxes $A_C$ and $A_G$ are shown in Figs. 3.4 and 3.5 together with the empirical values specified by the Central Bank of Russia [17]. Let us notice, however, that, the Gross Domestic Product $Y$ in Eq. (3.8), defined by Eq. (3.24), contains values of exogenous fluxes of money ($A_0$ included), so that the quantity $A_0$ cannot be estimated obviously from Eq. (3.8), this quantity is shown in Fig. 3.3 according to the Central Bank of Russia [17].

Let us pay attention to the conformity of two results of assessments of the flux $A_C$ (see Fig. 3.5). Results confirm the assumption made in Sect. 2.2 that commission payments and other gathering completely go for current maintenance and wages of the bank employees and, consequently, are not included in a balance parity (3.10). Some exception appears in some years, since 2007. The situation with a flux $A_G$ is more complex; actual variation of the international reserves of the government in 2005–2011 does not coincide with an assessment on a balance parity (3.10).

As it is possible to notice from Figs. 3.3, 3.4 and 3.5, sources of money $E_0$, $A_C$, $A_G$ appear strongly pulsating functions, which shows, that the relations (3.8), (3.10) and (3.16) should be the stochastic equations, but for the beginning we neglect the random contribution, assuming, that we operate with the average quantities. Further, turning to calculation of behaviour of variables, we shall use approximation of the sources of money with exponential functions

$$A_C = 160 \exp(0.0065t), \quad A_G = -300 \exp(0.008t). \tag{3.35}$$

### 3.5.3 Fundamental Characteristics of the System

A peculiarity of our approach to the description of monetary circulation, proposed in the work [10], is the introduction and use of fundamental characteristics of system, among which: coefficient of efficiency of the system $R$, which is the ratio of an assessment of average services of bank system to the pure output of production system (see Sect. 3.3.2), expenses for production and maintenance of circulation of one ruble in unit of time $\kappa$ and a measure of propagation of credit money $\xi^*$. The ratio $\kappa/R$ is 'velocity of circulation of money' in the known 'quantity theory of money' [3].

The coefficient of efficiency of system $R$ can be calculated from values of the Gross Domestic Product $Y$ and income of bank system $H$; these values can be found in Year Books of Rosstat (see, for example, [19]). In addition, it is necessary to know values of external fluxes $A_0$, $A_C$, $A_G$. As consequence of the Eq. (3.24), value $R$ is estimated according to the formula

$$R = \frac{H}{Y - H + A_0 + A_C + A_G} \tag{3.36}$$

The calculated values of $R$ are presented in Fig. 3.6. Values of efficiency increase from $R \approx 0.01$ in year 2000 up to $R \approx 0.04$ towards the end of the considered period. The figure also shows values of $\kappa$, which characterises expenses for maintenance of production and circulation of one ruble. The values of the quantity in the beginning and in the end of the period are estimated as $\kappa \approx 0.1$, or about ten kopecks on one circulating cash or non-cash ruble.

The norm of mandatory reservation $\xi$ is set by the central bank; its value can be estimated according to the known fraction of mandatory deposit and is equal $\xi \approx 0.1$

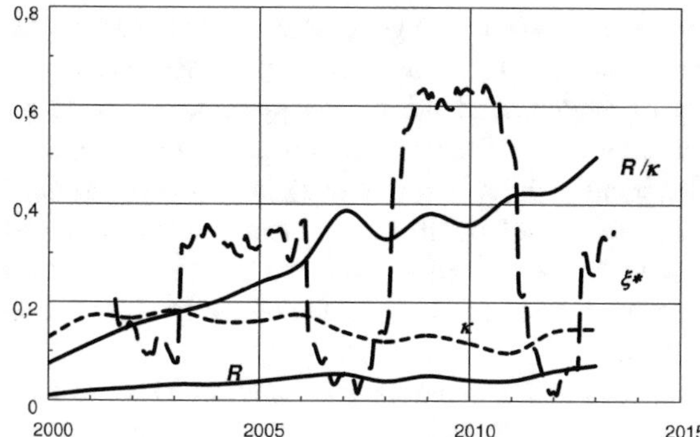


**Fig. 3.6** Fundamental characteristics of the system. The sharply changing dotted curve represents a measure of propagation of credit money. A solid bottom line presents a measure of participation of the money system in production activity (the efficiency ratio), the dashed line above—expenses of a society for maintenance in circulation one cash or non-cash ruble. The top solid line represents 'time of circulation of money' $R/\kappa$ in 'the quantity theory of money'

in the beginning of the considered period and decreases up to $\xi \approx 0.015$ towards the year 2012. Effective values $\xi^*$, characterising the true expansion of credit money, are calculated specifically under the formula (3.14) copied in the form

$$\xi^* = \frac{\mathrm{d}D_{\mathrm{B}}}{\mathrm{d}t} : \left(\frac{\mathrm{d}D}{\mathrm{d}t} + \frac{\mathrm{d}B}{\mathrm{d}t}\right). \tag{3.37}$$

The calculated values $\xi^*$ are strongly pulsating; average values of quantity, shown in Fig. 3.6, are increasing from the beginning of the considered period towards the end, reaching after the year 2010 the average value 0.3. The stationary values $\xi^*$, calculated under the formula (3.32) at current values of variables, appear much less than actual values and practically coincide with values of $\xi$.

### 3.5.4 *Trajectories of Evolution*

To show the consistency and adequacy of our concepts, we shall try to reproduce behaviour of the variables, described in Sect. 3.5.1, for the Russia money system within several years after 2000, being based on system of the Eq. (3.27). We believe, that the monetary system accompanies the production program with the rates of growth, which for sector output is equal to the rate of growth of the GDP; average value of GDP (in the current prices) for the considered first three years are equal $\sigma = 0.24$ per a year. The income of the government depends on GDP and the income of the population (see Eq. 3.21), but for calculation of evolution of monetary system, according to system (3.27), the knowledge of proficit of the budget $\Delta$ is needed only.

In the initial years, the income of the government exceed expenses; for calculation, the constant value $\Delta = 180 \cdot 10^9$ rubles per a year is accepted.

The trajectory of evolution of the money system, described by the system of Eq. (3.27), is essentially determined by values of emission of paper and credit money $E_C$, $E_0$, $E_G$, which include the fluxes of money into the system $A_0$, $A_C$ and $A_G$. The quantity $E_0$ and $E_G$ are defined by the government and the central bank, proceeding from the necessity to provide additional financing the budget area and government projects. The emission of credit money $E_C$ is defined by joint actions of the central and commercial banks.

Both the central bank and commercial bank have to take into account the production development. One can assume that the central bank should increase the deposits of the government and commercial bank, $D_G$ and $D_B$, according to the progress of production subsector, in the simplest case, with the rate of growth of production output $\sigma$. Then, assuming, for simplicity, that $E_G = 0$, the two other quantities can be estimated as

$$\begin{aligned} E_0 &= \sigma D_G - q_B B_B + r_B(D_B - \xi D) - \Delta, \\ E_C &= \sigma D_B - r_B(D_B - \xi D) + q_B B_B + \Delta + A_C + A_G. \end{aligned} \tag{3.38}$$

The external fluxes $A_C$ and $A_G$ are considered to be set; at calculations, we use the approximations (3.35) shown in Figs. 3.4 and 3.5.

According to the central bank of Russia [17], the rates of payments for credits and debts in central and commercial banks are pulsating quantities; average values in initial years after 2000: $q = 0.16$ and $r = 0.05$. In the same year's average value of the discount rate or the rate of refinancing (nowadays called the key rate) in the central bank is $q_B = 0.20$, the payment for deposits of commercial banks, apparently is small, it is possible to accept value $r_B = 0.02$.

Fundamental characteristics of system $R$ and $\kappa$ change slowly (see Fig. 3.6), but for simplicity of the description, they, as well as norm of mandatory deposit $\xi$, is assumed constant during all period of consideration. We use the values of characteristics of the system estimated in the previous section: value of coefficient of efficiency $R = 0.01$; factor of friction $\kappa = 0.1$ ruble for a year, norm of the mandatory deposit $\xi = 0.04$; value of a measure of propagation of credit money $\xi^* = 0.19$.

We accept values of variables for year 2000 as initial ones, that is: $D = 420$, $D_B = 150$, $D_G = 120$, $B = 680$, $B_B = 190$, $B_G = 550$ and $M_0 = 220$. We shall remind, that all quantities are estimated in $10^9$ rubles. The calculated and depicted on Fig. 3.2 with the thin lines trajectories of evolution reproduce initial behaviour of actual trajectories with comprehensible accuracy (recollect accuracy of estimated values of the quantities), and it is the first important step, which is necessary to show adequacy of our concepts. It is possible to calculate also the price index, which increases from the beginning of the period in correspondence with assessments of actual values of the quantities.

## 3.6 About the Unit of Measurement of Value

An arbitrary quantity of money units $M_0+D$ that is issued by the central and commercial banks is usually circulating in the production–distribution system. The value of monetary unit is determined eventually by amounts of products that can be contested to introduced money and is recognised by comparison of the quantity of money units $M_0+D$, circulating into the system, with the pure production output $\underline{I}+\underline{G}+\underline{C}$, measured by some 'physical' measure, that is by the quantity of products (see Eq. 3.25). The constancy of the monetary unit favours balanced development. For the monetary unit to be constant, the central bank has to support the correspondence of amount of money to production output, which is difficult to achieve, so that the monetary unit usually change its value. To exclude non-controllable changes, one uses the conventional monetary unit of constant purchasing capacity and price index $\rho$ (see Sect. 2.2.2). The requirement for the monetary unit of constant purchasing capacity to be constant can be found from Eq. (3.26) for dynamics of price index.

Throughout the centuries the role of 'monetary unit' was played by various products, but gold eventually achieved a special advantage, and the monetary unit before the first world war almost everywhere was defined as a quantity of gold (the gold standard). The 'gold' monetary unit is not a constant measure of value, but can be proxy of the unit of constant purchasing capacity. To be a proper unit of constant purchasing capacity, the changes in the efforts involved in producing a unit of gold have to correspond to the changes in effort for production of any other product. Apparently, this condition cannot be fulfilled in practice. The price of gold grows in the conventional monetary units, which testifies that the purchasing capacity of these units falls. Although the 'gold' monetary unit is not an ideal measure of purchasing capacity, it is better than any monetary unit not connected with 'a physical content'. Apparently, projects with a return to the gold standard will appear.

The modern continuously changing money units create severe difficulties, for both the functioning and the analysis of economic systems, though there are many powerful people, who receive income from a manipulation with the monetary units.[4] There is a question of whether some true scale of value similar to the kilogram or

[4]As known financier Lietaer [20, p. 254] writes: 'The world has been living without an international standard of value for decades, a situation which should be considered as inefficient as operating without standard of length or weight.' To demonstrate the indispensability of a constant unit of measure for production efficiency, we shall imagine a contractor who builds houses. It is possible to utilise any measure of length to build a house, a good house, which is pleasant not only to the customer, but also to another customer who will hasten to place an order. At the construction of the second house, the contractor does not notice that its measure of length has decreased a little bit; this will not prevent him from building precisely the same house, but a little bit smaller in size, which allows the contractor to save building materials. And what will occur, if the measure of length changes during construction of the house? Certainly, the cunning contractor has realised for a long time that the skilful manipulation of a measure of length brings good income, and he does not have any interest in changing that. And what do his clients think?.

metre for mass and length can be introduced. Is it possible to find an objective basis for the establishment of a monetary unit? The crisis situations of the past years have shown that the problem is worth thinking about. We shall return to this question in Chap. 11.

## References


1. Leontief, W.W.: Input-Output Economics, 2nd edn. Oxford University Press, New York (1986)
2. Blaug, M.: Economic Theory in Retrospect, 5th edn. Cambridge University Press, Cambridge (1997)
3. Fisher, I.: The Purchasing Power of Money: Its Determination and Relation to Credit, Interest, and Crises. Macmillan, New York (1911)
4. Keynes, J.M.: The General Theory of Employment, Interest and Money. Macmillan Cambridge University Press, New York (1936)
5. Graziani, A.: The Monetary Theory of Production. Cambridge University Press, Cambridge (2003)
6. Godley, W., Lavoie, M.: Monetary Macroeconomics: An Integrated Approach to Credit, Money, Income, Production and Wealth. Palgrave Macmillan, Basingstoke (2007)
7. Chiarella, C., Flaschel, P.: Keynesian monetary growth dynamics in open economies. Ann. Oper. Res. **89**, 35–59 (1999)
8. Keen, S.: The dynamics of the monetary circuit. In: Rossi, S., Ponsot, J.-F. (eds.) The Political Economy of Monetary Circuits: Tradition and Change, pp. 161–87. Palgrave Macmillan, London (2009)
9. Keen, S.: Endogenous money and effective demand. Rev. Keynes. Econ. **2**(3), 271–291 (2014)
10. Pokrovskii, V.N., Schinckus, Ch.: An elementary model of money circulation. Phys. A Stat. Mech. Appl. **643**, 111–122 (2016)
11. Werner, R.A.: Can banks individually create money out of nothing? - The theories and the empirical evidence. Int. Rev. Financ. Anal. **36**, 1–19 (2014)
12. McLeay, M., Radia, A., Thomas, R.: Money creation in the modern economy. Bank Engl. Q. Bull. **1**, 14–27 (2014)
13. Samuelson, P., Nordhaus, W.: Economics, 13th edn. McGraw-Hill Book Company, New York (1989)
14. Keynes, J.M.: A Treatise on Money. Macmillan, London (1930). Reprinted in: The collected writings of John Maynard Keynes, Vols. V and VI. Macmillan, London and Basingstoke (1971)
15. Harrod, R.F.: Money. Macmillan, London (1969)
16. Altuchov, Yu.A., Pokrovskii, V.N.: [A test of the elementary model of money circulation]. Ekonomika i matematicheskie metody. Economics and mathematical methods (2018, in press)
17. The Official Site of the Central Bank of Russia: www.cbr.ru. Accessed 17 Aug 2017
18. Andreev, M.J., Pilnik, N.P., Pospelov, I.G.: Ekonometricheskoe issledovanie i modelnoe opisanie deyatelnosti sovremennoy rossiyskoy bankovskoy sistemy (The Econometric Investigtion and Model Description of Activity of the Modern Russian Bank System). Dorodnitsyn Computer Center of the Russian Academy of Science, Moscow (2008). http://www.ccas.ru/mmes/AndreevPilnikPospelovRussianBankSystem.pdf. Accessed 17 Aug 2017
19. Rosstat: Russian statistical year-book 2013. Rosstat, Moscow (2013)
20. Lietaer, B.: The Future of Money. A New Way to Create Wealth, Work, and a Wiser World. Century, London (2001)

# Chapter 4
# Many-Sector Approach to Production System

**Abstract** In this chapter, the input-output model, already discussed in Chap. 2, is considered within the linear approximation. This allows us to formulate the static and dynamic equations for the gross output. The model determines the growth rate, which, in the considered case, when the only internal restrictions for development are taken into account, is the rate of potential growth. The restrictions imposed by labour and energy will be introduced in the following chapters, and we will return to many-sector dynamic equations in Chap. 9.

## 4.1 Linear Approximation

A method of production determines, first, what one needs to create this or that thing, which organizes the material side of production process. Exploiting, due to Leontief [1–3] and Sraffa [4], the many-sector model of the production system of the economy, one describes the transfer of products from one sector to another, which is reflected in the balance equations (2.3) and (2.4), that is,

$$X_i = \sum_{j=1}^{n} X_i^j + Y_i, \quad X^i = \sum_{j=1}^{n} X_j^i + Z^i \tag{4.1}$$

These equations contain the gross and final sector outputs $X_j$ and $Y_j$, intermediate production consumption $X_j^i$ and the sector production of value $Z^i$. The equations do not allow one to determine the output of the production system of the economy without extra information. A fortunate idea, originated by Leontief [1, 2], allows to reduce the number of variables by introducing quantities that are characteristics of the production system itself. Following to Leontief's works [1, 2], we will consider the simplest case of linear approximation to connect the intermediate production consumption $X_j^i$ and fixed production capital $K_j^i$ with the gross sector output $X_j$. This allows one to introduce fundamental technological characteristics of technology: technological matrices (4.4) and (4.13).



V. N. Pokrovskii, *Econodynamics*, New Economic Windows,
https://doi.org/10.1007/978-3-319-72074-6_4

### 4.1.1 The Input-Output Matrix

Apparently, the greater the gross output of a sector, the more intermediate products are needed for production. This observation allows one to reduce intermediate production consumption $X^i_j$ to gross output $X_i$ by introducing characteristics of technology, which can be done in various ways. In the original work [1, 2], it is assumed that the intermediate production consumption $X^i_j$ is proportional to the gross sector output,

$$X^i_j = a^i_j X_I \tag{4.2}$$

with coefficients of proportionality $a^i_j$ reflecting the exploited technology. The coefficients comprise a matrix—the matrix of intermediate consumption coefficients.

Otherwise, one can assume that a technological matrix can be introduced as the ratio of the velocities of the quantities

$$\frac{dX^i_j}{dt} = \tilde{a}^i_j \frac{dX_i}{dt}. \tag{4.3}$$

The last relation defines another matrix of intermediate consumption coefficients that are characteristics of the technology, introduced in a given moment of time, while Eq. (4.2) introduces the matrix of intermediate consumption coefficients that are average characteristics of all existing technology.

We shall follow Leontief [1, 2], who has chosen relation (4.2) to introduce the matrix of intermediate consumption coefficients, or input-output matrix that in the complete form can be written as

$$\mathsf{A} = \begin{Vmatrix} a^1_1 & a^2_1 & \dots & a^n_1 \\ a^1_2 & a^2_2 & \dots & a^n_2 \\ \dots & \dots & \dots & \dots \\ a^1_n & a^2_n & \dots & a^n_n \end{Vmatrix}. \tag{4.4}$$

Matrix $\mathsf{A}$ is a phenomenological characteristic of the technological organisation existing in the production system at the moment. It changes if the technology organisation changes during the time. The components of matrix $a^i_j$ represent a mixture of all technologies, old and new. In contrast, the components of the matrix $\tilde{a}^i_j$ represent characteristics of newly introduced technology.

It is easy to see that the matrixes introduced by the different methods are connected with each other

$$\tilde{a}^i_j = a^i_j + \frac{1}{\delta^i} \frac{da^i_j}{dt}, \tag{4.5}$$

where $\delta^i = \dfrac{1}{X_i} \dfrac{dX_i}{dt}$ is the growth rate of the gross output. Of course, in the case, when the technology does not depend on time, the components of matrixes introduced in alternative ways coincide. The difference can be negligible if the technology changes slowly.

### 4.1.2 Static Leontief Equation

One can use relation (4.2) to rewrite the balance equations (4.1) in the form

$$X_j = \sum_{i=1}^{n} a^i_j X_i + Y_j, \tag{4.6}$$

$$X_i = a^i X_i + Z^i, \quad a^i = \sum_{l=1}^{n} a^i_l. \tag{4.7}$$

The first of these equations is known as the Leontief equation. Equations (4.6) and (4.7) connect three vectors: the gross output $X_j$, the final output $Y_j$ and the sector production of value $Z^i$, so that only one of them can be considered to be independent. All the variables $X_j$, $Y_j$ and $Z_j$ in Eqs. (4.6) and (4.7) are referred to the same moment of time.

The first equation in (4.7) allows one to specify the properties of matrix $\mathsf{A}$. A requirement of productivity states that the sector production of value has to be non-negative,

$$Z^i \geq 0,$$

and taking into account Eq. (4.7), this is followed by the property

$$\sum_{l=1}^{n} a^i_l < 1, \quad i = 1, 2, \ldots, n.$$

It is also natural to suppose that all components of matrix $\mathsf{A}$ are non-negative, so that one has for each component of matrix $\mathsf{A}$

$$0 \leq a^i_j < 1, \quad i, j = 1, 2, \ldots, n. \tag{4.8}$$

### 4.1.3 Planning of Gross Output

To create a final product $Y_j$, one needs the products of, generally speaking, all sectors, so one has to consider production of the gross output in all sectors. Equation (4.6), which is known as the static Leontief equation, allows one to plan the gross output $X_j$, which is needed to get the final output $Y_j$ at a given technology. It is convenient to rewrite Eq. (4.6) in vector form,

$$\mathsf{X} = \mathsf{AX} + \mathsf{Y}. \tag{4.9}$$

The formal solution of this equation

$$\mathsf{X} = (\mathsf{E} - \mathsf{A})^{-1}\mathsf{Y}, \tag{4.10}$$

where $\mathsf{E}$ is the unity matrix, can be represented as

$$\mathsf{X} = \left(\mathsf{E} + \sum_{k=1}^{\infty} \mathsf{A}^k\right)\mathsf{Y}. \tag{4.11}$$

From property (4.8) of matrix $\mathsf{A}$,

$$\mathsf{A}^k \to 0,$$

so that the series in Eq. (4.11) converges. The non-negativity of components of the matrix ensures non-negativity of the gross output, when the final output is non-negative. It can be tested directly that expression (4.11) satisfies Eq. (4.9). Indeed,

$$\begin{aligned}
\mathsf{Y} &= (\mathsf{E} - \mathsf{A})\left(\mathsf{E} + \sum_{k=1}^{\infty} \mathsf{A}^k\right)\mathsf{Y} \\
&= \left(\mathsf{E} + \sum_{k=1}^{\infty} \mathsf{A}^k - \mathsf{A} - \sum_{k=1}^{\infty} \mathsf{A}^{k+1}\right)\mathsf{Y} \\
&= \left(\mathsf{E} + \sum_{k=1}^{\infty} \mathsf{A}^k - \sum_{i=1}^{\infty} \mathsf{A}^i\right)\mathsf{Y} \;=\; \mathsf{Y}.
\end{aligned}$$

Note that in representation (4.11), matrix $\mathsf{A}$ is called the matrix of direct input, matrix $\mathsf{A}^2$ is called the matrix of indirect input of the first order, and so on. The matrix $(\mathsf{E} - \mathsf{A})^{-1}$ is called the matrix of total input.

### 4.1.4 The Capital-Output Matrix

To produce something, one also needs in production equipment (production capital stock), the greater the amount of the stock, the more the scale of production. In linear approximation, the fixed production capital (value of basic production equipment) of type $j$ in sector $i$ is proportional to the gross sector output

$$K_j^i = b_j^i X_i. \tag{4.12}$$

The relation (4.12) defines the matrix of fixed capital coefficients (capital-output matrix)

$$\mathsf{B} = \left\| \begin{array}{cccc} b_1^1 & b_1^2 & \dots & b_1^n \\ b_2^1 & b_2^2 & \dots & b_2^n \\ \dots & \dots & \dots & \dots \\ b_n^1 & b_n^2 & \dots & b_n^n \end{array} \right\| . \tag{4.13}$$

In line with the input-output matrix $\mathsf{A}$, matrix $\mathsf{B}$ is also a characteristic of the technology used in the production system. The coefficient of proportionality $b_j^i$ represents a mixture of all technologies, old and new, if the technologies are changing during that time.[1]

It is easy to get a relation between the amount of production equipment of a certain kind $K_j$ and sector capital $K^i$. Indeed, from the definitions of the quantities in Sect. 2.3.2 and relation (4.12), one finds, that the quantities are connected with each other by means of components of the capital-output matrix

$$K_j = \sum_{i=1}^{n} \overline{b}_j^i K^i . \tag{4.14}$$

Here a non-dimensional matrix of capital coefficients is introduced

$$\overline{b}_j^i = (b^i)^{-1} b_j^i, \quad b^i = \sum_{l=1}^{n} b_l^i . \tag{4.15}$$

On the definition, components of this matrix are connected by the relations

$$\sum_{j=1}^{n} \overline{b}_j^i = 1, \quad i = 1, 2, \dots, n.$$

It is also possible to get a relation between the investment of product $j$ in all sectors $I_j$ and total investment $I^i$ in sector $i$. To obtain such a relation, one can refer to the equations for the dynamics of the total amount of equipment $K_j$ and for the dynamics of fixed sector capital $K^i$ from Sect. 2.3.2, that is, to the equations

$$\frac{dK_j}{dt} = I_j - \mu K_j, \quad j = 1, 2, \dots, n,$$

$$\frac{dK^i}{dt} = I^i - \mu K^i, \quad i = 1, 2, \dots, n. \tag{4.16}$$

To get the desired relation, one can use parity (4.14) and rewrite Eq. (4.16) in the form

[1] Instead of matrix $\mathsf{B}$, one can use matrix $\tilde{\mathsf{B}}$ which characterises currently introduced technology. The relations between components of matrixes $\mathsf{B}$ and $\tilde{\mathsf{B}}$ are the same as those between components of matrixes $\mathsf{A}$ and $\tilde{\mathsf{A}}$ in Eq. (4.5).

$$\sum_{i=1}^{n} \overline{b}_j^i \frac{dK^i}{dt} + \sum_{i=1}^{n} K^i \frac{d\overline{b}_j^i}{dt} = I_j - \mu \sum_{i=1}^{n} \overline{b}_j^i K^i, \quad j = 1, 2, \ldots, n,$$

$$\frac{dK^i}{dt} = I^i - \mu K^i, \quad i = 1, 2, \ldots, n. \tag{4.17}$$

Then, we can exclude the derivatives of sector capital from relations (4.17) to obtain

$$I_j = \sum_{i=1}^{n} \overline{b}_j^i I^i + \sum_{i=1}^{n} K^i \frac{d\overline{b}_j^i}{dt}. \tag{4.18}$$

The made earlier approximations concerned to representations of depreciated capital, so we ought to consider relations (4.18) to be valid within the first terms with respect to the growth rates.

## 4.2 Effects of Prices

Relations (4.6) and (4.7) are written with the implicit assumption that prices of products are given and constant. To obtain a law of transformation of matrix A when the prices are changing, we assume that there is a reference state with a set of the fixed prices, which are considered to be all equal to unity, and an arbitrary state with a given set of prices, so that components of the output in the reference (with sign $\hat{}$) and the arbitrary states are connected by relations

$$X_i = p_i \hat{X}_i, \quad Y_i = p_i \hat{Y}_i, \quad i = 1, 2, \ldots, n, \tag{4.19}$$

where $p_i$ is the price index in the arbitrary state. One can assume that the basic balance equations are valid at any system of prices.

### 4.2.1 Conditions of Consistency

After transition to the system of arbitrary prices, Eq. (4.6) can be rewritten in the form

$$\hat{X}_j = \sum_{i=1}^{n} a_j^i \frac{p_i}{p_j} \hat{X}_i + \hat{Y}_j.$$

The balance equation should have the same form for any system of prices, so that a new input-output matrix must be introduced

$$\hat{a}^i_j = a^i_j \frac{p_i}{p_j}, \quad a^i_j = \hat{a}^i_j \frac{p_j}{p_i}. \tag{4.20}$$

Certainly, the description also must be covariant for any way of description. In the case, if one accept the definition (4.3) and use matrix (4.5), one has, in line with Eq. (4.6), the equation

$$\frac{dX_j}{dt} = \sum_{i=1}^{n} \tilde{a}^i_j \frac{dX_i}{dt} + \frac{dY_j}{dt}, \tag{4.21}$$

which can be rewritten with the help of Eq. (4.19) in the form

$$p_j \frac{d\hat{X}_j}{dt} + \hat{X}_j \frac{dp_j}{dt} = \sum_{i=1}^{n} \tilde{a}^i_j \left( p_i \frac{d\hat{X}_i}{dt} + \hat{X}_i \frac{dp_i}{dt} \right) + p_j \frac{dY_j}{dt} + \hat{Y}_j \frac{dp_j}{dt}.$$

So as balance equations have the same form for any system of prices, one should write down two separate relations: a balance equation and an equation for prices,

$$\begin{aligned} \frac{d\hat{X}_j}{dt} &= \sum_{i=1}^{n} \hat{\tilde{a}}^i_j \frac{d\hat{X}_i}{dt} + \frac{d\hat{Y}_j}{dt}, \\ \hat{X}_j \frac{d \ln p_j}{dt} &= \sum_{i=1}^{n} \hat{\tilde{a}}^i_j \hat{X}_i \frac{d \ln p_i}{dt} + \hat{Y}_j \frac{d \ln p_j}{dt}, \end{aligned} \tag{4.22}$$

where, similar to relations (4.20), one has

$$\hat{\tilde{a}}^i_j = \tilde{a}^i_j \frac{p_i}{p_j}, \quad \tilde{a}^i_j = \hat{\tilde{a}}^i_j \frac{p_j}{p_i}.$$

Relation (4.22) can be considered as a set of equations for quantities

$$\frac{d \ln p_i}{dt}, \quad i = 1, 2, \ldots, n$$

which can have a non-trivial solution, if the following condition is fulfilled

$$\left| (\delta^i_j - \tilde{a}^i_j) X_i - Y_j \delta^i_j \right| = 0. \tag{4.23}$$

If $\tilde{a}^i_j$ does not depend on time, relation (4.23) is always valid; otherwise, condition (4.23) presents an equation for the growth rates of components $\tilde{a}^i_j$.

One can see that the prices of products cannot be quite arbitrary quantities. A non-trivial solution of set (4.22) determines a relation between prices of different products. However, the prices are appointed independently, and it is possible to imagine a situation in which the prices are chosen in such a way that Eq. (4.22) is not satisfied. This means an infringement of value balance of products, which cannot last long.

Note that it is convenient to rewrite relations (4.19) in vector form

$$\mathsf{X} = \mathsf{P}\hat{\mathsf{X}}, \quad \hat{\mathsf{X}} = \mathsf{P}^{-1}\mathsf{X}, \qquad \mathsf{Y} = \mathsf{P}\hat{\mathsf{Y}}, \quad \hat{\mathsf{Y}} = \mathsf{P}^{-1}\mathsf{Y}, \tag{4.24}$$

where the transformation matrixes are introduced,

$$\mathsf{P} = \begin{Vmatrix} p_1 & 0 & \dots 0 \\ 0 & p_2 & \dots 0 \\ \dots & \dots & \dots \\ 0 & 0 & \dots p_n \end{Vmatrix}, \quad \mathsf{P}^{-1} = \begin{Vmatrix} p_1^{-1} & 0 & \dots 0 \\ 0 & p_2^{-1} & \dots 0 \\ \dots & \dots & \dots \\ 0 & 0 & \dots p_n^{-1} \end{Vmatrix}. \tag{4.25}$$

Then, the rules (4.20) for transformation of the technological matrix can be written as

$$\mathsf{A} = \mathsf{P}\hat{\mathsf{A}}\mathsf{P}^{-1}, \quad \hat{\mathsf{A}} = \mathsf{P}^{-1}\mathsf{A}\mathsf{P}. \tag{4.26}$$

It is easy to see that analogous relations are valid for the matrix of capital coefficients,

$$\mathsf{B} = \mathsf{P}\hat{\mathsf{B}}\mathsf{P}^{-1}, \quad \hat{\mathsf{B}} = \mathsf{P}^{-1}\mathsf{B}\mathsf{P}.$$

One calls matrices $\hat{\mathsf{A}}$ and $\mathsf{A}$, and also matrices $\hat{\mathsf{B}}$ and $\mathsf{B}$, similar matrices [5]. To separate changes of matrices that are connected with technology changes from the changes influenced by price changes, one has to consider some invariant combinations of components of the matrices. It is known that similar matrices have the same eigenvalues that, thus, do not depend on the prices.

### 4.2.2 Dynamics of Sector Production of Value

In line with relation (4.7), an equation for the derivatives of the quantities can be written

$$\frac{dZ^i}{dt} = (1 - \tilde{a}^i)\frac{dX_i}{dt}. \tag{4.27}$$

This relation, taking the formulae (4.19) into account, can be rewritten in the form

$$\frac{dZ^i}{dt} = (1 - \tilde{a}^i)\left(p_i\frac{d\hat{X}_i}{dt} + \hat{X}_i\frac{dp_i}{dt}\right). \tag{4.28}$$

The growth rate of the sector production of value in a new set of prices is broken into two parts: a rate connected with a change of gross output in the 'natural' units and a rate connected with a change of the price.

One can consider the sector as an economic agent that plans its activity and can make decisions about the amount of gross output. The aim of the sector is to obtain a bigger amount of production of value $Z^j$. One can assume that increase of output is stimulated by increase of production of value in the sector. In the simplest case,

$$\frac{d\hat{X}_i}{dt} = k_i \frac{dZ^i}{dt}, \tag{4.29}$$

where $k_i$ $(i = 1, 2, \dots, n)$ are sensibility coefficients. The coefficient $k_i$ shows how sector $i$ reacts with an increase of production of value. We consider $k_i$ to be non-negative and limited due to the possibilities of production

$$k_i < \frac{1}{(1 - \tilde{a}^i)p_i}. \tag{4.30}$$

Relations (4.28) and (4.29) determine the rate of sector production of value,

$$\frac{dZ^i}{dt} = \frac{(1 - \tilde{a}^i)X_i}{[1 - k_i(1 - \tilde{a}^i)p_i]p_i} \frac{dp_i}{dt}. \tag{4.31}$$

From this equation, using relations (4.21) and (4.27), one can obtain formulae for the growth rates of the gross and final products of the sector,

$$\frac{dX_i}{dt} = \frac{X_i}{[1 - k_i(1 - \tilde{a}^i)p_i]p_i} \frac{dp_i}{dt}, \quad i = 1, 2, \dots, n, \tag{4.32}$$

$$\frac{dY_j}{dt} = \sum_{i=1}^{n} \frac{(\delta^i_j - \tilde{a}^i_j)X_i}{[1 - k_i(1 - \tilde{a}^i)p_i]p_i} \frac{dp_i}{dt}, \quad j = 1, 2, \dots, n. \tag{4.33}$$

The last expression defines the partial derivatives of the function $Y_j = Y_j(p_1, p_2, \dots, p_n)$, which is called the supply function, as

$$\frac{\partial Y_j}{\partial p_i} = \frac{(\delta^i_j - \tilde{a}^i_j)X_i}{[1 - k_i(1 - \tilde{a}^i)p_i]p_i}. \tag{4.34}$$

One can see that the final output of a sector is an increasing function of the price of its own product, and a decreasing function of the prices of all other products.

## 4.3 Dynamics of Output

The relations given in the previous sections allow one to write equations for gross and final outputs at given matrixes $\mathsf{A}$ and $\mathsf{B}$ as functions of time. To find a dynamic equation for gross output, we differentiate relation (4.12), obtaining

$$\frac{dK^i_j}{dt} = b^i_j \frac{dX_i}{dt} + X_i \frac{db^i_j}{dt}, \quad i, j = 1, 2, \ldots, n,$$

and refer to Eq. (2.22) for dynamics of value of basic production equipment $K^i_j$ of the type $j$ in sector $i$, that is to the equations

$$\frac{dK^i_j}{dt} = I^i_j - \mu K^i_j, \quad i, j = 1, 2, \ldots, n. \tag{4.35}$$

A comparing of the above equations determines a set of dynamic equations for gross output $X_i$ in the form

$$b^i_j \frac{dX_i}{dt} + X_i \frac{db^i_j}{dt} = I^i_j - \mu b^i_j X_i, \quad i, j = 1, 2, \ldots, n. \tag{4.36}$$

It is assumed that matrix $\mathsf{B}$ and gross investment $I^i_j$ are given as functions of time.

### 4.3.1 Dynamic Leontief Equation

The situation becomes simpler, if one assumes that time dependence of the matrix $\mathsf{B}$ can be neglected. In this case, one can rewrite Eq. (4.36) as

$$\sum_{i=1}^{n} b^i_j \frac{dX_i}{dt} = I_j - \mu \sum_{i=1}^{n} b^i_j X_i, \quad j = 1, 2, \ldots, n. \tag{4.37}$$

The gross investment of product $j$ in all sectors $I_j$ is determined, according to relation (2.9), as

$$I_j = Y_j - C_j - G_j, \quad j = 1, 2, \ldots, n, \tag{4.38}$$

where $C_j$ is the total personal consumption of product $j$, and $G_j$ is investment in storage of intermediate and fundamental products.

Because of the non-negativity of the quantities $C_j$ and $G_j$, the investments $I_j$ are certain parts of the final output, so they can be conveniently written as

$$I_j = s_j Y_j, \quad 0 < s_j < 1, \quad j = 1, 2, \ldots, n, \tag{4.39}$$

and, then, from Eq. (4.6), as

$$I_j = s_j \sum_{i=1}^{n} (\delta^i_j - a^i_j) X_i, \quad 0 < s_j < 1, \quad j = 1, 2, \ldots, n, \tag{4.40}$$

The quantity $s_j$ is introduced here as the ratio of investment products to final product of sector $j$.

The last relation allows us to rewrite Eq. (4.37) in the following form

$$\sum_{i=1}^{n} b_j^i \frac{dX_i}{dt} = s_j \sum_{i=1}^{n} (\delta_j^i - a_j^i) X_i - \mu \sum_{i=1}^{n} b_j^i X_i, \quad j = 1, 2, \ldots, n. \tag{4.41}$$

This is a dynamic equation for gross sector output in a form that was originally derived by Leontief [2]. The equation can be conveniently rewritten in vector form

$$\mathsf{B}\frac{d\mathsf{X}}{dt} = \big[\, \mathsf{S}(\mathsf{E} - \mathsf{A}) - \mu \mathsf{B} \,\big]\mathsf{X}, \tag{4.42}$$

where $\mathsf{S}$ is a symbol for the diagonal matrix with values $s_j$ on the diagonal.

### 4.3.2 *Balanced Growth*

Equation (4.41) is applicable to a real system, if changes of matrices $\mathsf{A}$ and $\mathsf{B}$ due to technological changes can be neglected. In the considered case, when all parameters in Eq. (4.41) are constant, the gross output can be found in the form

$$X_j(t) = X_j(0) e^{\sigma t}, \tag{4.43}$$

where $X_j(0)$ satisfy the set of algebraic equations

$$[\, \mathsf{S}(\mathsf{E} - \mathsf{A}) - (\mu + \sigma)\mathsf{B} \,]\mathsf{X}(0) = 0. \tag{4.44}$$

The solution (4.43) presents a trajectory of the balanced (homothetic) growth, when the growth rates of the gross output in all sectors are the same.

There are non-trivial solutions of Eq. (4.44), if the determinator of the system (4.44) is equal to zero, that is,

$$|\, \mathsf{S}(\mathsf{E} - \mathsf{A}) - (\mu + \sigma)\mathsf{B} \,| = 0. \tag{4.45}$$

This is an equation for the growth rate $\sigma$. However, it is easy to see that, if there is a sector in the production system which does not create products for investment (that is the case!), the determinant (4.45) is identically equal to zero at any value of $\sigma$.

However, the growth rate of the output is restricted. Specially developed methods [6] are used (see, for example, [7, 8]) to calculate the growth rate for the homothetic trajectories. To apply these methods, one can consider Eq. (4.41) in moments of time $t = 0, 1, 2, \ldots$ and assume

$$X_i = X_i(t), \quad \frac{dX_i}{dt} = X_i(t+1) - X_i(t),$$

which implies

$$\sum_{i=1}^{n} b^i_j X_i(t+1) = \sum_{i=1}^{n} \left[ s_j(\delta^i_j - a^i_j) + (1-\mu)b^i_j \right] X_i, \quad j = 1, 2, \ldots, n.$$

This equations can be rewritten in the form of the inequality

$$\alpha \mathsf{B}\mathsf{X} \leq (\mathsf{E} - \mathsf{A} - (1-\mu)\mathsf{B})\mathsf{X}, \tag{4.46}$$

if one can introduce the ratio of growth

$$\alpha = \min_{i=1\div n} \frac{X_i(t+1)}{X_i(t)}.$$

This quantity depends on the vector $\mathsf{X}$ which can be chosen in such a way that $\alpha$ would take the greatest value,

$$\hat{\alpha} = \max_{\mathsf{X}} \alpha(\mathsf{X}).$$

This value of $\alpha$ determines the optional balance trajectory of the quickest growth,

$$X_i(t) = \hat{X}_i(0)\hat{\alpha}^t = X_i(0)e^{\hat{\sigma}t}. \tag{4.47}$$

### 4.3.3 *Potential Investment*

One can avoid the restriction of the previous section and formulate dynamic equations for the general case, when components of technological matrices $\mathsf{A}$ and $\mathsf{B}$ assumed to be functions of time. By combining (4.6) and (4.12), the final output can be defined as a linear function of the sector capital stock,

$$Y_j = \sum_{i=1}^{n} \xi^i_j K^i, \quad \xi^i_j = \frac{\delta^i_j - a^i_j}{b^i}, \tag{4.48}$$

where a matrix of marginal productivities of sector capital with components $\xi^i_j$ is introduced. Capital stock $K^i$ is governed by the dynamic Eq. (2.26), that is, by the equation

$$\frac{dK^i}{dt} = I^i - \mu K^i, \tag{4.49}$$

where sector investments $I^i$ obey, according to formulae (4.18) and (4.39), a system of algebraic equations

$$\sum_{i=1}^{n} \overline{b^i_j} I^i = s_j Y_j - \sum_{i=1}^{n} K^i \frac{d\overline{b}^i_j}{dt}. \tag{4.50}$$

Equations (4.50) do not determine the sector investment $I^i$ in a unique way: the number of equations is less than the number of variables. For given values of parameters, Eqs. (4.48)–(4.50) determine a set of dynamic trajectories for a system with $n$ sectors. To separate a unique trajectory of evolution of the dynamic system, one has to complete the system of Eq. (4.50) or use an extra condition. One can say about any extra condition that one needs to determine a trajectory of evolution as about the principle of development.

#### 4.3.3.1 Trajectory of the Quickest Growth

It is convenient to require some criterion to be optimal. For example, the criterion can be connected with the growth rates of the final output,

$$\frac{dY_j}{dt} = \sum_{i=1}^{n} \xi^i_j \frac{dK^i}{dt} + \sum_{i=1}^{n} K^i \frac{d\xi^i_j}{dt}. \tag{4.51}$$

We can consider the rate of the total of the final products and require it to have a maximum value at given technology, that is,

$$\max \frac{dY}{dt} = \max \sum_{i=1}^{n} \xi^i (I^i - \mu K^i), \quad \xi^i = \sum_{j=1}^{n} \xi^i_j, \tag{4.52}$$

while relations (4.50) ought to be considered as restrictions that must be rewritten in the form

$$\sum_{i=1}^{n} \overline{b^i_j} I^i \leq s_j Y_j. \tag{4.53}$$

In this way, investments are specified as a function of the parameters of the problem

$$\tilde{I}^i = \tilde{I}^i(\xi^i, b^i_j, s_j, Y_j).$$

In the considered case, the principle of development means that the society manages its resources in the best way to secure the fastest growth.

When investments are known, dynamic Eqs. (4.48) and (4.49) determine a unique trajectory: the trajectory of the largest potential growth. One can imagine that the trajectory of evolution will be sought by numerical methods, while the maximisation problem (4.52) and (4.53) has to be solved by standard methods of linear programming in every step of the solution of the Cauchy problem.

#### 4.3.3.2 The Balanced Growth

Instead of the described procedure for calculating the investment, other procedures can be invented. In the simplest way, the potential investment can be specified as an investment of homothetic trajectory. In this case, the growth rate of all sectors are equal to the growth rate of the entire system, and the potential investment, due to Eqs. (4.39) and (4.48), can be written as

$$I = \sum_{i,j=1}^{n} s_j \, \xi_j^i \, K^i, \quad \xi_j^i = \frac{\delta_j^i - a_j^i}{b^i}.$$

This allows one to define the potential growth rate of the gross output in any sector $l$ as

$$\tilde{\delta}^l = -\mu + \frac{1}{K} \sum_{i,j=1}^{n} s_j \, \xi_j^i \, K^i, \quad l = 1, 2, \ldots, n \tag{4.54}$$

#### 4.3.3.3 Final Remarks

Thus, a procedure for calculating investment can be invented, but in any case, it is assumed that total personal consumption of the product $C_j$ and the investments in stored intermediate and fundamental products $G_j$ are given; in this case, the investments, defined by that or this way on the basis of the recorded equations, ought to be considered as *potential investments* $\tilde{I}^i$. To find actual investments, we have to take into account other restrictions for progress of the system. The availability of labour and energy impose some strong restrictions on the development of the production system and should be included in the theory. In any case, actual investments do not exceed potential investments, and actual amounts of final personal consumption and streams of intermediate products appear to be consequences of calculation of progress of the system. We shall discuss the problem of real investment in the following chapters and return to the many-sector model in Chap. 9.

## 4.4 Enterprise and Basic Technological Processes

From a microeconomic point of view, a production system consists of numerous enterprises, each of them including one or more *basic technological processes*. The latter can be considered as *atoms* of the production system. Methods to describe the resources of an enterprise consisting of a finite or an infinite set of basic technological processes were proposed by von Neumann [6] and Gale [9], respectively. We shall consider a finite set of basic technological processes, namely, the von Neumann model of enterprise.

Every enterprise produces one or several products, and it consumes some products. In other words, the enterprise transforms a set of the input products $x_j, x_i, \ldots$, where the labels of products $j,\ i, \ldots$ are fixed, into an output set: $x_l, x_m, \ldots$, where the labels of products $l,\ m, \ldots$ are also fixed. It is convenient to introduce the input and output vectors $\mathsf{u}$ and $\mathsf{v}$ with non-negative components $u_k$ and $v_k$, $k = 1, 2, \ldots, n$. Some of the components are equal to zero.

We assume that the components of the vectors are measured in units of value, so the final output of the enterprise is

$$y = \sum_{j=1}^{n} (v_j - u_j). \tag{4.55}$$

This is a contribution of the enterprise to the gross national product.

A couple of vectors ($\mathsf{u}$,$\mathsf{v}$) characterise the technology used in production or, to put it differently, the technological process is represented as a couple of vectors ($\mathsf{u}$,$\mathsf{v}$). The technological process can be depicted as a point in Euclidean space of dimension $2n$ (Fig. 4.1).

One can assume that there are basic technological processes, each of which cannot be divided or changed, for example, a car assembly line, the characteristics of which are constant. The only action the manager can perform is to switch the line on and off. One can assume that the enterprise consists of a set of basic technological processes, which can be used in different combinations, so that the technological process of the enterprise can be represented as an expansion over the basic technological processes

$$(\mathsf{u}, \mathsf{v}) = \sum_{j=1}^{r} (\mathsf{a}^j, \mathsf{h}^j) z_j, \tag{4.56}$$

where $z_j \geq 0$ is the intensity of the use of the basic technological process labelled $j$.

For given basic technological processes and an arbitrary vector $\mathsf{z} = (z_1, z_2, \ldots, z_r)$, relation (4.56) determines a set of possible technological processes of the enterprise, that is, a technological set that is a specific characteristic of the enterprise. The aim of the enterprise and of its investigator is to find a value of $\mathsf{z}$ such that the final output (4.55) takes the greatest value.

It is convenient to use the following representation for the components of the input and output vectors

$$u_i = \sum_{j=1}^{r} a_i^j z_j, \quad v_i = \sum_{j=1}^{r} h_i^j z_j, \tag{4.57}$$

where $a_i^j$ and $h_i^j$ are components of input and output matrices $\mathsf{A}$ and $\mathsf{H}$, respectively. One can say that the von Neumann model is given if matrices $\mathsf{A}$ and $\mathsf{H}$ are given.

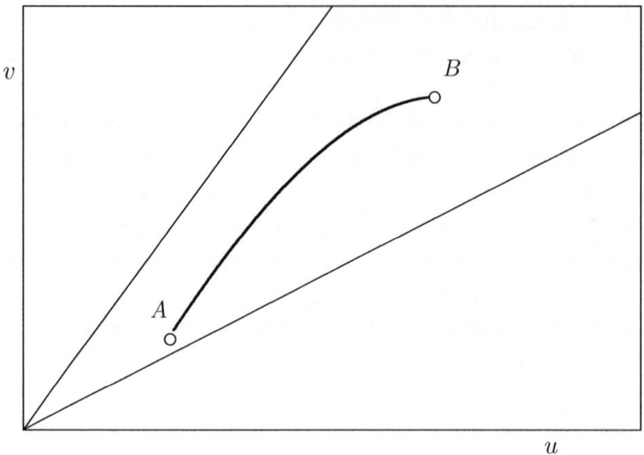


**Fig. 4.1** Input-output space. All trajectories are situated (similar to the trajectory *AB*) inside the sector restricted by straight lines which represent basic technological processes

Then, the final output of the enterprise can be written as

$$y(\mathsf{z}) = \sum_{i=1}^{n} \sum_{j=1}^{r} (h_i^j - a_i^j) z_j. \tag{4.58}$$

Considering the dynamics of the process, we are looking for the greatest growth of output (4.58). The problem can be reduced to the problem considered in Sect. 4.3.2. For this aim, one can introduce a potential output-input ratio

$$\alpha(\mathsf{z}) = \min_{j=1,2,\ldots,n;\ v_i \neq 0} \frac{v_i}{u_i} \tag{4.59}$$

which depends on the intensity $\mathsf{z} = (z_1, z_2, \ldots, z_r)$.

Because the final output is non-negative, it follows from relation (4.58) that

$$\mathsf{Az} \leq \mathsf{Hz}$$

or, using definition (4.59),

$$\alpha \mathsf{Az} \leq \mathsf{Hz}. \tag{4.60}$$

It remains to find the greatest rate of growth $\alpha$. One can see that Eq. (4.60) is quite similar to Eq. (4.46) from Sect. 4.3.2.

If the basic technological processes are unchanged during time, vector $\mathsf{z} = constant$ and the production trajectory is a straight line in technological space—the von Neumann ray (Fig. 4.1). Otherwise, the problem appears more complicated one. One can assume that intensity $\mathsf{z}$ is a function of time, so that the von Neumann model can describe complex dynamic processes or, in geometrical terms, the complex trajectory in input-output space [7]. Hundreds of papers have been written about the properties of von Neumann-type growth models, the golden rules of capital accumulation, turnpike theorems and so on. However, all this is concerned with potential trajectories of a production system.

## References


1. Leontief, W.W.: Quantitative input and output relations in the economic system of the United States. Rev. Econ. Stat. **18**, 105–125 (1936)
2. Leontief, W.W.: The Structure of the American Economy 1919–1939. Harvard University Press, Cambridge (1941)
3. Leontief, W.W.: Input-Output Economics, 2nd edn. Oxford University Press, New York (1986)
4. Sraffa, P.: Production of Commodities by Means of Commodities: Prelude to a Critique of Economic Theory. Cambridge University Press, Cambridge (1975)
5. Korn, G.A., Korn, T.M.: Mathematical Handbook for Scientists and Engineers. McGraw-Hill, New York (1968)
6. Von Neumann, J.: Über ein okonomisches Gleichungssystem und eine Verallgemeinerung des brouwerschen Fixpunktsatzes. Ergebnisse eines Mathematischen Kolloqueiums **8**, 73–83 (1937)
7. Makarov, V.L., Rubinov, A.M.: Matematicheskaya teoriya eklnomicheskoy dinamiki i ravnovesiya (The Mathematical Theory of Economic Dynamics and Equilibrium). Nauka, Moscow (1973)
8. Cheremnykh, Y.N.: Analiz povedenija trajectorii dinamiki narodno-khozjaistvennykh modelei (Analysis of Trajectories of Economic Models). Nauka, Moscow (1982)
9. Gale, D.: The Theory of Linear Economic Models. McGraw-Hill, Auckland (1960)

# Chapter 5
# Production Factors and Technology

**Abstract** The description of production processes in the previous chapter assumes that something can be made from something else and that there are tools for making things. Now we are going to look at the production process from another side, considering efforts of humans and work of production equipment. We are introducing two sets of quantities, which, in the macroeconomic approach, reflect the level of technology in the economy: (1) the technological coefficients $\lambda$ and $\varepsilon$ that show how much labour and substitutive work are needed to introduce a unit of investment in the production system, and (2) the potential rates of growth of production factors $\tilde{\nu}$ and $\tilde{\eta}$. The first set of quantities includes inherent characteristics of production equipment, whereas the second set, in this chapter, will be considered as a set of exogenous factors. The technological coefficients appear to be appropriate and convenient characteristics of technology used in an economy and can be easily estimated considering the performance of production equipment.

## 5.1 Dynamics of Production Factors

To describe the process of production in a proper way, one should to take into account fixed production equipment (fixed production capital stock) and two factors: manpower and the work of production equipment substituting manpower. We know that it is a good approximation to consider the production system as consisting of $n$ sectors, while each of them is characterised by the basic production equipment $K^i$ of the sector, which has to be animated by the corresponding production factors $L^i$ and $P^i$. In this chapter, we proceed considering the production system as a whole, or, in other words, as consisting of only one sector that is characterised by aggregate amounts of production factors

$$K = \sum_{i=1}^{n} K^i, \quad L = \sum_{i=1}^{n} L^i, \quad P = \sum_{i=1}^{n} P^i. \tag{5.1}$$



V. N. Pokrovskii, *Econodynamics*, New Economic Windows,
https://doi.org/10.1007/978-3-319-72074-6_5

### 5.1.1 Dynamics of Production Capital

To develop a proper description of a technological process, it is necessary to take into account that equipment with a variety of technological characteristics is introduced into the production process at different moments in time. Therefore, the age of existing production capital is various and, similar to formula (2.19), we can write an expression for the structure of fixed production capital

$$K(t) = \int_{-\infty}^{t} k(t,s)ds, \tag{5.2}$$

where $k(t,s)$ is the part of the fixed capital existing at time $t$ which was introduced at time $s$ during a unit of time. This part of the capital stock is depreciating according to the law

$$\frac{\partial k(t,s)}{\partial t} = -\mu(t,s)k(t,s) \tag{5.3}$$

from the initial amount

$$k(s,s) = I(s). \tag{5.4}$$

Equation (5.3) can be integrated with respect to $s$, which gives an expression

$$\frac{dK}{dt} = I - \int_{-\infty}^{t} \mu(t,s)k(t,s)ds. \tag{5.5}$$

One can introduce an effective coefficient of depreciation

$$\mu(t) = \frac{1}{K}\int_{-\infty}^{t} \mu(t,s)k(t,s)ds$$

to rewrite Eq. (5.5) in the conventional [1] form

$$\frac{dK}{dt} = I - \mu K. \tag{5.6}$$

One can see that a solution of this equation, if $\mu = const$, can be written, in accordance with (5.2), as

$$K(t) = \int_{-\infty}^{t} k(t,s)ds, \quad k(t,s) = e^{-\mu(t-s)}I(s). \tag{5.7}$$

### 5.1.2 Dynamics of Substitutive Work and Labour

The existing technology determines how many workers' efforts and how much extra energy is needed for the production mechanism to be in action; therefore, the dynamic equations for labour and substitutive work must contain technological characteristics of capital equipment. Further, we reconsider derivation developed in works [2, 3].

Similar to the presentation of capital (5.2), the amounts of production factors can be represented by the formulae

$$L(t) = \int_{-\infty}^{t} l(t,s)ds, \quad P(t) = \int_{-\infty}^{t} e(t,s)ds, \tag{5.8}$$

where $l(t,s)$ and $e(t,s)$ are the labour and substitutive work which are necessary for part of the fixed capital $k(t,s)$ to be in action, so the following relations can be written:

$$l(t,s) = \lambda(t,s)k(t,s), \quad e(t,s) = \varepsilon(t,s)k(t,s). \tag{5.9}$$

From Eq. (5.4), one has the relations

$$\begin{aligned} l(s,s) &= \lambda(s)I(s), \quad \lambda(s) = \lambda(s,s), \\ e(s,s) &= \varepsilon(s)I(s), \quad \varepsilon(s) = \varepsilon(s,s), \end{aligned} \tag{5.10}$$

where $I(s)$ is the gross investment at time $s$.

The coefficients $\lambda(s) > 0$ and $\varepsilon(s) > 0$ determine the required amount of labour and substitutive work, delivered by external energy sources, per unit of increase in capital; therefore, they can be denominated as labour requirement and energy requirement, respectively. The values of these coefficients are determined by the applied technology, and we can call them the technological coefficients.

One can see that, from relation (5.3), equations for the dynamics of quantities (5.9) can be written as follows:

$$\begin{aligned} \frac{\partial l(t,s)}{\partial t} &= -\mu_{\mathsf{L}}(t,s)l(t,s), \quad \mu_{\mathsf{L}}(t,s) = \mu(t,s) - \frac{1}{\lambda(t,s)}\frac{\partial \lambda(t,s)}{\partial t}, \\ \frac{\partial e(t,s)}{\partial t} &= -\mu_{\mathsf{P}}(t,s)e(t,s), \quad \mu_{\mathsf{P}}(t,s) = \mu(t,s) - \frac{1}{\varepsilon(t,s)}\frac{\partial \varepsilon(t,s)}{\partial t}. \end{aligned} \tag{5.11}$$

The last terms in the definitions of the depreciation coefficients $\mu_{\mathsf{L}}(t,s)$ and $\mu_{\mathsf{P}}(t,s)$ are connected with the change of quality of the production equipment after it has been installed at time $s$.

One can use definitions (5.8) to determine the required amount of the production factors,

$$\frac{dL}{dt} = \lambda I - \int_{-\infty}^{t} \mu_{\mathrm{L}}(t,s) l(t,s) ds, \quad \frac{dP}{dt} = \varepsilon I - \int_{-\infty}^{t} \mu_{\mathrm{P}}(t,s) e(t,s) ds. \quad (5.12)$$

The first terms on the right side of these relationships describe the increase in the quantities of interest, caused by gross investments $I$; the second terms reflect the decrease of the corresponding quantities due to both the change of the quality of production equipment after installation and the removal of a part of the production equipment from the service.

We can rewrite Eq. (5.12), introducing a special notation for the last parts of the equations, as

$$\frac{dL}{dt} = \lambda I - (\mu + \nu')L, \quad \frac{dP}{dt} = \varepsilon I - (\mu + \eta')P. \quad (5.13)$$

The quantities $\mu + \nu'$ and $\mu + \eta'$ are effective depreciation coefficients of the production factors. If, for example, the installed technological equipment requires more labour during ageing, then $\nu' < 0$, which means a decrease in the effective depreciation coefficient. If the technological equipment does not change its quality over time, that is the technological coefficients in Eq. (5.9) do not depend on the argument $t$, the quantities $\nu' = 0$ and $\eta' = 0$ and all the depreciation coefficients in Eqs. (5.6) and (5.13) will appear to be the same. The quantities $\nu'$ and $\eta'$ could be proper corrections for the growth rates of labour and substitutive work, if any discrepancy in empirical data is observed. These quantities could also compensate possible errors.

## 5.2 Macroeconomic Characteristics of Production Equipment

Technology is commonly understood as methods of production, so characteristics of tools, machines, materials, techniques and sources of power are needed to describe the technology of production of useful things. Below we consider the introduced phenomenological characteristics of production equipment.

### *5.2.1 Technological Coefficients*

To discuss the meaning and properties of the technological coefficients $\lambda(t)$ and $\varepsilon(t)$ in the dynamic equation (5.13), one can neglect the quantities $\nu'$ and $\eta'$ for simplicity. It is easy to see from dynamic equations (5.6) and (5.13) that the constant technological coefficients can be expressed as

$$\lambda = \frac{L}{K}, \quad \varepsilon = \frac{P}{K}.$$

It is convenient to introduce the non-dimensional technological quantities

$$\overline{\lambda} = \lambda \frac{K}{L}, \qquad \overline{\varepsilon} = \varepsilon \frac{K}{P}, \tag{5.14}$$

so that at constant technology $\overline{\lambda} = 1$ and $\overline{\varepsilon} = 1$.

The requirement of positivity of the marginal productivities (Chap. 6, Sect. 6.1) puts some restrictions (formula 6.10) on the values of the technological coefficients, namely,

$$\overline{\lambda} < 1 < \overline{\varepsilon} \quad or \quad \overline{\lambda} > 1 > \overline{\varepsilon}. \tag{5.15}$$

Apart from this, it is natural to consider the technological coefficients to be non-negative; one can hardly imagine a situation where one of them would be negative.

The technological coefficients play an essential role in the description of the production system; they ought to be considered as independent variables in the set of equations of evolution of the system. Variation of technological coefficients specifies variation in consumption of workers' efforts and substitutive work. A case, for example, when

$$\lambda < \frac{L}{K}, \qquad \varepsilon < \frac{P}{K}$$

means, that labour-saving and energy-saving technologies are being introduced. Indeed, the Eq. (5.13) allows us to estimate reduction of consumption of the production factors at introduction of new technologies. For example, at variation of technological coefficient from initial value 1 up to smaller number $\overline{\lambda}$ the reduction rate of consumption of workers' efforts is specified as

$$\frac{1}{L}\frac{d(L_0 - L)}{dt} = (1 - \overline{\lambda})\frac{I}{K} \tag{5.16}$$

Note that the quantities reciprocal to the technological coefficients can be interpreted as the cost of the equipment needed to introduce a unit of labour or a unit of substitutive work into the production process or, in other words, *the prices of introduction* of the corresponding production factors

$$\frac{1}{\lambda} = \frac{K}{\overline{\lambda} L}, \quad \frac{1}{\varepsilon} = \frac{K}{\overline{\varepsilon} P}.$$

For the unchangeable technology, the costs of introduction are

$$\frac{1}{\lambda} = \frac{K}{L}, \quad \frac{1}{\varepsilon} = \frac{K}{P}.$$

It is essential to note that the price of introduction of substitutive work as a production factor is different from the price of use of substitutive work as a production factor, defined by Eq. (2.35). The latter is not the price of an energy career with the corresponding energy content, but the price of equipment (capital) which allows energy to be converted usefully into other forms.

### 5.2.2 Technological Index

Equation (5.13) can be considered as relations, which determine a demand of the production factors for a given technology and investment. The values of investment and production factors cannot be quite arbitrary and have to correspond to the values of technological coefficients, which are subjects of the restrictions (5.15).

One can consider the quantities $\frac{1}{L}\frac{dL}{dt} + \nu'$ and $\frac{1}{P}\frac{dP}{dt} + \eta'$ in Eqs. (5.6) and (5.13) to be the effective growth rates of labour and substitutive work. It is convenient to introduce special notation for these quantities

$$\delta = \frac{1}{K}\frac{dK}{dt}, \quad \nu = \nu' + \frac{1}{L}\frac{dL}{dt}, \quad \eta = \eta' + \frac{1}{P}\frac{dP}{dt}. \tag{5.17}$$

This allows us to rewrite Eqs. (5.6) and (5.13) in the following form

$$\delta = \frac{I}{K} - \mu, \qquad \nu = \overline{\lambda}\frac{I}{K} - \mu, \qquad \eta = \overline{\varepsilon}\frac{I}{K} - \mu, \tag{5.18}$$

where the symbols for the non-dimensional technological quantities (5.14) are used.

Equation (5.18) describe exact relations between the effective rates of growth of production factors $\delta(t)$, $\nu(t)$ and $\eta(t)$ and technological coefficients in the form

$$\overline{\lambda} = \frac{\nu + \mu}{\delta + \mu}, \qquad \overline{\varepsilon} = \frac{\eta + \mu}{\delta + \mu}. \tag{5.19}$$

The depreciation coefficient $\mu$ can be excluded from relations (5.19). Therefore, we can obtain a relation between the different effective rates of growth,

$$\delta = \nu + \alpha(\eta - \nu), \quad \alpha = \frac{1 - \overline{\lambda}}{\overline{\varepsilon} - \overline{\lambda}}. \tag{5.20}$$

The technological index $\alpha$ is considered as a given quantity.

The condition of productivity of the production factors (5.15) imposes a certain restriction on the values of the technological index,

$$0 < \alpha < 1. \tag{5.21}$$

Moreover, some other estimates of this quantity can be made. It will be shown in Chap. 6 that the technological index $\alpha$ has the meaning of the share of expenses for maintenance of substitutive work as a production factor in the total expenses for maintenance of production factors (labour $L$ and substitutive work $P$).

Let us recall that the quantities $\nu$ and $\eta$ represent the effective growth rates of production factors and can coincide with the rates of real growth, if corrections $\nu'$ and $\eta'$ can be neglected.

## 5.3 Investment and Dynamics of Technology

### 5.3.1 *Investment and Three Modes of Development*

To find out investment, one must take into account the restriction imposed by internal (the limiting output and necessary level of consumption) and external circumstances. According to Eq. (4.54), if no restrictions are imposed by the availability of labour and substitutive work, the potential growth rate of capital can be written in terms of a many-sector model as

$$\tilde{\delta} = -\mu + \frac{1}{K} \sum_{j,l=1}^{n} s_j K^l \xi_j^l. \tag{5.22}$$

Because the quantity $s_j$, which is a share of the investment product in the final product of the sector $j$, changes from 0 to 1 in every sector, the rate of potential growth of investment $\tilde{\delta}$ is restricted by the inequalities

$$0 < \tilde{\delta} < \frac{Y}{K} - \mu.$$

The other restrictions emerge from the non-availability of other production factors. One can assume here that there are external sources of labour and energy, so that the amounts of available labour $\tilde{L}$ and substitutive work $\tilde{P}$ are known. It is convenient to consider them solutions of the equations

$$\frac{d\tilde{L}}{dt} = \tilde{\nu}\,\tilde{L}, \quad \frac{d\tilde{P}}{dt} = \tilde{\eta}\,\tilde{P} \tag{5.23}$$

Though the rates of potential growth $\tilde{\nu}$ and $\tilde{\eta}$ can be, in principle, calculated as was discussed in Chap. 2 (see Sects. 2.4.2 and 2.5.5), later on they are assumed to be given as functions of time.

Thus, we assume the rates of potential growth of production factors to be known as functions of time,

$$\tilde{\delta} = \tilde{\delta}(t), \qquad \tilde{\nu} = \tilde{\nu}(t), \qquad \tilde{\eta} = \tilde{\eta}(t).$$

In any case, the effective rates of growth $\delta$, $\nu$ and $\eta$, defined by Eq. (5.18), do not exceed the rates of potential growth $\tilde{\delta}$, $\tilde{\nu}$ and $\tilde{\eta}$, that is,

$$\delta \leq \tilde{\delta}, \quad \nu \leq \tilde{\nu}, \quad \eta \leq \tilde{\eta}.$$

This determines the restrictions for investments in the production system

$$I \leq (\mu + \tilde{\delta})K, \quad I \leq \frac{\mu + \tilde{\nu}}{\lambda} L, \quad I \leq \frac{\mu + \tilde{\eta}}{\varepsilon} P. \tag{5.24}$$

The real investments are determined by a competition between potential investments from one side and labour and energy supplies from the other side. One can assume that the production system tries to swallow up all available production factors. In this case, for investments we should write

$$I = (\delta + \mu)K = \min \begin{cases} (\tilde{\delta} + \mu)K \\ (\tilde{\nu} + \mu)K/\overline{\lambda} \\ (\tilde{\eta} + \mu)K/\overline{\varepsilon} \end{cases}. \tag{5.25}$$

The effective rates of growth of production factors $\delta$, $\nu$ and $\eta$ are different from the rates of potential growth. According to the three lines of relation (5.25), one can define three modes of economic development for which we have different formulae for calculation. From Eqs. (5.18) and (5.25), the effective rates of growth can be calculated for the three modes as

$$\begin{array}{lll} \delta = \tilde{\delta}, & \nu = (\tilde{\delta} + \mu)\overline{\lambda} - \mu, & \eta = (\tilde{\delta} + \mu)\overline{\varepsilon} - \mu, \\ \delta = (\tilde{\nu} + \mu)\frac{1}{\overline{\lambda}} - \mu, & \nu = \tilde{\nu}, & \eta = (\tilde{\nu} + \mu)\frac{\overline{\varepsilon}}{\overline{\lambda}} - \mu, \\ \delta = (\tilde{\eta} + \mu)\frac{1}{\overline{\varepsilon}} - \mu, & \nu = (\tilde{\eta} + \mu)\frac{\overline{\lambda}}{\overline{\varepsilon}} - \mu, & \eta = \tilde{\eta}. \end{array} \tag{5.26}$$

The first set of equations is valid in the case of lack of investment, and abundance of labour, substitutive work and raw materials. The second line is valid in the case of lack of labour, and abundance of investment, substitutive work and raw materials. The last line of equations is valid in the case of lack of substitutive work, and abundance of investment, labour and raw materials.

### 5.3.2 Unemployment and Principle of Development

According to the preceding speculation, the rates of real growth of production factors are not bigger than the potential ones. If, indeed, the production system is trying to devour all available production factors, the growth of one of the production factors coincides with potential growth. This means that the gaps between real and potential amounts of production factors, for example, a gap between labour supply $\tilde{L}$ and labour demand $L$, will increase. Therefore, one can see that the index of unemployment $u = (\tilde{L} - L)/\tilde{L}$, for example, cannot shrink in a 'natural' way. Considering the situation in a one-sector approach, in order to decrease the gaps between real and potential amounts of production factors, an intervention in the form of governmental investments is needed, and to take it into account, one can rewrite relation (5.26) in the form

$$I = (\delta + \mu)K = \chi(u)\,K + \min \begin{cases} (\tilde{\delta} + \mu)K \\ (\tilde{\nu} + \mu)K/\overline{\lambda}\ , \\ (\tilde{\eta} + \mu)K/\overline{\varepsilon} \end{cases} \tag{5.27}$$

where quantity $\chi(u)$ is designed to regulate the gaps between supply and demand of production factors. One has to reserve some amount of products for investments to be regulated. In this case, three modes of economic development exist as well. In the many-sector approach, the intervention is defined in similar way, but the interpretation can be different (see Sect. 9.1.2).

### 5.3.3 Dynamics of Technological Coefficients

To find out the law of evolution of the technological coefficients in time, one has to consider again the restrictions on investments. According to the three modes defined by relations (5.26), one can obtain three sets of relations for the non-dimensional technological quantities $\overline{\lambda}$, $\overline{\varepsilon}$ and their ratio $\Theta = \overline{\varepsilon}/\overline{\lambda}$

$$1 = \frac{\tilde{\delta} + \mu}{\delta + \mu}, \quad \overline{\lambda} \leq \frac{\tilde{\nu} + \mu}{\delta + \mu}, \quad \overline{\lambda} \leq \frac{\tilde{\nu} + \mu}{\tilde{\delta} + \mu}, \quad \overline{\varepsilon} \leq \frac{\tilde{\eta} + \mu}{\delta + \mu}, \quad \overline{\varepsilon} \leq \frac{\tilde{\eta} + \mu}{\tilde{\delta} + \mu},$$

$$1 \leq \frac{\tilde{\delta} + \mu}{\delta + \mu}, \quad \overline{\lambda} = \frac{\tilde{\nu} + \mu}{\delta + \mu}, \quad \overline{\lambda} \geq \frac{\tilde{\nu} + \mu}{\tilde{\delta} + \mu}, \quad \overline{\varepsilon} \leq \frac{\tilde{\eta} + \mu}{\delta + \mu}, \quad \Theta \leq \frac{\tilde{\eta} + \mu}{\tilde{\nu} + \mu},$$

$$1 \leq \frac{\tilde{\delta} + \mu}{\delta + \mu}, \quad \overline{\lambda} \leq \frac{\tilde{\nu} + \mu}{\delta + \mu}, \quad \overline{\varepsilon} = \frac{\tilde{\eta} + \mu}{\delta + \mu}, \quad \overline{\varepsilon} \geq \frac{\tilde{\eta} + \mu}{\tilde{\delta} + \mu}, \quad \Theta \geq \frac{\tilde{\eta} + \mu}{\tilde{\nu} + \mu}. \tag{5.28}$$

In the first case, there is an internal restriction to growth. In the last two lines, one of the production factors, $L$ or $P$, is limited.

One can assume that there are internal technological changes, which lead to alteration of the technological coefficients, whereas the production system tries to use all available resources. This means that the technological coefficients have tendencies to change such that the inequalities in conditions (5.28) are trending to turn into equalities. These processes are connected with the invention of new technologies and propagation of known ones.

One can consider the rates of growth

$$\frac{d\Theta}{dt}, \quad \frac{d\overline{\lambda}}{dt}, \quad \frac{d\overline{\varepsilon}}{dt}$$

to be functions of differences

$$\Theta - \frac{\tilde{\eta}+\mu}{\tilde{\nu}+\mu}, \quad \overline{\lambda} - \frac{\tilde{\nu}+\mu}{\tilde{\delta}+\mu}, \quad \overline{\varepsilon} - \frac{\tilde{\eta}+\mu}{\tilde{\delta}+\mu}.$$

In the first approximation, the tendencies to changes can be described by equations for the quantities

$$\frac{d\Theta}{dt} = -\frac{1}{\tau_\theta}\left(\Theta - \frac{\tilde{\eta}+\mu}{\tilde{\nu}+\mu}\right), \tag{5.29}$$

$$\frac{d\overline{\lambda}}{dt} = -\frac{1}{\tau_\lambda}\left(\overline{\lambda} - \frac{\tilde{\nu}+\mu}{\tilde{\delta}+\mu}\right), \tag{5.30}$$

$$\frac{d\overline{\varepsilon}}{dt} = -\frac{1}{\tau_\varepsilon}\left(\overline{\varepsilon} - \frac{\tilde{\eta}+\mu}{\tilde{\delta}+\mu}\right). \tag{5.31}$$

So as $\Theta = \overline{\varepsilon}/\overline{\lambda}$, only two of Eqs. (5.29)–(5.31) are independent. For Eqs. (5.29)–(5.31) to be consistent, relaxation times $\tau_\lambda$ and $\tau_\varepsilon$ should be equated and to be connected with relaxation time $\tau_\theta$ as follows:

$$\tau_\lambda = \tau_\varepsilon = \frac{1}{\overline{\lambda}}\frac{\tilde{\nu}+\mu}{\tilde{\delta}+\mu}\tau_\theta.$$

Equations (5.29)–(5.31) are the first-order approximations to the relaxation equations with respect to the quantities in the brackets, so one should take the zero-order terms of the relaxation times in these equations. This means that, in the considered approximation, all relaxation times are equal to each other, namely,

$$\tau_\lambda = \tau_\varepsilon = \tau_\theta = \tau, \tag{5.32}$$

so that the subscripts to the relaxation times in Eqs. (5.29)–(5.31) can be omitted in the following exposition. The meaning of $\tau$ is the time of crossover from one technological situation to another, when external parameters $\tilde{\nu}$ and $\tilde{\eta}$ change. It is determined by internal processes of developing and attracting of the proper technology.

It is easy to see that, if the rates of growth are constant, Eq. (5.29), for example, at initial value

$$\Theta(0) = (1-\Delta)\frac{\tilde{\eta}+\mu}{\tilde{\nu}+\mu}$$

has a simple solution

$$\Theta(t) = \frac{\tilde{\eta} + \mu}{\tilde{\nu} + \mu} \left[ 1 - \Delta \exp\left( -\frac{t}{\tau} \right) \right]. \tag{5.33}$$

### 5.3.4 *Dynamics of the Technological Index*

Now we can directly calculate changes of the technological index

$$\alpha = \frac{1 - \overline{\lambda}}{\overline{\varepsilon} - \overline{\lambda}}.$$

After differentiating the quantity and making use of Eqs. (5.30) and (5.31), one gets the relation

$$\frac{d\alpha}{dt} = \frac{\tilde{\delta} - \tilde{\nu} - \alpha\,(\tilde{\eta} - \tilde{\nu})}{\tau(\overline{\varepsilon} - \overline{\lambda})(\tilde{\delta} + \mu)} \tag{5.34}$$

To specify the change of the technological index, we have to compare the rate of capital potential growth $\tilde{\delta}$ with the combination of the rates of potential growth of the other factors $\tilde{\nu} + \alpha\,(\tilde{\eta} - \tilde{\nu})$. If the first quantity is bigger than the second, the technological index grows. One can assume that, in a steady-state situation, there is a relation

$$\tilde{\delta} = \tilde{\nu} + \alpha\,(\tilde{\eta} - \tilde{\nu}), \tag{5.35}$$

which is similar to relation (5.20) for the rates of real growth of production factors. In this case, the technological index $\alpha$ appears to be constant during evolution, in other words, the technological index appears to be the first integral of evolution. One can expect that there are social mechanisms which ensure the validity of Eq. (5.35). Of course, this equation should be considered as an approximate equality, which can be violated by disturbances in the social life.

## 5.4 Evolution of Production System

The recorded parities allow to formulate a system of equations describing evolution of characteristics of the production system, which in this chapter is considered in the most rough approximation as a sole sector. It is natural, that such description is possible to be formulated only at some simplifications and assumptions which have been described earlier in this chapter. Then, we shall discuss the results of application of the description (the system of equations) for reconstruction of the development of the U.S.A. economy to estimate adequacy of our previous speculation.

### 5.4.1 The System of Equations

First of all, we are collecting together the equations for the production factors, that is, Eqs. (5.6), (5.13), (5.25), (5.30) and (5.31)

$$\frac{dK}{dt} = I - \mu K, \quad \frac{dL}{dt} = \left(\overline{\lambda}\,\frac{I}{K} - \mu\right) L, \quad \frac{dP}{dt} = \left(\overline{\varepsilon}\,\frac{I}{K} - \mu\right) P,$$

$$\frac{I}{K} = \min\left\{(\tilde{\delta} + \mu), \quad (\tilde{\nu} + \mu)\frac{1}{\overline{\lambda}}, \quad (\tilde{\eta} + \mu)\frac{1}{\overline{\varepsilon}}\right\},$$

$$\frac{d\overline{\lambda}}{dt} = -\frac{1}{\tau}\left(\overline{\lambda} - \frac{\tilde{\nu} + \mu}{\tilde{\delta} + \mu}\right), \quad \frac{d\overline{\varepsilon}}{dt} = -\frac{1}{\tau}\left(\overline{\varepsilon} - \frac{\tilde{\eta} + \mu}{\tilde{\delta} + \mu}\right), \quad \alpha = \frac{1 - \overline{\lambda}}{\overline{\varepsilon} - \overline{\lambda}} \tag{5.36}$$

This is a set of five independent differential equations for seven variables $K$, $L$, $P$, $I$, $\overline{\lambda}$, $\overline{\varepsilon}$ and $\alpha$, so one can choose five variables to be independent. It is assumed that the initial values of the five independent variables as well as the rates of potential growth of production factors

$$\tilde{\delta} = \tilde{\delta}(t), \quad \tilde{\nu} = \tilde{\nu}(t), \quad \tilde{\eta} = \tilde{\eta}(t)$$

together with the time of crossover from one technological situation to another $\tau$ and the coefficient of amortisation $\mu$ are given. These equations allow us to analyse the evolution of an economic system.

In the general case, the system (5.36) describes development of the production system at the rates of potential growth of the capital, expenditures of labour and substitutive work, $\tilde{\delta}$, $\tilde{\nu}$ and $\tilde{\eta}$, as functions of time. At the given initial values of the variables, the problem reduces to a Cauchy problem, which can be solved by numerical methods. However, the greatest difficulty is that the rates of potential growth remain unknown and should be the object of special research.

### 5.4.2 Dynamics of Development

To test the applicability of the theory to reality, we have considered [3] the development of the United States economy for the period 1900–2000 (see Appendix B). Scenarios of development can be obtained, if one sets the rates of potential growth of capital, labour and substitutive work $\tilde{\delta}$, $\tilde{\nu}$ and $\tilde{\eta}$ as function of time. Considering the past development, the rates of potential growth can be estimated empirically. Before the year 2000, the rates of potential growth of labour and productive energy $\tilde{\nu}$ and $\tilde{\eta}$ are taken to be a little bit more than the rates of real growth for such amounts, that the calculated trajectories of evolution of the production factors $L$ and $P$ would

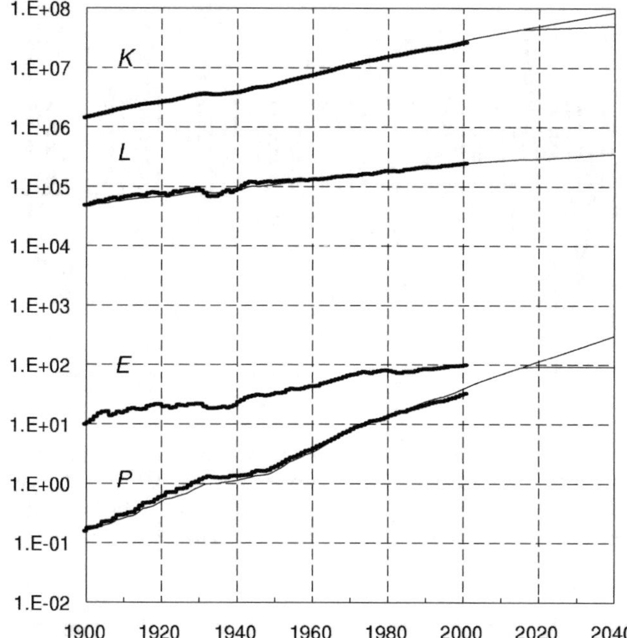


**Fig. 5.1** Production factors in the U.S. economy. The basic production equipment (capital stock) $K$ in million dollars for year 1996; consumption of labour $L$ in million man-hours per year; primary energy $E$ and substitutive work $P$ in quads (1 quad $= 10^{18}$ J) per year. The thick solid lines represent empirical values, while the thin lines show the results of calculation. Adapted from [3]

correspond to the empirical values of production factors $K$, $L$ and $P$. This allows us to determine trajectories of the other variables.

One can see on the chart of Fig. 5.1 that the calculated trajectory for production factors approximates the real time dependence for the period 1900–2000 (see Appendix B), though some details, for example, the behaviour in the turmoil years 1930–1940, are not reproduced correctly. Beyond the year 2000, we explore two scenarios of development that were calculated in year 2003 [3] to demonstrate effect of change in the substitutive work usage. In both cases, the rate of growth of labour $\tilde{\nu}$ coincides with the rate of population growth, namely, $\tilde{\nu} = 0.01$ for the U.S. The first scenario corresponds to the value $\tilde{\eta} = 0.05$ for all years. The second one shows the effect of diminishing the energy supply for the substitutive work in the economy: the value of $\tilde{\eta} = 0.05$ in year 2000 decreases to zero in year 2010. The thin lines in Fig. 5.1 show the results of calculation of the production factors at values of the depreciation coefficient $\mu$ calculated according to cited statistical data ($\mu \approx 0.02$ before year 1925 and increases from 0.026 to 0.068 over years 1925–2000) and time of technological rearrangement $\tau = 1$ year.

The initial values of all variables, apart from the technological variables, are known from empirical data. The initial values of the technological variables can

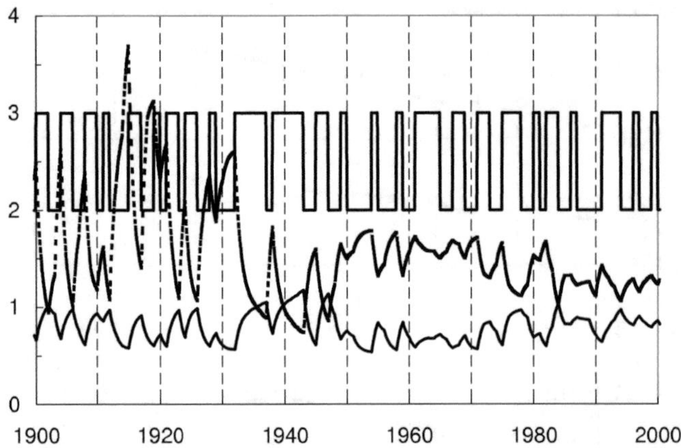


**Fig. 5.2** Technological coefficients in economy of the USA. The labour requirement (the bottom solid curve) and energy requirement (the dashed curve) oscillate with the period of pulsations about 4 years. The top step curve shows change of modes of development from the deficiency of human work (value 2) to the deficiency of substitutive work (value 3) and back

be chosen arbitrarily, because, due to the relaxation equations from the set (5.36), the initial values of the technological variables are being forgotten in $\tau \approx 1$ year. However, the choice of the technological variables must correspond to the value of the technological index $\alpha$.

The rates of potential growth for the past development are chosen such that calculated trajectories of the production factors conform with the actual ones, while the trajectories of the technological variables $\overline{\lambda}$ and $\overline{\varepsilon}$ are determined by the equations. The research detects the pulsation of the technological coefficients and change of mode of development (Fig. 5.2), which is connected with existence of various modes of functioning of the system of production that have been described in Sect. 5.3. In the considered period, in production system of the USA was realised the second and third case, and there is a change of modes of progress over the period of time about 4 years. The production proceeds under the abundance of investments and raw material, but under deficiency of workers' efforts, when $\frac{d\overline{\lambda}}{dt} < 0$, or under deficiency of substitutive work, when $\frac{d\overline{\lambda}}{dt} > 0$. The change of modes leads to that the rates of actual growth of production factors appear less than the corresponding rates of potential growth. The restrictions for growth are interchangeable between lack of humans' work and lack of substitutive work (but not investments). It is possible to assume that this is a typical situation for every economy.

One can see that the system of Eq. (5.36) allows us to draw scenarios of evolution of national economies for possible development of available production factors. However, the potential rates of growth should be considered as endogenous quantities in the problem of evolution of the human production system. This problem was discussed in Chap. 2, but there is no solution yet.

## 5.5 Mechanism of Evolution of Production System

In this section, we turn to the microeconomic approach and consider the production system to consist of numerous enterprises, each of them including one or more *technological processes*, as was discussed in Sect. 4.4. Each technological process consumes some products in order to output other ones. In other words, the enterprise transforms the input set of products $x_j, x_i, \ldots$, where the labels of products $j,\ i, \ldots$ are fixed, into an output set: $x_l, x_m, \ldots$, where the labels of products $l,\ m, \ldots$ are also fixed. This side of the technological process was described in Sect. 4.4. It is convenient to use, as before, the input and output vectors $\mathsf{u}$ and $\mathsf{v}$ with non-negative components $u_k$ and $v_k$, $k = 1, 2, \ldots, n$.

In addition, each technological process is characterised by equipment with value $K$ and the production factors: the labour $L$ and substitutive work $P$ needed to animate the production equipment. So, the technological process or the enterprise can be given by a set of five quantities

$$\mathsf{u},\ \mathsf{v},\ K,\ L,\ P$$

or, considering the scale of production as non-essential, by four quantities

$$\frac{\mathsf{u}}{K},\ \frac{\mathsf{v}}{K},\ r_1 = \frac{L}{K},\ r_2 = \frac{P}{K}. \tag{5.37}$$

The technological process can be depicted by a point in a many-dimensional space which consists of input-output space (Fig. 4.1) and production-factor space (Fig. 5.3).

According to Schumpeter [4], the mechanism of evolution of the production system can be imagined as the emergence, growth and disappearance of technological processes. An essential moment in this scheme is the emergence of a new enterprise, which can use a known technology or create a new one. In the first case, one speaks of diffusion of the known technology. The values of quantities (5.37) are the same for all enterprises with similar technology. One says that the production expands extensively. In the second case new technological processes appear. The new enterprise

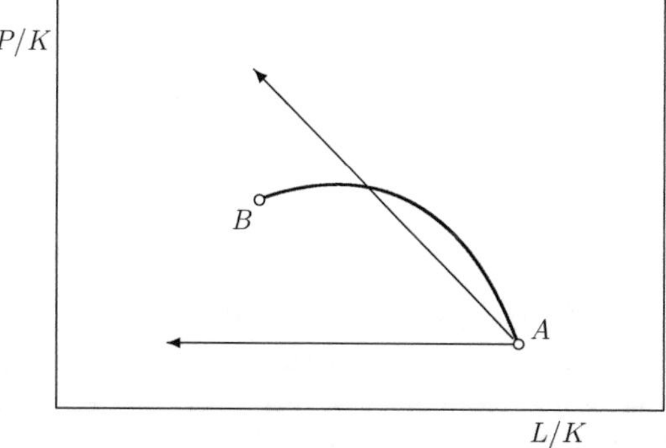


**Fig. 5.3** Space of production factors. The area inside the sector presents a set of more effective technologies in comparison with initial technology *A*

(technological process) produces either known products by new methods (process innovation), or completely new products (product innovation). In any case, the values of quantities (5.37) have been arising in new combinations. Some people take the risk of investigating new combinations of production factors and setting up a new technological process. The people running such businesses—the *neo-entrepreneurs* according to Schumpeter [4]—are central figures of economic development.

One can imagine a simple scheme of development of a production system consisting of many enterprises. We do not discuss stages of infancy, adolescence, maturity and senescence of the enterprise. For simplicity, we assume that a technological process remains unchanged until the moment of disappearance and consider elementary acts of evolution of production system to be emergence and reproduction of a new technological process. One can enumerate all technological processes according to the time of their emergence by a label $\alpha$ ($\alpha = 1, 2, \ldots$). An enterprise (technological process) with characteristics $K^{\alpha},\ L^{\alpha},\ P^{\alpha}$ emerges at moment $t^{\alpha}$ and disappears at moment $t^{\alpha}+\tau$. For simplicity, the time of existence of the enterprises $\tau$ is assumed to be equal for all enterprises. The quantities $K^{\alpha},\ L^{\alpha},\ P^{\alpha}$ and $t^{\alpha}$ are random ones, so *an emergence distribution function* should be introduced. It is convenient to use variables $t^{\alpha},\ r_1^{\alpha},\ r_2^{\alpha}$ ($\alpha = 1, 2, \ldots$) as arguments of the emergence function.

Then, one can define the technological progress in terms of microeconomic variables $r_1$ and $r_2$. The empirical data for the progressively developing U.S. economy, demonstrated with formulae (2.29), (2.30) and (2.37), show that the rate of capital growth exceeds the rates of growth of labour, whereas the rate of substitutive work growth exceeds the rate of capital growth; so, in order for an enterprise to be a partner in the technological progress, the relations between the variables should be

$$r_1^{\alpha} < r_1^0, \quad r_2^{\alpha} > r_2^0, \quad r_2^{\alpha}/r_1^{\alpha} > r_2^0/r_1^0, \tag{5.38}$$

where the superscript 'zero' denotes values of quantities of the foregoing (initial) technology. It can be seen in Fig. 5.3 that a point corresponding to a progressively new technology falls in a sector of more productive technologies.

Now, expressions for the production factors can be defined. One can neglect deterioration of the equipment and write

$$\begin{aligned}
K(t) &= \sum_{\alpha} \left\langle K^{\alpha} \left[\theta(t - t^{\alpha}) - \theta(t - \tau - t^{\alpha})\right]\right\rangle, \\
L(t) &= \sum_{\alpha} \left\langle L^{\alpha} \left[\theta(t - t^{\alpha}) - \theta(t - \tau - t^{\alpha})\right]\right\rangle, \\
P(t) &= \sum_{\alpha} \left\langle P^{\alpha} \left[\theta(t - t^{\alpha}) - \theta(t - \tau - t^{\alpha})\right]\right\rangle,
\end{aligned} \tag{5.39}$$

where the symmetric step function $\theta(x)$ and, later, the Dirac delta function $\delta(x)$ are used (see, for example, Korn and Korn [5] for explanation of the properties of these functions). The angle brackets in formula (5.39) denote averaging with respect to the emergence function.

To calculate the technological coefficient on the basis of relations (5.39), it is necessary to know the emergence distribution function or, in other words, the specific mechanism of evolution of production system should be estimated. Without discussing the specific mechanism, one can assume that the emergence function is steady state; that is, it does not depend explicitly on time. This assumption allows one to calculate the derivative of (5.39) and to find expressions for the investment and technological coefficients,

$$
\begin{aligned}
I(t) &= \sum_{\alpha} \left\langle K^{\alpha} \left[ \delta(t - t^{\alpha}) - \delta(t - \tau - t^{\alpha}) \right] \right\rangle , \\
\lambda(t) &= \frac{1}{I(t)} \sum_{\alpha} \left\langle r_1^{\alpha} K^{\alpha} \left[ \delta(t - t^{\alpha}) - \delta(t - \tau - t^{\alpha}) \right] \right\rangle , \\
\varepsilon(t) &= \frac{1}{I(t)} \sum_{\alpha} \left\langle r_2^{\alpha} K^{\alpha} \left[ \delta(t - t^{\alpha}) - \delta(t - \tau - t^{\alpha}) \right] \right\rangle .
\end{aligned}
\tag{5.40}
$$

Time dependence of the technological coefficients is determined by the dependence of variables $r_1$ and $r_2$ on index $\alpha$. To estimate the behaviour of the technological coefficients, one can introduce two non-dimensional functions of the index,

$$
r_1^{\alpha} = r_1^0 (1 - \varphi_1(\alpha)), \quad r_2^{\alpha} = r_2^0 (1 + \varphi_2(\alpha)), \tag{5.41}
$$

where $\varphi_1(\alpha)$ and $\varphi_2(\alpha)$ are assumed to be small positive quantities in the case of progressive evolution of a production system.

This allows one to see an obvious result: the technological coefficient $\lambda(t)$ is a diminishing function of time in this case. This approach allows us to investigate details of the mechanism of evolution of the production system.

## References


1. Blanchard, O.J., Fisher, S.: Lectures on Macroeconomics. MIT Press, Cambridge (1989)
2. Pokrovski, V.N.: Physical Principles in the Theory of Economic Growth. Ashgate Publishing, Aldershot (1999)
3. Pokrovski, V.N.: Energy in the theory of production. Energy **28**(8), 769–788 (2003)
4. Schumpeter, J.A.: Theorie der wirtschaftlichen Entwicklung: Eine Untersuchung über Unternehmergewinn, Kapital, Kredit, Zins und den Kojunkturzyklus. Dunker und Humblot, Berlin (1911)
5. Korn, G.A., Korn, T.M.: Mathematical Handbook for Scientists and Engineers. McGraw-Hill, New York (1968)

# Chapter 6
# Production of Value

**Abstract** In this chapter, the relationship between the production of value and the original (primary) sources of value, the production factors, is considered. The increase in production of value is connected, as we know (Chap. 4), with an increase in production equipment (capital stock), and the capital stock is conventionally considered a production factor. On the other hand, production of value can be associated with an increase in technological work, that is, with an increase in efforts of humans and work of production equipment (Chap. 1). In all, the production of value is determined by three production factors, but the roles of production factors are different: humans' efforts and substitutive work are the true sources of value, while capital stock presents the means by which the workmen and energy resources are attracted to the production, allowing workers' efforts to be substituted by machine's work.

## 6.1 Output and Production Factors

We consider the final output $Y$ that represents the value of the products created by the production system per unit of time. It is assumed that such scale of value is chosen (see Sect. 2.2.3), so that purchasing capacity of the monetary unit remains identical at all times. Otherwise, the price index $\rho$ appears, and the production of value in current money units has to be written as

$$d\hat{Y} = \rho\,(dY)_\rho + \hat{Y}\,d\ln\rho,$$

where $(dY)_\rho$ is the production of value at constant prices.

Our task now is to express the production of value $Y$ through original (primary) sources of value, the production factors, the history of the choice of which was discussed briefly in Sect. 1.3.



V. N. Pokrovskii, *Econodynamics*, New Economic Windows,
https://doi.org/10.1007/978-3-319-72074-6_6

### 6.1.1 Specification of the Production Function

One can proceed from the conviction that functioning of the production system, in the most simple approximation, can be correctly described by means of three variables: fixed capital stock $K$, expenditures of work force $L$ and substitutive work $P$. The output $Y$ must be considered to depend on the three production factors

$$Y = Y(K, L, P). \tag{6.1}$$

To specify this general form[1] of the law of production of value, one has to take into account two issues. First, as far as there is a relation (5.20) among the growth rates of the production factors, the variables $K$, $L$ and $P$ appear to be interdependent: only two of the production factors are independent.[2] Then, the technological description assumes that the machine's work and humans' input act as substitutes to each other, and the amount of production equipment, universally measured by its value $K$, has to be considered to be a complement to technological work ($L$ and $P$) of the production equipment.[3] All this motivates one to write the relation between output and production factors in the form of two alternative lines

$$Y = \begin{cases} Y(K) \\ Y(L, P) \end{cases}, \qquad dY - \Delta dt = \begin{cases} \xi(K)\, dK \\ \beta(L, P)\, dL + \gamma(L, P)\, dP \end{cases}, \tag{6.2}$$

where $\Delta dt$ is a part of the increment of production of value which is connected with the change of characteristics of the production system (the structural and technological change). When $\Delta = 0$, the marginal productivities $\xi$, $\beta$ and $\gamma$ correspond to the value produced by the addition of a unit of capital, or by the addition of a unit of labour input at constant external energy consumption or by the addition of a unit of energy at constant labour input, respectively. In line with the existing practice, these quantities can be labelled as marginal productivities of the corresponding production factors. The written relation can be considered an expression of a substitution law.

The forms (6.2) seem to be consistent with some different approaches to the theory of production of value [4, 5]. The present theory keeps the main attributes of the

[1]Note that in the more general case, it is possible to admit, that production of value $Y(t)$ is function of the arguments $L(t-s)$ and $P(t-s)$ taken at the previous points in time, that is, that production of value is a function of a trajectory of evolution. This possibility was considered earlier [1], though, as it is clear from the analysis, there is no indispensability to use this concept of production function.

[2]One can see that Eq. (6.8) is an inexplicit relation for the production factors.

[3]The relationship of complementarity and substitutability among capital $K$, labour $L$ and primary energy $E$ has been analysed by some researchers. For example, Berndt and Wood [2] remark on p. 351 '...that $E - K$ complementarity and $E - L$ substitutability are consistent with the recent high-employment, low-investment recovery path of the U.S. economy.' Patterson ([3], p. 382) found 'for New Zealand (1960–1985) that energy and labour inputs acted as mild substitutes to each other, and energy and capital inputs were mild complements to each other'. These research works dealt with the total primary consumption of energy $E$ and could not discover the relationship of exact substitution between labour $L$ and substitutive work $P$, which is a part of the primary energy.

neoclassical approach, i.e. the concept of value produced by production factors (donor value) and the concept of the production factors themselves, and can be considered as a generalisation and extension of the conventional neo-classical approach. In the conventional, neoclassical theory, capital as a variable played two distinctive roles: capital stock as value of production equipment and capital service as a substitute for labour. We consider capital stock to be the means of attracting labour and energy services to the production, while human work and the work of external energy sources are considered to be the true sources of value.[4] Human work is replaced by the work of external energy sources using different sophisticated appliances. In contrast to the conventional theory, the perfect substitution of labour and energy does not lead to any discrepancies. One can imagine a factory working without energy or without labour, but one cannot imagine a factory without production equipment.

We shall use formulae (6.2) as a starting point for the productivity theory to obtain relations between marginal productivities of capital on one side and of labour and energy on the other side (see formulae (6.12)). Note that, just as capital consists of many parts with their own productivity, labour and energy can be divided into separate parts according to their qualities, so that Eq. (6.2) can be generalised. In this chapter, however, we shall use the simplest approach.

### 6.1.2 Principle of Productivity

The production factors are used to create things and services, and it is natural to believe that an addition of any of the production factors has to increase production. Therefore, one has to consider all marginal productivities to be positive; this statement is known as the *principle of productivity*.

It was discovered by Marx [8], that the labour force is such a commodity that gives surplus value. Taking the Marx's statement into account, the principle of productivity can be written as

$$(\beta - w)\, dL > 0, \tag{6.3}$$

where $w$ is a price of labour (see Sect. 2.4.2).

To explain the modern surplus product in industrial societies, one should take the other production factor, substitutive work $P$, into account and write

$$(\beta - w)\, dL + (\gamma - p)\, dP > 0, \tag{6.4}$$

[4]Some argue that, in this case, labour can be reduced to energy, and one has the only argument: energy as a source of value [6, 7]. However, labour and energy are measured in different units, and nobody knows how to calculate the real work provided by these production factors and compare them. Besides, even if possible, such a comparison could not be universal, so it is better to deal with the two separate arguments.

where $p$ is the price of production–consumption of energy providing the substitutive work (see Eq. 2.35). Both the first and the second terms in formula (6.4) are expected to be positive. This statement can be considered as the *strong principle of productivity*.

## 6.2 Productivities and Technological Coefficients

To link the marginal productivities with technological coefficients, one can refer to differential expressions (6.2) for the production of value. We can rewrite them in the following form

$$\frac{dY}{dt} - \Delta = \begin{cases} \xi \dfrac{dK}{dt}, \\ \beta \dfrac{dL}{dt} + \gamma \dfrac{dP}{dt} \end{cases}, \tag{6.5}$$

where $\xi$, $\beta$ and $\gamma$ are marginal productivities of the corresponding production factors. The derivatives of the production factors can be written on the basis of Eqs. (5.6) and (5.13) in the form

$$\frac{dK}{dt} = \left(\frac{I}{K} - \mu\right) K, \quad \frac{dL}{dt} = \left(\overline{\lambda}\frac{I}{K} - \nu' - \mu\right) L, \quad \frac{dP}{dt} = \left(\overline{\varepsilon}\frac{I}{K} - \eta' - \mu\right) P, \tag{6.6}$$

where $\overline{\lambda} = \lambda K/L$ and $\overline{\varepsilon} = \varepsilon K/P$ are the non-dimensional technological variables which characterise the quality of introduced equipment. Both the technological coefficients $\lambda$, $\varepsilon$ and the non-dimensional technological variables $\overline{\lambda}$, $\overline{\varepsilon}$ are functions of time.

Combining Eqs. (6.5) and (6.6) and assuming, for the start, that the quantities $\nu'$ and $\eta'$ in Eq. (6.6) can be neglected, one rewrites production of value in the form

$$\frac{dY}{dt} - \Delta = \begin{cases} \xi \left(\dfrac{I}{K} - \mu\right) K, \\ (\beta \overline{\lambda} L + \gamma \overline{\varepsilon} P) \dfrac{I}{K} - \mu (\beta L + \gamma P) \end{cases} \tag{6.7}$$

It is assumed that the investment $I$ is independent variable, so that, comparing the right-hand sides of Eq. (6.7), one finds relations between the characteristic parameters of the system

$$\overline{\lambda} \beta L + \overline{\varepsilon} \gamma P = \xi K,$$
$$\beta L + \gamma P = \xi K.$$

The relations ought to be considered as a set of equations for the marginal productivities $\beta$ and $\gamma$. Then, one can find, following the earlier work [9], the expressions of the marginal productivities through the technological coefficients

$$\xi = \beta \frac{L}{K} + \gamma \frac{P}{K}, \tag{6.8}$$

$$\beta = \xi \frac{\overline{\varepsilon} - 1}{\overline{\varepsilon} - \overline{\lambda}} \frac{K}{L}, \quad \gamma = \xi \frac{1 - \overline{\lambda}}{\overline{\varepsilon} - \overline{\lambda}} \frac{K}{P}. \tag{6.9}$$

If the technological variables $\overline{\lambda}$ and $\overline{\varepsilon}$ are unrestricted, it is possible that one of the marginal productivities is negative, but they cannot both be negative. One can see that the marginal productivities are positive, if

$$\overline{\lambda} < 1 < \overline{\varepsilon} \quad or \quad \overline{\lambda} > 1 > \overline{\varepsilon}. \tag{6.10}$$

It is natural to expect that the marginal productivities must be positive for some span of time, that is one of the sets of requirements (6.10) is fulfilled, which was discussed already in Sect. 5.2.2. One can consider inequalities (6.10) as a formulation of the productivity principle.

In the general case, when $\nu' \neq 0$, $\eta' \neq 0$, that is the change of technology during the time of exploitation is taken into account, the relations between marginal productivities (6.8) and (6.9) can be generalised as

$$\xi = \beta \left(1 + \frac{\nu'}{\mu}\right) \frac{L}{K} + \gamma \left(1 + \frac{\eta'}{\mu}\right) \frac{P}{K},$$

$$\beta = \xi \frac{\overline{\varepsilon} - \left(1 + \frac{\eta'}{\mu}\right)}{\overline{\varepsilon}\left(1 + \frac{\nu'}{\mu}\right) - \overline{\lambda}\left(1 + \frac{\eta'}{\mu}\right)} \frac{K}{L}, \quad \gamma = \xi \frac{\left(1 + \frac{\nu'}{\mu}\right) - \overline{\lambda}}{\overline{\varepsilon}\left(1 + \frac{\nu'}{\mu}\right) - \overline{\lambda}\left(1 + \frac{\eta'}{\mu}\right)} \frac{K}{P}.$$

## 6.3 Approximation of Marginal Productivities

The production function is assumed to satisfy some requirements. One of them—the principle of *productivity*—has been discussed in the previous sections. Then, one should take into account that the arguments of the production function must be non-dimensional quantities, so the production function can be written as

$$Y = Y_0 \, f\left(\frac{L}{L_0}, \frac{P}{P_0}\right),$$

where $L_0$ and $P_0$ are values of labour and capital services in the base year. In this section, we shall consider restrictions imposed on the production function by the requirements of *uniformity* and *universality*.

### 6.3.1 Principle of Universality

This requirement means that a proposed production function can be used not only for a given case, but for many various situations. For example, the initial point can be chosen arbitrarily, but the form of production function must not be affected by this arbitrariness. If, for example, two initial points, $t_0$ and $t_1 > t_0$, are chosen, one must write for production of value

$$Y = Y_1\, f\left(\frac{L}{L_1}, \frac{P}{P_1}\right), \quad Y_1 = Y_0\, f\left(\frac{L_1}{L_0}, \frac{P_1}{P_0}\right).$$

This can be rewritten as

$$Y = Y_0\, f\left(\frac{L_1}{L_0}, \frac{P_1}{P_0}\right)\, f\left(\frac{L}{L_1}, \frac{P}{P_1}\right),$$

so that, in order for the description to be universal, that is, independent of the arbitrary choice of initial point (to be consistent), the production function must satisfy the following relation:

$$Y_0\, f\left(\frac{L}{L_0}, \frac{P}{P_0}\right) = Y_0\, f\left(\frac{L_1}{L_0}, \frac{P_1}{P_0}\right)\, f\left(\frac{L}{L_1}, \frac{P}{P_1}\right).$$

One must choose a form of function $f(\cdot)$ such that values of $L_1,\ P_1$ on the right-hand side of the last formula would disappear. This requirement puts some restrictions in the form of the production function. One can see that production function of the form

$$Y = Y_0 \left(\frac{L}{L_0}\right)^{\alpha} \left(\frac{P}{P_0}\right)^{\beta} \tag{6.11}$$

obeys the above requirement. The parameters $\alpha$ and $\beta$ do not depend on the initial point and can be considered as characteristics of the production system.

### 6.3.2 Principle of Uniformity

One expects the production output for a large system to be proportional to the scale of production. This means that the production function has to be a homogenous, uniform function of first order, that is

$$Y(\lambda L, \lambda P) = \lambda Y(L, P).$$

Under this condition function (6.11) can be written as

$$Y = Y_0 \frac{L}{L_0} \left( \frac{L_0}{L} \frac{P}{P_0} \right)^{\alpha}. \tag{6.12}$$

The constant $Y_0$ is determined by the initial conditions, so that the only parameter which remains unknown in the expression for the production function, to say nothing of the initial values of the variables, is the quantity $\alpha = \gamma_0 = 1 - \beta_0$. It is a characteristic of the production system, which, as shown below, coincides with the technological index introduced in Sect. 5.2.2. The productivity principle restricts values of the technological index, $0 < \alpha < 1$.

Of course, one can consider the index $\alpha$ to be a function of production factors, which can be represented by an expansion series in the powers of quantities $\ln \frac{L}{L_0}$, $\ln \frac{P}{P_0}$. For example, in linear approximation

$$\alpha = \alpha_0 + a \ln \frac{L}{L_0} + b \ln \frac{P}{P_0},$$

where the parameters $a, b$ are also characteristics of the production system.[5] As for any other forms of the production function known in the literature, it is worth testing them for consistency before using them.

### 6.3.3 Marginal Productivities

Function (6.12) has the exact form of the neoclassical, i.e. Cobb–Douglas production function (1.3), in which capital $K$ stands for substitutive work $P$. This function, in accordance with Eq. (6.5), provides the following expressions for marginal productivities and the contribution from technological and structural change:

$$\beta = Y_0 \frac{1-\alpha}{L_0} \left( \frac{L_0}{L} \frac{P}{P_0} \right)^{\alpha}, \quad \gamma = Y_0 \frac{\alpha}{P_0} \left( \frac{L_0}{L} \frac{P}{P_0} \right)^{\alpha-1}, \tag{6.13}$$

---

[5]The first-order terms of expansion determine the first-order function

$$Y = Y_0 \left( \frac{L}{L_0} \right)^{\beta_0} \left( \frac{P}{P_0} \right)^{\gamma_0} \exp \frac{1}{2} \left( \beta_l \ln^2 \frac{L}{L_0} + (\beta_e + \gamma_l) \ln \frac{L}{L_0} \ln \frac{P}{P_0} + \gamma_e \ln^2 \frac{P}{P_0} \right),$$

which under the condition of uniformity takes the following form:

$$Y = Y_0 \frac{L}{L_0} \left( \frac{L_0}{L} \frac{P}{P_0} \right)^{\alpha} \exp \left[ \frac{1}{2} b \ln^2 \left( \frac{L_0}{L} \frac{P}{P_0} \right) \right], \quad b = -\beta_l = \beta_e = \gamma_l = -\gamma_e.$$

$$\Delta = Y \ln \left( \frac{L_0}{L} \frac{P}{P_0} \right) \frac{d\alpha}{dt} \tag{6.14}$$

where $L_0$ and $P_0$ are values of labour and capital services in the base year.

Comparing expressions (6.9) and (6.13) for the marginal productivities, one obtains

$$\xi = Y_0 \frac{L}{L_0 K} \left( \frac{L_0}{L} \frac{P}{P_0} \right)^{\alpha}, \qquad \alpha = \frac{1 - \overline{\lambda}}{\overline{\varepsilon} - \overline{\lambda}} \tag{6.15}$$

Thus, the index $\alpha$ in Eq. (6.12) is, indeed, the same quantity as the technological index introduced by Eq. (5.20). In addition, all available information about the technological performance could be introduced by estimating this quantity. Moreover, a condition regarding the optimal use of production factors enables us to establish a relation between the parameter $\alpha$ on one hand and the shared costs of production factors on the other (see Sect. 6.7). This provides a different means of estimating the technological index.

The requirement of productivity $0 < \alpha < 1$ means that the marginal productivities are increasing functions of the ratio of substitutive work to labour, that is,

$$\frac{d\beta}{d(P/L)} > 0, \quad \frac{d\gamma}{d(P/L)} > 0.$$

## 6.4 Production Function and Equations of Growth

### *6.4.1 Fundamental Equations of Evolution*

The preceding results allow us to record output $Y$ as a function of production factors $K$, $L$ and $P$, whereas properties of the production system itself are fixed by the internal characteristics $\alpha$ and $\xi$ connected with Eq. (6.15). From these findings, formula (6.2) can be specified as

$$Y = \begin{cases} \xi K \\ Y_0 \dfrac{L}{L_0} \left( \dfrac{L_0}{L} \dfrac{P}{P_0} \right)^{\alpha} \end{cases} \tag{6.16}$$

As was discussed in Sect. 5.4, trajectories of production factors $K$, $L$ and $P$ are determined by available resources and used technology and obey the system of Eq. (5.36) that are reproduced here

$$\frac{dK}{dt} = I - \mu K, \quad \frac{dL}{dt} = \left( \overline{\lambda} \frac{I}{K} - \mu \right) L, \quad \frac{dP}{dt} = \left( \overline{\varepsilon} \frac{I}{K} - \mu \right) P,$$

$$\frac{I}{K} = \min \left\{ (\tilde{\delta} + \mu), \quad (\tilde{\nu} + \mu) \frac{1}{\overline{\lambda}}, \quad (\tilde{\eta} + \mu) \frac{1}{\overline{\varepsilon}} \right\},$$

$$\frac{d\overline{\lambda}}{dt} = -\frac{1}{\tau}\left(\overline{\lambda} - \frac{\tilde{\nu}+\mu}{\tilde{\delta}+\mu}\right), \quad \frac{d\overline{\varepsilon}}{dt} = -\frac{1}{\tau}\left(\overline{\varepsilon} - \frac{\tilde{\eta}+\mu}{\tilde{\delta}+\mu}\right), \quad \alpha = \frac{1-\overline{\lambda}}{\overline{\varepsilon}-\overline{\lambda}}$$

The symbols were explained in Chap. 5. Note that the quantities $\alpha$ in equations for production factors and (6.16) are identical ones. The calculation of the trajectory of production factors was done for the U.S. economy; the example can be found in Sect. 5.4.2.

The above written relations compose the system of evolutionary equations for output of the production system. However, we have to remind the main trouble: to compute the evolution of production factors, according to the system of Eq. (5.36), one has to know the rates of potential growth of production factors $\tilde{\delta},\ \tilde{\nu},\ \tilde{\eta}$, which is easier said than done. Due to these difficulties one can try to formulate the problem in another way, which will be considered further. Nevertheless, the system of Eqs. (5.36) and (6.16) remains to be the fundamental system of equations of evolution.

### 6.4.2 Decomposition of the Growth Rate of Output

The growth rate of output in terms of the present theory can be written in the form of two alternative expressions

$$\frac{1}{Y}\frac{dY}{dt} = \begin{cases} \dfrac{1}{K}\dfrac{dK}{dt} + \dfrac{1}{\xi}\dfrac{d\xi}{dt}, \\ (1-\alpha)\dfrac{1}{L}\dfrac{dL}{dt} + \alpha\dfrac{1}{P}\dfrac{dP}{dt} + \ln\left(\dfrac{L_0}{L}\dfrac{P}{P_0}\right)\dfrac{d\alpha}{dt} \end{cases} \tag{6.17}$$

The first terms in the first and second lines on the right-hand side of this formula represent the contribution to growth due to the growth of production factors: capital, labour and substitutive work. The last ones are due to the change of the production system itself; changes of quantities $\xi$ and $\alpha$ are connected with technological and structural changes. The last terms cannot be reduced to any function of production factors.

Let us note that variations in characteristics of production system $\xi$ and $\alpha$ are connected with each other. To find a formula for a change of capital marginal productivity, we differentiate relation (6.15)

$$\frac{1}{\xi}\frac{d\xi}{dt} = -\frac{1}{K}\frac{dK}{dt} + (1-\alpha)\frac{1}{L}\frac{dL}{dt} + \alpha\frac{1}{P}\frac{dP}{dt} + \ln\left(\frac{L_0}{L}\frac{P}{P_0}\right)\frac{d\alpha}{dt}$$

To simplify this relation, one can use the dynamic equations for production factors (6.6), on the assumption $\nu' = 0,\ \eta' = 0$, and obtain

$$\frac{1}{\xi}\frac{d\xi}{dt} = \ln\left(\frac{L_0}{L}\frac{P}{P_0}\right)\frac{d\alpha}{dt}, \tag{6.18}$$

so that, in line with Eq. (6.14), one has

$$\Delta = Y \frac{1}{\xi}\frac{d\xi}{dt}. \tag{6.19}$$

Returning now to the expression for the output growth rate (6.17) and using Eq. (5.26) for the growth rates of production factors in three possible cases, one can record the following

$$\frac{1}{Y}\frac{dY}{dt} = \frac{1}{\xi}\frac{d\xi}{dt} + \begin{cases} \delta, & \text{if } \frac{d\overline{\lambda}}{dt} > 0 \text{ and } \frac{d\overline{\varepsilon}}{dt} > 0, \\ (\nu+\mu)\frac{1}{\overline{\lambda}} - \mu, & \text{if } \frac{d\overline{\lambda}}{dt} < 0 \text{ and } \frac{d\overline{\varepsilon}}{dt} > 0, \\ (\eta+\mu)\frac{1}{\overline{\varepsilon}} - \mu, & \text{if } \frac{d\overline{\lambda}}{dt} > 0 \text{ and } \frac{d\overline{\varepsilon}}{dt} < 0 \end{cases} \tag{6.20}$$

It is assumed that the characteristics of the equipment do not change after its installation, that is $\nu' = 0$ and $\eta' = 0$; otherwise, the expression for the growth rate takes a more complicated form.

The growth rate of the output is expressed by three different relations for three modes of development, which were described in Sect. 5.3.1 (Chap. 5). To simplify relations (6.20), one can use Eq. (5.20) to express the energy requirement and the growth rate of substitutive work through the technological index as

$$\overline{\varepsilon} = \frac{1-(1-\alpha)\overline{\lambda}}{\alpha}, \quad \eta = \frac{\delta-(1-\alpha)\nu}{\alpha}, \quad 0 < \alpha < 1.$$

These relations allow one to identify the expressions in the second and the third lines of Eq. (6.20). Moreover, one can see, that these relations practically exclude the first line, so that Eq. (6.20) reduce to a universal expression for the growth rate of output,

$$\frac{1}{Y}\frac{dY}{dt} = \frac{\nu+(1-\overline{\lambda})\mu}{\overline{\lambda}} + \frac{1}{\xi}\frac{d\xi}{dt} \tag{6.21}$$

So, the growth rate of the output is determined by four quantities:

- Productivity of capital stock $\xi$. This is a fundamental quantity connected with the fundamental technological matrixes (see Eq. 4.48), when one refers to the many-sector approach. Technological and structural changes are introduced through this quantity.
- The non-dimensional technological coefficient $\overline{\lambda}$. If the quantity $\overline{\lambda} < 1$, the consumption (for unit of capital stock) of labour decreases and consumption of productive energy (substitutive work of production equipment) increases. The situation is opposite, if the quantity $\overline{\lambda} > 1$.

- The coefficient of depreciation $\mu$. This quantity does not affect the growth rate of output, if $\overline{\lambda} = 1$.
- The growth rate of labour $\nu$. The growth rate of output coincides with this quantity, if $\overline{\lambda} = 1$.

All the quantities are characteristics of the method of production, that is characteristics of technology. If the technological coefficient $\overline{\lambda}$ is equal to unity, labour productivity is constant and all addition of a product is connected only with an increase in the number of workers. Under the condition, when the labour requirement $\overline{\lambda}$ is less than unity, efforts of workers are partially replaced with the work of the machines movable by outer energy sources.

### *6.4.3 Equations of the Programmed Development*

The results of the previous section allow us to develop a different approach to the description of growth. One can see that the growth rate of output, according to Eq. (6.21), can be calculated on the assumption that the four quantities: capital stock productivity $\xi(t)$, the rate of labour growth $\nu(t)$, the depreciation coefficient $\mu(t)$ and the non-dimensional technological coefficient $\overline{\lambda}(t)$, are given as functions of time. To find the time dependence of the output, there is no need to know the time dependence of the production factors, though it is convenient to formulate a simple scheme which allows one to calculate trajectories of the production factors as well. The above results allow one to write the system of equations

$$\begin{aligned}
Y &= \xi K, \quad \xi = \xi(t), \\
\frac{dK}{dt} &= \delta K, \quad \delta = \frac{\nu + (1-\overline{\lambda})\mu}{\overline{\lambda}}, \quad \mu = \mu(t), \quad \overline{\lambda} = \overline{\lambda}(t), \\
\frac{dL}{dt} &= \nu L, \quad \nu = \nu(t), \\
\frac{dP}{dt} &= \eta P, \quad \eta = \frac{\delta - (1-\alpha)\nu}{\alpha}, \quad \alpha = \frac{\ln\left(\frac{Y}{Y_0}\frac{L_0}{L}\right)}{\ln\left(\frac{L_0}{L}\frac{P}{P_0}\right)}
\end{aligned} \tag{6.22}$$

The system determine the time dependence of the variables $Y$, $K$, $L$, $P$ and $\alpha$, if the four quantities $\xi$, $\mu$, $\nu$ and $\overline{\lambda}$ are given as functions of time. Initial values of all five variables have to be given, and the initial value of capital stock must correspond to the initial values of output and marginal productivity $K_0 = Y_0/\xi(0)$, while the initial values of labour $L_0$ and substitutive work $P_0$ can be chosen independently. Note also that these quantities are connected by Eq. (6.15), which shows that one has to require the consistency of the solution to the given quantity $\xi$.

## 6.5 Exponential Growth

Considering applications of the evolution equations, it is natural to address, first of all, to cases of growth, when it is assumed that the time dependence of production function is given. The simplest illustration is the example, when all variables can be described by exponential functions; the case is known as the 'stylised' economic growth.

### 6.5.1 *Empirical Facts*

One can see on the plot in Fig. 5.1 that the time dependence of production factors $K$, $L$ and $P$ for the relatively calm period of years 1950–2000 can be approximately depicted by straight lines, so that, for these years, the growth of production factors can be described by exponential functions, as was demonstrated in Chap. 2 (see formulae (2.29), (2.30) and (2.36)), with the rates of growth (in units of $year^{-1}$)

$$\frac{1}{K}\frac{dK}{dt} = 0.0316, \quad \frac{1}{L}\frac{dL}{dt} = 0.0147, \quad \frac{1}{P}\frac{dP}{dt} = 0.0585. \tag{6.23}$$

The empirical growth of the GDP of the U.S. for these years can be also considered to be approximately exponential (see formula (2.14))

$$Y = 1.69 \times 10^{12} \cdot e^{0.0326\,t} \ \text{dollar}(1996)/\text{year}.$$

Time $t$ is measured in years, and $t = 0$ corresponds to year 1950.

The fundamental empirical fact is that the average growth rate of output 0.0326 is approximately equal to the growth rate of a fixed capital $\delta = 0.0316$, which was recorded for various countries by many researchers [10, 11].

### 6.5.2 *Asymptotic Solution*

Turning to the system of Eq. (5.36), we can note that, in the simplest case, when all rates of growth are given as constant:

$$\delta = \tilde{\delta}, \quad \nu = \tilde{\nu}, \quad \eta = \tilde{\eta},$$

the equations have a simple asymptotic solution

$$K = K_0 e^{\delta t}, \quad L = L_0 e^{\nu t}, \quad P = P_0 e^{\eta t},$$

$$\frac{I}{K} = \delta + \mu, \quad \overline{\lambda} = \frac{\nu + \mu}{\delta + \mu}, \quad \overline{\varepsilon} = \frac{\eta + \mu}{\delta + \mu}, \quad \alpha = \frac{\delta - \nu}{\eta - \nu}. \tag{6.24}$$

In this case, the rates of potential growth of production factors coincide with the real ones.

For exponential growth of production factors, an expression for output follows immediately relations (6.16) and (6.24) and can be written in the following form

$$Y = Y_0 e^{[\nu + \alpha(\eta - \nu)]t} = Y_0 e^{\delta t}. \tag{6.25}$$

One can see that the growth rate of output is equal to the growth rate of capital and is connected with the growth rates of labour and substitutive work.[6] The theory describes the 'stylised' facts of economic growth, described by exponential functions.

Returning to the empirical data for the U.S. economy, one can consider the rates of growth in the period of 1950–2000 as approximately constant and equal to the values of the growth rates of production factors given in Sect. 6.6.1, that is,

$$\delta = 0.0316, \quad \nu = 0.0147, \quad \eta = 0.0585.$$

Formulae (6.24) allows one to estimate the technological coefficients as well as the technological index, so that one can find a value of the technological index as $\alpha = 0.39$. This estimate is naturally close to the other one (see Sect. 7.1), which confirms that the rates of real growth of production factors can be considered approximately constant in the described period.

In accordance with relation (6.25), the empirical averaged growth rate of output 0.0326 is approximately equal to the growth rate of capital $\delta = 0.0316$, which is confirmed by numerous observations [10, 11]. The difference between the growth rates of capital and output seems to be quite unreliable, given the rough estimate of the parameters of the problem, although under more detailed consideration, the difference can be attributed to an intrasectoral technological change and to the difference of the growth rates of sector outputs (see Sect. 9.3.2, formula (9.19)).

---

[6]Here the reader can be pertinently reminded of the contradictions arising if one uses for an assessment of output the neoclassical production function in the Cobb–Douglas form,

$$Y = Y_0 \frac{L}{L_0} \left( \frac{L_0}{L} \frac{K}{K_0} \right)^{\alpha'}, \quad 0 < \alpha' < 1.$$

Considering exponential growth (6.27), the expression for output is determined in the form of

$$Y = Y_0 e^{[(1-\alpha')\nu + \alpha' \delta]t}.$$

It is easy to see that the Cobb–Douglas production function describes empirical data for the U.S. for the years 1950–2000 at $\alpha' \approx 1$, which excludes the influence of labour. Moreover, the index $\alpha'$ can be interpreted as a share of capital in total expenses for maintenance of production factors —the quantity that is equal to 0.3–0.4 for the U.S. economy. This is a well-known fact ([11], p. 4), which has led to the introduction of *the total factor of productivity* [12] and to numerous modifications of the neoclassical production function [13, 14].

### 6.5.3 The Factor of Total Productivity

The rate of growth of output (6.25) can be broken into two parts. For the U.S. economy, on average a fraction of the rate $(1-\alpha)\nu \approx 0.0112$ is connected with the growth of expenditures of labour, and the other part $\alpha\eta \approx 0.0235$ – with the growth of substitutive work. The capital is the means of attracting the production factors to production, thus an increase in consumption of the production factors is connected with an increase in capital. One can formally separate the growth rate of capital $\delta$ within the growth rate of substitutive work $\eta$ to get a breakdown of the growth rate of output in conventional terms: the contribution from the labour growth $(1-\alpha)\nu \approx 0.0112$ and the contribution from the capital growth $\alpha\delta \approx 0.0126$. One can see that the Solow residual (total factor productivity) can be expressed through the technological index and the growth rates as

$$Solow\ \ Residual = \alpha(\eta-\delta) = (1-\alpha)(\delta-\nu) \approx 0.0109. \tag{6.26}$$

Structural and technological changes, if they exist, compensate each other in this simple case of exponential growth. A detailed decomposition of the Solow residual for the U.S. economy is shown in Fig. 6.1.

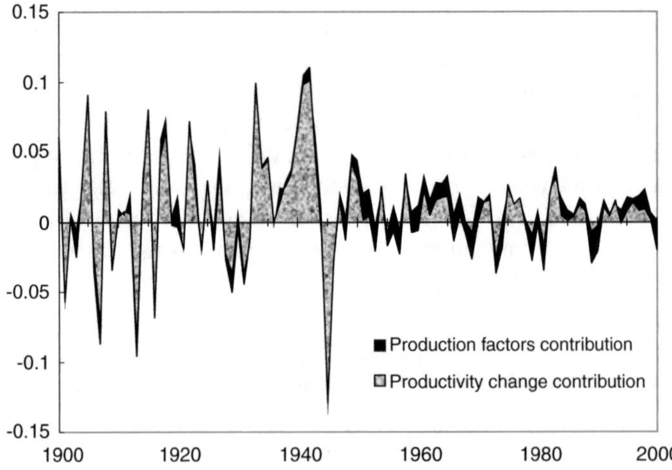


**Fig. 6.1** Decomposition of the Solow residual. The conventional 'total factor productivity growth' is broken into two parts: the difference between the growth rates of capital services and capital stock (black area) and the residual connected with changes of the production system itself (the technological and structural changes)

## 6.6 Dynamics of the U.S. Economy

### 6.6.1 Trajectory of Output

To illustrate the procedure of drawing a scenario and deviations arising due to approximations, we refer to the U.S. data for the years 1900–2000. Earlier in Chap. 5, we have reproduced historical empirical dependence of the production factors for this case (Fig. 5.1), and this allows us to restore time dependence of the Gross Domestic Product (GDP) for the U.S. economy for those years according to relation (6.16). At the known time dependence of the production factors, the calculated time dependence of output (the thin lines) approximates, as one can see in Fig. 6.2, the real-time dependence of the GDP (the thick solid line) before the year 2000 in all details [9], which confirms the consistency of the theory.

It was assumed [9] that a development will continue beyond year 2000. The outputs of two scenarios of development of the U.S. economy for the years 2000–2040 are presented in Fig. 6.2 by thin lines. As was described in Sect. 5.4.2, the growth rate of labour $\nu$ for either scenario coincides with the rate of population growth, namely, $\nu = 0.01$ for the U.S. economy. The first scenario corresponds to the values of $\mu$ and $\overline{\lambda}$ in the year 2000, that is $\mu = 0.68$ and $\overline{\lambda} = 0.78$ for all years. The second one shows the effect of diminishing the energy supply in the economy: the value of $\overline{\lambda}$ increases to unity in the year 2010. One can see a decrease in the growth rate of the output in the case when the growth rate of productive consumption of energy is decreasing. Of course, these results should be considered as an illustration of a

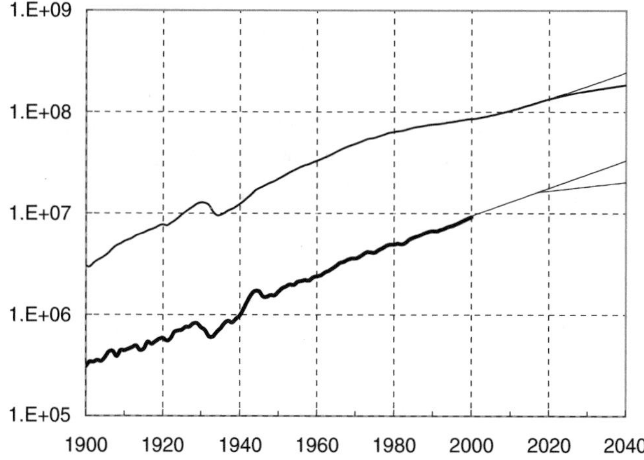


**Fig. 6.2** Output and national wealth in the U.S. economy. The empirical (solid shorter line for GDP) and calculated values of the total national wealth $W$ and GDP in millions of dollars for the year 1996. Two scenarios are shown; the lower line after the year 2010 corresponds to a diminishing supply of energy

method of forecasting rather than as the forecast itself. One needs to know the future availability of labour and substitutive work to do a real prediction.

The results of investigation [9] show that such values of variables of the system (5.36) and (6.16) can be found that description of situation corresponds to reality. The output can also be reproduced under application of the system of Eq. (6.22), if empirical values of the capital stock productivity $\xi(t)$, the rate of labour growth $\nu(t)$, the depreciation coefficient $\mu(t)$ and the non-dimensional technological coefficient $\overline{\lambda}$ are given as function of time. One can get the picture of development for any economy, if one set the values of variables $Y$, $L$ and $\alpha$ in the initial year and the program of development for the future in terms of the four quantities $\xi$, $\mu$, $\nu$ and $\overline{\lambda}$. No procedure of adjustment is needed, but, to set initial values and to design a program of development, we must know something about the investigated economy.

### *6.6.2 Pulsations in the Production Development*

The numerous empirical data demonstrates the pulsating character of development of production. The growth rate of the Gross Domestic Product of the U.S. economy (see Fig. 2.2 in Sect. 2.2.3) fluctuates with the period of pulsations about 4 years. These pulsations can be naturally connected with the pulsations of the technological coefficients depicted on the charts of Fig. 5.2 for the same period of time [15].

In simple approximation, when characteristics of the equipment do not change after its installation, the rate of growth of the Gross Domestic Product, according to Eq. (6.21) is recorded as

$$\frac{1}{Y}\frac{dY}{dt} = \frac{\nu + (1-\overline{\lambda})\mu}{\overline{\lambda}} + \frac{1}{\xi}\frac{d\xi}{dt}. \tag{6.27}$$

The rate of growth of output is connected with four quantities, while the change of the growth rate of the labour demand is connected with the coefficient of labour requirement $\overline{\lambda}$, which is a strongly pulsating quantity, as is possible to see in the charts of Fig. 5.2. The existence of pulsations of output can be connected with the existence of the three alternative types of functioning of the production system, which were described in Sect. 5.3. In the considered period, the second and third cases (see Sect. 5.4.2) are realised for production of the U.S., that is the processes are running at deficiency of labour and abundance of investments, substitutive work and raw materials, when $\frac{d\overline{\lambda}}{dt} < 0$; or at deficiency of substitutive work and abundance of investments, work and raw materials, when $\frac{d\overline{\lambda}}{dt} > 0$.

To describe an ideal cycle, one can start from the point, where the coefficient of labour requirement has its minimum value. In the case, when $\frac{d\overline{\lambda}}{dt} > 0$, the production system is experiencing deficiency of substitutive work, whereas the creation of working places is restricted, and, at small values of the coefficient of labour requirement, the index of unemployment starts to increase. The coefficient of labour requirement is also growing, according to Eq. (5.30); the production system attracts more labour, but it appears insufficient to decrease the index of unemployment immediately, and the index grows simultaneously with the technological coefficient. The growth of unemployment stops when the production system succeeds in using all the extra supply of labour, and the technological coefficient reaches its potential value at $\frac{d\overline{\lambda}}{dt} = 0$. The situation is being balanced at the peak of unemployment, and, at this point, the change of the growth mode has occurred. Further, when $\frac{d\overline{\lambda}}{dt} < 0$, the production system is functioning at a deficiency of labour and is able to use all available resources of workers: the index of unemployment decreases simultaneously with decrease in labour demand, until at some point in time a new balance occurs at $\frac{d\overline{\lambda}}{dt} = 0$, where the type of functioning of the production system is changing again: a new cycle begins.

There remain many reasons for the observable real cycles not to be ideal. However, empirical data for the U.S. economy confirm the general patterns of business cycle phenomena: the changes of the coefficient of labour requirement and the index of unemployment correlate with a coefficient of correlation close to unity [15]. The period of a cycle is connected with the mechanism of propagation of exploited technologies and for the US economy is equal to approximately four years.

The considered approximation of a national economy as a uniform sector allows us to describe the dynamics of short cycles in the U.S. economy in the last century from the point of view of production functioning. In reality, the national economy consists of many sectors, and each sector is characterised by the technological coefficients with different times of propagations of technology. Thus, it appears to be possible to have a wider set of modes of development leading to the occurrence of cycles of various durations; those were detected empirically when the functioning of the world economy was investigated [16]. The special interest is caused by very long cycles—the Kondratyev's cycles of duration of 40–60 years, the nature of which is not understood properly. Probably, the long waves of production can be connected with pulsation of the state of social mind between two ways of managing that are dubbed by Shcherbakov [17] as productive and appropriative. Under the *productive* character of managing, the role and place of a person is appreciated according to its working contribution (both physical and mental) into the social production. Under the *appropriative* character—according to his wealth; the most honourable employment is 'to do money'; to work 'is a shame'. To construct a mathematical model of the phenomenon and analyse the problem, it is necessary to consider simultaneously dynamics of production and monetary circulation.

### 6.6.3 National Wealth

In accordance with definitions of Chap. 2, the total national wealth $W$ (see Sect. 2.3) consists of the capital stock $K$ and the storage of products, which are material and non-material products of value $R$,

$$W = K + R.$$

The separate parts of the national wealth, in accordance with Eqs. (2.27) and (2.28) can be estimated from the equations

$$\frac{dK}{dt} = I - \mu K, \quad \frac{dR}{dt} = Y - I - C - \mu R,$$

where $I$ is the investment in the stock of production capital and $C$ is the current consumption which can be calculated, according to Eq. (6.32), as the cost of labour,

$$C = cL = \frac{1-\alpha}{\alpha} \mu K.$$

It is assumed that the depreciation coefficient has the same value for both tangible and non-tangible stocks.

Then, from the above relations, the total national wealth is determined by the equation

$$\frac{dW}{dt} = Y - C - \mu W. \tag{6.28}$$

Results of calculations of national wealth $W$ for the U.S., in line with gross domestic product $Y$, for the two scenarios described in Sect. 6.6.1 are shown in Fig. 6.2 by thin lines.

## 6.7 The Best Utilisation of Production Factors

One can assume that the production factors $L$ and $P$ are chosen in such amounts to be the most effective in production, that is their values maximise the production function at given total expenses for production factors. In this case, the question of the choice of production factors can be interpreted as a problem of finding the maximum value of function $Y(L, P)$ at condition

$$cL + pP = V,$$

where $c$ and $p$ are prices of 'consumption' of production factors, and $V$ is a part of the gross output, which goes for the maintenance of the production factors. The discussion of the prices $c$ and $p$ is given in Sects. 2.4.2 and 2.5.3 (Eq. 2.35).

One can follow a general method of searching a conditional extremum, so that we look for the unconditional maximum of the Lagrange function

$$Y(L, P) - \kappa(cL + pP - V), \tag{6.29}$$

where $\kappa$ is a Lagrange multiplier.

Therefore, we obtain equations for a point of the conditional extremum

$$\frac{\partial Y}{\partial L} = \kappa c, \quad \frac{\partial Y}{\partial P} = \kappa p,$$

which can be rewritten, remembering the definitions of the marginal productivities, as

$$\beta = \kappa c, \quad \gamma = \kappa p.$$

From the last relations, it follows that the ratio of the prices of production factors is equal to the ratio of marginal productivities or referring to Eq. (6.9), is inversely proportional to the ratio of production factors

$$\frac{c}{p} = \frac{\beta}{\gamma} = \frac{\overline{\varepsilon} - 1}{1 - \overline{\lambda}} \frac{P}{L}.$$

The relation between the prices of production factors, taking definition (5.20) for the technological index into account, can also be written as

$$\frac{c}{p} = \frac{\beta}{\gamma} = \frac{1 - \alpha}{\alpha} \frac{P}{L}.$$

Therefore, the index $\alpha$ can be expressed through the prices and the amounts of production factors,

$$\alpha = \frac{pP}{cL + pP}. \tag{6.30}$$

This relation means that the technological index $\alpha$ in the equilibrium situation can be interpreted as follows: it is a share of the expenses, needed for utilisation of substitutive work as a production factor, within the total expenses for production factors. If the production factors are chosen as optimal, then

$$0 < \alpha < 1,$$

which implies the known restrictions on values of technological variables

$$\overline{\lambda} < 1 < \overline{\varepsilon} \quad or \quad \overline{\lambda} > 1 > \overline{\varepsilon}. \tag{6.31}$$

These conditions coincide with conditions of positivity of marginal productivities (see formulae (6.10) in Sect. 6.2).

Expression (6.30) allows one to estimate the technological index $\alpha$ due to the estimates of the cost of consumption of production factors. Remembering the definition of the cost of substitutive work (Eq. 2.35), one can define the cost of labour as

$$c = \mu \frac{\overline{\varepsilon} - 1}{1 - \overline{\lambda}} \frac{K}{L} = \mu \frac{1 - \alpha}{\alpha} \frac{K}{L}. \tag{6.32}$$

Note that the cost of labour $c$, which is the value of products needed to compensate current living expenses, does not include, in contrast to the price of labour (wage), any accumulation. It is the value of the minimum amount of products, which are needed for humans to subsist.

## 6.8 On the Choice Between Consumption and Saving

The production system of the economy is driven by the desires of economic subjects, first of all, by the desires of producers to produce as much as they can. The production system tries to swallow all available resources, and three modes of economic development, for which we have different formulae for calculation, are possible (see Sect. 5.3.1). In the case of abundance of labour and energy, the desires of producers can meet restrictions from the side of consumers. In this case, a nation must decide how much it should save or consume, taking into account present and future consumption. According to Blanchard and Fisher [11], there are two basic models for solving this problem: the infinite horizon optimising model [18] and the overlapping generations model with finite horizon [19–21].

Frank Ramsay [18] used a simple model consisting of the neoclassical production function (1.2) and the equation for capital dynamics (5.6). According to the conventional conviction of that time, he supposed that a trajectory of evolution can be chosen in such a way that the consumption for the time $T$ is the biggest one. In other words, one ought to maximise the function

$$U(T) = \int_0^T u(c)\, e^{-\theta s}\, ds, \tag{6.33}$$

where consumption $c$ is a function of time $s$, $u(c)$ is a concave objective function and $T$ is the horizon of planning.

The second model—the overlapping generation model [19–21]—assumes that at any time individuals of different generations are alive and may be interacting with one another. The investment is generated by individuals who save during their lives to ensure their consumption during retirement. In this way, preferences of individuals determine investment into the production system.

These two models can be very useful in estimating potential investment and the potential rate of capital growth. However, it is necessary to take the restriction on the availability of labour and energy into account to calculate a real trajectory of evolution. Further, we shall demonstrate how to formulate the Ramsay problem (the solution of the original problem can be found, for example in the monograph by Blanchard and Fisher [11]) in more realistic way. We can consider the problem applying to evolution Eqs. (5.6) and (5.13), which were formulated in Sect. 5.1. To introduce the consumption into equations, one can take the relation

$$I = Y - cL \tag{6.34}$$

and rewrite the equation of the second line in formula (6.7) for production of value in another form to get the system of equations

$$\frac{dY}{dt} = (\beta\lambda + \gamma\varepsilon)\, Y - (\beta\lambda + \gamma\varepsilon)\, cL - \mu\, (\beta\, L + \gamma\, P),$$

$$\frac{dK}{dt} = Y - cL - \mu\, K, \quad \frac{dL}{dt} = \lambda\, Y - (\lambda\, c + \mu) L, \quad \frac{dP}{dt} = \varepsilon\, Y - \varepsilon\, cL - \mu P. \tag{6.35}$$

Further, it is convenient to use the variables

$$y = \frac{Y}{L}, \quad \epsilon = \frac{P}{L}, \quad \ell = \frac{L}{K}$$

to write a system of evolutionary equations in the form

$$\begin{aligned}
\frac{dy}{dt} &= (\beta(\epsilon) - y)[\overline{\lambda}\, (y - c)\, \ell - \mu] + \gamma(\epsilon)[\overline{\varepsilon}\, (y - c)\, \ell - \mu]\, \epsilon, \\
\frac{d\epsilon}{dt} &= (\overline{\varepsilon} - \overline{\lambda})\, (y - c)\, \epsilon\, \ell, \\
\frac{d\ell}{dt} &= (\overline{\lambda} - 1)\, (y - c)\, \ell^2.
\end{aligned} \tag{6.36}$$

The technological variables $\overline{\lambda} = \lambda K/L,\ \overline{\varepsilon} = \varepsilon K/S$ are determined by Eqs. (5.30) and (5.31). It is assumed that the marginal productivities are given, for example, by formula (6.13).

One can use the standard procedure to find a trajectory which maximises function (6.33) under restrictions (6.36). The result is a differential equation for consumption as a function of time.

## References

1. Pokrovski, V.N.: Physical Principles in the Theory of Economic Growth. Ashgate Publishing, Aldershot (1999)
2. Berndt, E.R., Wood, D.O.: Engineering and econometric interpretations of energy - capital complementarity. Am. Econ. Rev. **69**, 342–354 (1979)
3. Patterson, M.G.: What is energy efficiency? concepts, indicators and methodological issues. Energy Policy **24**, 377–390 (1996)
4. Odum, H.T.: Environmental Accounting. Emergy and Environmental Decision Making. John Wiley & Sons, New York (1996)
5. Valero, A.: Thermoeconomics as a conceptual basis for energy-ecological analysis. In: Ulgiati, S. (ed.) Advances in Energy Studies Workshop: Energy Flows in Ecology and Economy, Porto Venere, Italy 1998, pp. 415–444. MUSIS, Rome (1998)
6. Costanza, R.: Embodied energy and economic valuation. Science **210**, 1219–1224 (1980)
7. Cleveland, C.J., Costanza, R., Hall, C.A.S., Kaufmann, R.: Energy and the U.S. economy: A biophysical perspective. Science **225**, 890–897 (1984)
8. Marx, K.: Capital. Encyclopaedia Britannica, Chicago (1952). English translation of Karl Marx, Das Kapital. Kritik der politischen Oekonomie, Otto Meissner, Hamburg (1867)
9. Pokrovski, V.N.: Energy in the theory of production. Energy **28**, 769–788 (2003)
10. Scott, M.F.G.: A New View of Economic Growth. Clarendon Press, Oxford (1989)
11. Blanchard, O.J., Fisher, S.: Lectures on Macroeconomics. MIT Press, Cambridge (1989)
12. Solow, R.: Technical change and the aggregate production Function. Rev. Econ. Stud. **39**, 312–330 (1957)
13. Ferguson, C.E.: The Neo-Classical Theory of Production and Distribution. Cambridge University Press, Cambridge (1969)
14. Brown, M.: On the Theory and Measurement of Technological Change. Cambridge University Press, Cambridge (1966)
15. Pokrovskii, V.N.: Pulsation of the growth rate of output and technology. Phys. A Stat. Mech. Appl. **390**(23–24), 4347–4354 (2011)
16. Korotayev, A.V., Tsirel, S.V.: A spectral analysis of World GDP dynamics: Kondratiev waves, Kuznets swings, Juglar and Kitchin cycles in global economic development, and the 2008–2009 economic crisis. Struct. Dyn. **4**, 3–57 (2010)
17. Shcherbakov, A.V.: Upravlenie krizisami v ekonomike (The Management of Crises in Economy). In: Akaev, A.A., Korotayev, A.V., Malkov, SYu. (eds.) Mirovaya dinamika: Zakonomernosti, tendentsii, perspektivy (The World Dynamics: Laws, Tendencies, Pospects), pp. 362–394. URSS, KRASAND, Moscow (2013)
18. Ramsey, F.P.: A mathematical theory of saving. Econ. J. **38**, 543–559 (1928)
19. Allais, M.: Economie et interet. Imprimerie Nationale, Paris (1947)
20. Samuelson, P.A.: An exact consumption-loan model of interest with or without the social contrivance of money. J. Polit. Econ. **66**, 467–482 (1958)
21. Diamond, P.A.: National debt in a neo-classical growth model. Am. Econ. Rev. **55**, 1126–1150 (1965)

# Chapter 7
# Estimation of Parameters of Production Processes

**Abstract** In this chapter, to illustrate and test the theory, we refer to time series for the U.S. economy for the years 1890–2009 collected in Appendix 2. The choice of this case is justified by the availability and reliability of the data, which can be easily found on web pages of the U.S. Census Bureau and the U.S. Bureau of Economic Analysis. These organisations have been permanently improving the methods of estimation of time series, and the data has been permanently revised in order for the numbers to be as accurate as possible. We have used the latest available series to illustrate the methods of estimation of some quantities: substitutive work, technological index, marginal productivities, technological coefficients and bulk productivity of workman.

## 7.1 Technological Characteristics

The empirical time series of output $Y$, capital $K$ and labour $L$ are usually known, and, for the U.S. economy, are collected in Appendix 2. Methods of direct estimates of the third production factor—substitutive work $P$—are feasible (see Sect. 2.5.2), but are not developed sufficiently to make up a time series, so that a method of indirect estimation of substitutive work for the past periods is developed. In Sect. 7.1.2, will be shown that the substitutive work $P$ and also simultaneously the technological index $\alpha$ can be calculated according to the time series of the GDP $Y$ and production factors $K$ and $L$. When time series for all production factors are known, the technological characteristics of the production system can be estimated.

### *7.1.1 Personal Consumption and the Technological Index*

The technological index $\alpha$ appears to be a very important characteristic of the production system. An estimate of the technological index can be obtained when the optimal use of production factors is considered. According to Sect. 6.7 (formula 6.30), the technological index $\alpha$ can be expressed through prices and amounts of production factors as


V. N. Pokrovskii, *Econodynamics*, New Economic Windows,
https://doi.org/10.1007/978-3-319-72074-6_7

$$\alpha = \frac{pP}{cL + pP}. \tag{7.1}$$

This means that the technological index $\alpha$ represents the share of expenses needed for utilisation of capital services as a production factor (substitutive work) within the total expenses for production factors. Expression (7.1) allows one to estimate the technological index $\alpha$ from estimates of the cost of consumption of production factors.

The current consumption $C = cL$ is defined as the value of the minimum amount of products which are needed for the labour force to maintain. The cost of maintenance of labour can be estimated (see Sect. 2.2.4) through the poverty threshold $c^*$ as $cL = c^*N$, where $N$ is the number of population. Measured in this way, the consumption in year 1996 was $C = 2,120$ billion dollars compared with the expenses for maintenance of consumption of capital services $pP = \mu K = 1,378$ billion dollars (1996). So, for the latest decade of the twentieth century, one can obtain $\alpha \approx 0.4$, which means that about 40% of the total expenses for production factors take energy as a substitute of labour.

### 7.1.2 Substitutive Work and the Technological Index

A simple method allows us to build time series for both substitutive work $P$ and the technological index $\alpha$, if the empirical time series for output $Y$, capital $K$ and labour $L$ are known.

The value of the technological index $\alpha$ can be represented, due to Eq. (6.12), as

$$\alpha = \frac{\ln\left(\frac{Y}{Y_0}\frac{L_0}{L}\right)}{\ln\left(\frac{L_0}{L}\frac{P}{P_0}\right)} \tag{7.2}$$

However, the amount of substitutive work $P$ itself depends on the value of the technological index $\alpha$. The growth rate of substitutive work, from Eq. (5.20), is calculated as

$$\eta = \frac{\delta - (1-\alpha)\nu}{\alpha}, \quad 0 < \alpha < 1. \tag{7.3}$$

where $\nu$ is defined by Eq. (5.17) and includes possible correction determined by relations (7.8). Then, the time dependence of substitutive work can be restored by solving the equation

$$\frac{dP}{dt} = \eta(\alpha)P. \tag{7.4}$$

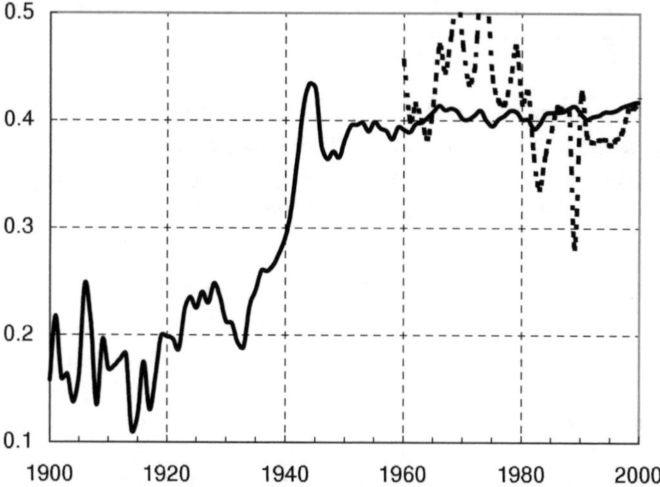


**Fig. 7.1** Technological Index. The solid line represents values of $\alpha$ found according to Eqs. (7.2)–(7.4). The dashed line represents values calculated due to formula (7.1). Adapted from [1]

Equations (7.2)–(7.4) allow to estimate the technological index $\alpha$ and substitutive work $P$ at given time series for $Y$, $K$ and $L$.

The results of calculation for $\alpha$ are depicted on the plot of Fig. 7.1 in line with the values of $\alpha$ calculated due to formula (7.1). Note that the choice of initial value of the technological index allows us to move the whole curve of $\alpha$ up and down, so it is important to have at least one point where absolute value of $\alpha$ is known, which, according to the estimation in Sect. 7.1.1, is taken as $\alpha \approx 0.4$ in year 1997. The calculated values of the technological index are used to estimate the total personal consumption. The results for the U.S. in twentieth century are shown in Fig. 2.4 in Sect. 2.2.4.2. The estimated values of the technological index allow us to calculate the growth rate of substitutive work $\eta$ and to restore the time dependence of the quantity $P$. The results are shown in Fig. 2.8 in Sect. 2.5.1 in line with the total (primary) consumption of energy in the U.S. economy. One can see that the substitutive work grew on average faster than the total consumption of energy in years 1900–2000; however, there are some years of recession.

### *7.1.3 Estimation of the Technological Coefficients*

From formula (5.20), the technological index $\alpha$ can also be calculated through the technological coefficients $\overline{\lambda}$ and $\overline{\varepsilon}$, which can be estimated independently. If one has time series of the production factors and investment, values of the coefficients of labour and substitutive work requirement can be found from Eq. (5.13), which can be rewritten in terms of the non-dimensional technological coefficients as

$$\frac{dL}{dt} = \left(\overline{\lambda}\frac{I}{K} - \nu' - \mu\right) L, \qquad \frac{dP}{dt} = \left(\overline{\varepsilon}\frac{I}{K} - \eta' - \mu\right) P. \tag{7.5}$$

These equations allow one to develop methods of estimation of the technological coefficients $\overline{\lambda}$ and $\overline{\varepsilon}$, whereas one has to take into account the conditions (6.10) of non-negativity of the marginal productivities (principle of productivity), which, in terms of the technological index, can be rewritten as

$$0 < \overline{\lambda} < \frac{1}{1-\alpha}, \quad 0 < \overline{\varepsilon} < \frac{1}{\alpha}, \quad 0 < \alpha < 1. \tag{7.6}$$

The depreciation coefficient $\mu$ in Eq. (7.5) is estimated from the time series of $I$ and $K$; these empirical values are determined in Sect. 2.3.3 and shown in Fig. 2.6. The conditions (7.6) allow one to separate the extra depreciation rate of labour and substitutive work $\nu'$ and $\eta'$ on the basis of time series of quantities $I/K$, $L$ and $P$.

One can assume that all consumption of labour $L$ is productive, thus values of the labour requirement $\overline{\lambda}$ can be calculated directly from the first equation from the set (7.5). An attempt to exploit this procedure, considering $\nu' = 0$, determines negative values of the technological coefficients (see the solid lines on the top plot in Fig. 7.2), which can be connected with errors in estimating the amount of labour and must be corrected by the quantity $\nu'$. To ensure the fulfillment of relations (7.6), one has to set the amendment $\nu'$ as

$$\nu' = -\frac{1}{L}\frac{dL}{dt} - \mu + \begin{cases} \overline{\lambda}_0 \dfrac{I}{K}, & \dfrac{1}{L}\dfrac{dL}{dt} + \mu < \overline{\lambda}_0 \dfrac{I}{K}, \quad \overline{\lambda}_0 < \dfrac{1}{1-\alpha}, \\[2ex] \dfrac{1-\alpha\,\overline{\varepsilon}_0}{1-\alpha}\dfrac{I}{K}, & \dfrac{1}{L}\dfrac{dL}{dt} + \mu > \dfrac{1-\alpha\,\overline{\varepsilon}_0}{1-\alpha}\dfrac{I}{K}, \quad \overline{\varepsilon}_0 < \dfrac{1}{\alpha} \end{cases} \tag{7.7}$$

where the positive quantities $\overline{\lambda}_0$ and $\overline{\varepsilon}_0$ are the prescribed bottom values of the technological coefficients.

The corrected results of the technological coefficients $\overline{\lambda}$ are shown by the solid lines on the bottom plot in Fig. 7.2. Values of the extra rate of depreciation $\nu'$ appear to be noticeable in the first half of the century, but quite insignificant after year 1950. It is known that estimates of the economic quantities for the first half of the century are less reliable than those for the second half, so the deviation of the quantity $\nu'$ from zero can be connected not only with performance of technological equipment, but with some possible errors in estimating economic quantities.

Then, one can turn to the estimation of the second technological coefficient – substitutive work requirement $\overline{\varepsilon}$. The dotted lines on the top plot in Fig. 7.2 shows values of primary energy requirement, calculated on the basis of the second equation from the set (7.5), in which primary energy $E$ stands for substitutive work $P$. The dashed line on the bottom plot in Fig. 7.2 shows values of substitutive work requirement, calculated, at known values of the labour requirement and technological index $\alpha$ (see the previous section), according to the formula

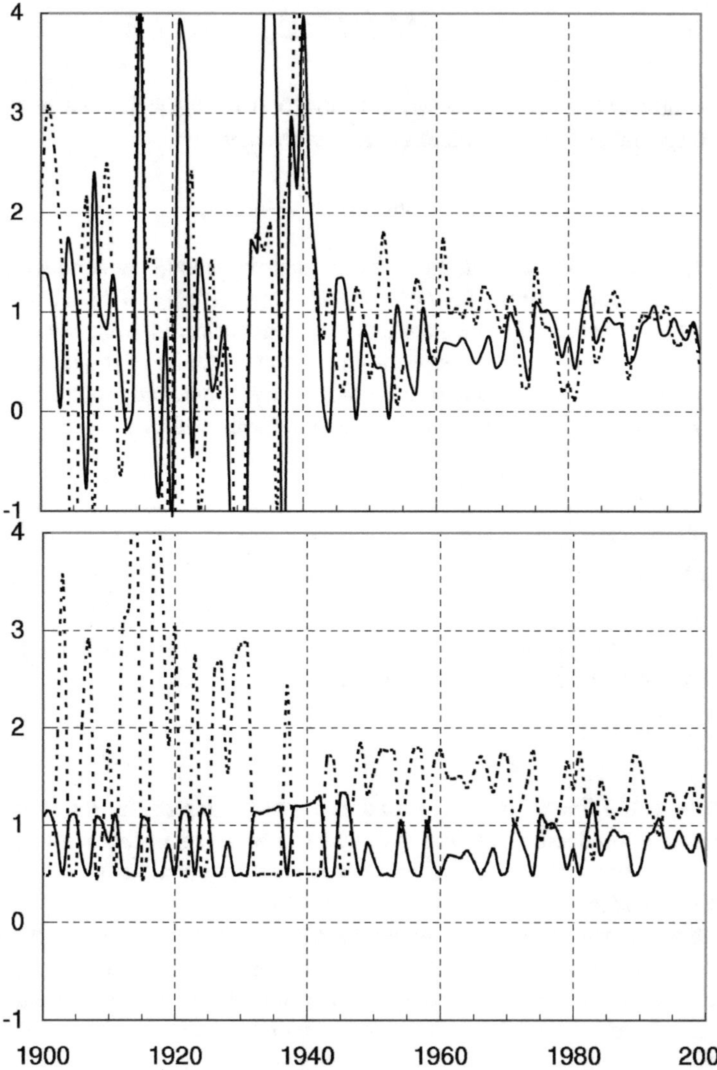


**Fig. 7.2** Technological coefficients for the U.S. economy. Top: Labour (solid lines) and primary energy (dotted lines) requirements are calculated according to relations from (7.5) (for labour at $\nu' = 0$) on the basis of time series for capital stock, labour, primary energy and investment. One can see that sometimes inequalities (7.6) are not fulfilled. Bottom: Corrected labour (solid lines) and substitutive work (dotted lines) requirements are calculated due to relations (7.5) and (7.8) (at $\nu' \neq 0$). Values of $\nu'$ are estimated due to Eq. (7.7) at $\overline{\lambda}_0 = \overline{\varepsilon}_0 = 0.5$

$$\overline{\varepsilon} = \frac{1 - (1 - \alpha)\overline{\lambda}}{\alpha}. \tag{7.8}$$

For a relatively reliable data of period 1950–2000, the technological coefficients do not change much, thus one can estimate mean values

$$\overline{\lambda} = 0.758, \quad \overline{\varepsilon} = 1.367.$$

The first quantity is less than unity. It means that on average labour-saving technologies are introduced during this span of time. One can see that the technological coefficients are determined by two controversial tendencies. In order to save labour and energy, they both must be less than unity. However, one of the technological coefficients must be greater than unity in order for the marginal productivity to be positive. As a result, it appears that the technological coefficients pulsate around unity.

### 7.1.4 Decomposition of Primary Energy

One can assume that, in accordance with the speculations of Sect. 2.5.1, the total primary energy $E$ can be broken down into two parts,

$$E = E_{\mathrm{C}} + E_{\mathrm{P}}. \tag{7.9}$$

The last part is the amount of energy carriers providing, after some transformation, the pure work of production equipment $P$, which is only a small part of total primary energy $E$. The two constituents of primary energy, as functions of time, behave differently with respect to labour $L$ as function of time. The property of production factor $P$ to be a substitute for labour allows us to state that an increase in consumption of substitutive work, which corresponds to an increase in consumption of primary substitutive work, can lead to a decrease in consumption of labour and otherwise. One expects the change of the first part $E_{\mathrm{C}}$ to correlate with the change of labour, and the change of the second part $E_{\mathrm{P}}$ to anti-correlate.

To analyse the situation, following [2], one has to consider the growth rates of the production factors, which, as was shown earlier, are connected with the investment $I$, depreciation coefficient $\mu$, and technological characteristics, $\overline{\lambda}$ and $\overline{\varepsilon}$, of a production system by Eq. (7.5). The non-dimensional technological coefficients $\overline{\lambda}$ and $\overline{\varepsilon}$ are characteristics of production equipment, which denote the required amount of labour and substitutive work per unit of introduced equipment (measured in units of total amount of capital $K$), respectively. The technological coefficients apparently have to be considered for characterisation of the process of substitution. From the definition of substitutive work and empirical investigation of these quantities (see Fig. 7.2), one has to define the correlation[1] of the technological coefficients as

$$\text{corr}\,(\overline{\lambda}, \overline{\varepsilon}) = -1. \tag{7.10}$$

Additionally, one has to consider changes of primary energy $E$ and its parts $E_{\mathrm{C}}$ and $E_{\mathrm{P}}$, the last being primary substitutive work. We assume that each quantity is also characterised by its own technological coefficients, so that, in line with Eq. (7.5), one can write three more balance equations,

$$\frac{dE}{dt} = \left(\overline{\varepsilon}_{\mathrm{E}}\frac{I}{K} - \mu\right) E, \quad \frac{dE_{\mathrm{C}}}{dt} = \left(\overline{\varepsilon}_{\mathrm{C}}\frac{I}{K} - \mu\right) E_{\mathrm{C}}, \quad \frac{dE_{\mathrm{P}}}{dt} = \left(\overline{\varepsilon}_{\mathrm{P}}\frac{I}{K} - \mu\right) E_{\mathrm{P}}. \tag{7.11}$$

The second terms on the right sides of these equations reflect the decrease in the production factors due to the removal of a part of the production equipment from service. For simplicity, it is assumed that the depreciation coefficients of all quantities in Eq. (7.11) are equal to the depreciation coefficient $\mu$ of production equipment (capital stock). This is true for the case when installed technological equipment does not change its quality during the time of service, which is assumed in the above equations.

One can presume the quantity $\overline{\varepsilon}$ to be a proxy of the quantity $\overline{\varepsilon}_{\mathrm{P}}$, and the quantity $\overline{\varepsilon}_{\mathrm{C}}$ to be proportional to quantity $\overline{\lambda}$

$$\overline{\varepsilon}_{\mathrm{C}} = \frac{\langle \overline{\varepsilon}_{\mathrm{C}} \rangle}{\langle \overline{\lambda} \rangle} \overline{\lambda}, \quad \overline{\varepsilon}_{\mathrm{P}} = \overline{\varepsilon}, \tag{7.12}$$

so that some of the correlations of the technological coefficients have to be defined as

$$\text{corr}\,(\overline{\lambda}, \overline{\varepsilon}_{\mathrm{C}}) = 1, \quad \text{corr}\,(\overline{\lambda}, \overline{\varepsilon}_{\mathrm{P}}) = -1, \quad \text{corr}\,(\overline{\varepsilon}, \overline{\varepsilon}_{\mathrm{P}}) = 1. \tag{7.13}$$

Note that these relations are the consequences of assumptions (7.12) and, in contrast to relation (7.10), have to be considered as approximate ones.

One can see that, due to Eqs. (7.9) and (7.11), some of the technological coefficients are connected by the relation

$$\overline{\varepsilon}_{\mathrm{E}} = (1 - x)\,\overline{\varepsilon}_{\mathrm{C}} + x\,\overline{\varepsilon}_{\mathrm{P}}, \quad x = \frac{E_{\mathrm{P}}}{E}. \tag{7.14}$$

---

[1]The correlation and covariance of two quantities $a$ and $b$ are defined as

$$\text{corr}\,(a, b) = \frac{\text{cov}\,(a, b)}{\Delta a\, \Delta b}, \quad (\Delta a)^2 = \frac{1}{n}\sum_{j=1}^{n}(a_j - \langle a \rangle)^2,$$

$$\text{cov}\,(a, b) = \frac{1}{n}\sum_{j=1}^{n}(a_j - \langle a \rangle)(b_j - \langle b \rangle), \quad \langle a \rangle = \frac{1}{n}\sum_{j=1}^{n} a_j.$$

This equation can be easily obtained, if one sums the last two equations from (7.11) and compares the result with the first equation from the same set.

To find an equation for the ratio $x$, we consider statistical characteristics of the technological coefficients. Relation (7.14) is followed by the relations for mean values, covariances and correlations, respectively,

$$\langle\overline{\varepsilon}_{\rm E}\rangle = (1-x)\langle\overline{\varepsilon}_{\rm C}\rangle + x\langle\overline{\varepsilon}_{\rm P}\rangle, \tag{7.15}$$

$$\mathrm{cov}\,(\overline{\lambda}, \overline{\varepsilon}_{\rm E}) = (1-x)\,\mathrm{cov}\,(\overline{\lambda}, \overline{\varepsilon}_{\rm C}) + x\,\mathrm{cov}\,(\overline{\lambda}, \overline{\varepsilon}_{\rm P}) \tag{7.16}$$

$$\mathrm{corr}\,(\overline{\lambda}, \overline{\varepsilon}_{\rm E})\Delta\overline{\varepsilon}_{\rm E} = (1-x)\,\mathrm{corr}\,(\overline{\lambda}, \overline{\varepsilon}_{\rm C})\,\Delta\overline{\varepsilon}_{\rm C} + x\,\mathrm{corr}\,(\overline{\lambda}, \overline{\varepsilon}_{\rm P})\Delta\overline{\varepsilon}_{\rm P}. \tag{7.17}$$

The last relation, taking Eq. (7.13) into account, can be rewritten as

$$\mathrm{corr}\,(\overline{\lambda}, \overline{\varepsilon}_{\rm E})\Delta\overline{\varepsilon}_{\rm E} = (1-x)\,\Delta\overline{\varepsilon}_{\rm C} - x\,\Delta\overline{\varepsilon}_{\rm P}. \tag{7.18}$$

One can use relations (7.12) and (7.15) to find the deviations of the quantities

$$\Delta\overline{\varepsilon}_{\rm C} = \frac{\langle\overline{\varepsilon}_{\rm C}\rangle}{\langle\overline{\lambda}\rangle}\,\Delta\overline{\lambda}, \quad \Delta\overline{\varepsilon}_{\rm P} = \Delta\overline{\varepsilon}, \quad \langle\overline{\varepsilon}_{\rm C}\rangle = \frac{\langle\overline{\varepsilon}_{\rm E}\rangle - x\,\langle\overline{\varepsilon}_{\rm P}\rangle}{1-x} \tag{7.19}$$

Equations (7.18) and (7.19) determine a formula for calculation of the ratio of primary substitutive work to total primary energy

$$\frac{E_{\rm P}}{E} = \frac{\langle\overline{\varepsilon}_{\rm E}\rangle\,\Delta\overline{\lambda} - \langle\overline{\lambda}\rangle\,\mathrm{corr}\,(\overline{\lambda}, \overline{\varepsilon}_{\rm E})\,\Delta\overline{\varepsilon}_{\rm E}}{\langle\overline{\varepsilon}\rangle\,\Delta\overline{\lambda} + \langle\overline{\lambda}\rangle\,\Delta\overline{\varepsilon}}. \tag{7.20}$$

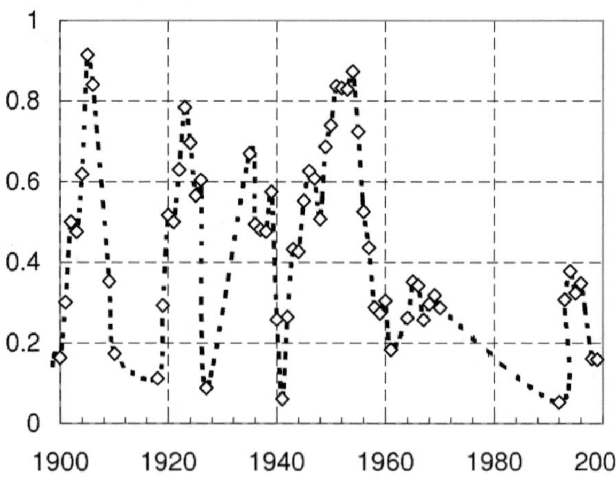


**Fig. 7.3** Share of primary substitutive work The ratio of primary substitutive work to total primary work $E_{\rm P}/E$ is calculated according to Eq. (7.20)

The formula contains statistical characteristics of the quantities $\overline{\lambda}$, $\overline{\varepsilon}$ and $\overline{\varepsilon}_E$, which can be estimated directly according to Eqs. (7.9) and (7.11) on the basis of values of substitutive work and time series for capital $K$, labour $L$, primary energy $E$ and investment from Table of Appendix 2. The values of the ratio for the U.S. economy are represented in Fig. 7.3. According to these results, absolute values of primary substitutive work are shown in Fig. 2.8 of Sect. 2.5.1. The results are realistic (close to the direct estimates of this quantity) and show ups and downs of the quantity in contrast to oversimplified direct estimates. The deviations of the calculated values of primary substitutive work from empirical ones are quite understandable, considering the rather arbitrary assumptions made in the empirical estimation of the quantity. For years 1911–1917, 1927–1934, 1962–1963 and 1971–1988, the calculated values are unrealistically small; one can suppose, assumptions (7.12), which are consequences of assumptions about the rates of depreciation of quality of production equipment, are too coarse in these cases.

## 7.2 Marginal Productivities

The differential formulae (6.2) and (6.14), that is

$$dY - \Delta dt = \begin{cases} \xi\, dK \\ \beta\, dL + \gamma\, dP \end{cases}, \qquad \Delta = Y \ln\left(\frac{L_0}{L}\frac{P}{P_0}\right)\frac{d\alpha}{dt} \tag{7.21}$$

allow one to estimate directly the marginal productivities $\xi$, $\beta$ and $\gamma$ due to empirical data. We assume that the time series for output $Y$, production factors $K$, $L$ and $P$ and the technological index $\alpha$ are known (the last two quantities obtained by exploiting the method of calculation described in Sect. 7.1.2).

From expressions (6.8), (6.13) and (6.15), the marginal productivities are connected with each other, that is

$$\xi = \beta\frac{L}{K} + \gamma\frac{P}{K}, \quad \beta = \xi\,(1-\alpha)\,\frac{K}{L}, \quad \gamma = \xi\,\alpha\,\frac{K}{P} \tag{7.22}$$

which allows us to test the theory using alternative estimates of the marginal productivities.

### 7.2.1 *Productivity of Capital Stock*

One way, to calculate the capital marginal productivity $\xi$, is a direct use of the formula

$$dY - \Delta dt = \xi\, dK. \tag{7.23}$$

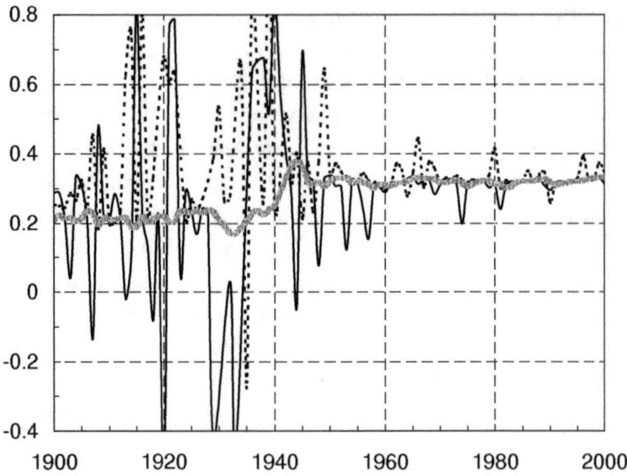


**Fig. 7.4** Marginal productivity of capital stock. The solid line represents direct estimates of $\xi$ from the empirical data and the equation $dY - \Delta dt = \xi\, dK$. The dashed line shows the marginal productivity calculated according to Eq. (7.22), while $\beta$ and $\gamma$ are estimated directly due to the empirical data and the equation $dY - \Delta dt = \beta\, dL + \gamma\, dP$. The dotted line represents the ratio $Y/K$

Another way is to use the bulk productivity of capital $\Xi$ defined by the relation

$$Y = \Xi\ K \tag{7.24}$$

and calculate the marginal productivity via the bulk productivity as

$$\xi = \Xi + K\,\frac{d\Xi}{dK}. \tag{7.25}$$

The two methods of calculation give almost identical results, which are shown in Fig. 7.4 by a solid line. For years 1950–2000, the mean values of the marginal productivity and its standard deviation can be estimated as

$$\xi = (0.307 \pm 0.044)\ \text{year}^{-1}. \tag{7.26}$$

Note that the differences $dY$ and $dK$ cannot be determined with great accuracy, thus negative and very large values of the marginal productivity are excluded as erroneous. It is not out of place to remember here the words of Morgenstern ([3], p. 4): 'There are many reasons why one should be deeply concerned with the "accuracy" of quantitative economic data and observations'.

One more way to estimate the capital marginal productivity is to use the first of relations (7.22), assuming that labour and energy marginal productivities, $\beta$ and $\gamma$, respectively, are directly calculated from empirical data, as will be demonstrated in the next subsection. The quantity $\xi$, calculated in this way, is shown in Fig. 7.4 by the dashed line. For the years 1950–2000, in this case, the mean values of the marginal productivity and its standard deviation can be estimated as

$$\xi = (0.337 \pm 0.039)\ \text{year}^{-1}. \tag{7.27}$$

The values of the marginal productivity (7.26) and (7.27) practically coincide with the averaged bulk productivity $Y/K$ that is $(0.321 \pm 0.009)$ year$^{-1}$; this is evidence that the capital marginal productivity does not depend on argument $K$. According to numerous observations [4, 5], the growth of $Y$ is approximately equal to the rate of growth of capital $K$, which can be confirmed by comparing formulae (2.14) and (2.29), so that the bulk productivity of capital $Y/K$ is approximately constant in the U.S. economy in the second half of the twentieth century.

### 7.2.2 Productivity of Labour and Substitutive Work

The direct way to calculate the marginal productivities of labour and substitutive work is to use the formula

$$dY - \Delta dt = \beta\, dL + \gamma\, dP. \tag{7.28}$$

Otherwise, the marginal productivities $\beta$ and $\gamma$ can be expressed through the bulk productivities of labour and substitutive work, $B$ and $\Gamma$, which are considered to be functions of the ratio $P/L$ and are defined by the relations

$$Y = B(P/L)\, L, \quad Y = \Gamma(P/L)\, P. \tag{7.29}$$

Indeed, having calculated the total differential of the output from the first equation, one obtains

$$\beta = B + \frac{P}{L}\frac{d\,B}{d(P/L)}, \quad \gamma = \frac{d\,B}{d(P/L)}, \tag{7.30}$$

Similarly, one can obtain, having calculated the total differential of the output from the second equation of the set (7.29),

$$\beta = -\left(\frac{P}{L}\right)^2 \frac{d\,\Gamma}{d(P/L)}, \quad \gamma = \Gamma + \frac{P}{L}\frac{d\,\Gamma}{d(P/L)}. \tag{7.31}$$

Calculations, according to relations (7.30) or relations (7.31), give slightly different values for the marginal productivities: one can use mean values of the two calculations. In comparison with Eqs. (7.28), (7.30) and (7.31) give an alternative method of direct estimation of the marginal productivities.

The non-dimensional marginal productivities $\beta L/K$ and $\gamma E/K$, estimated directly according to formula (7.28), are shown on the plot of Fig. 7.5 by the solid lines. The

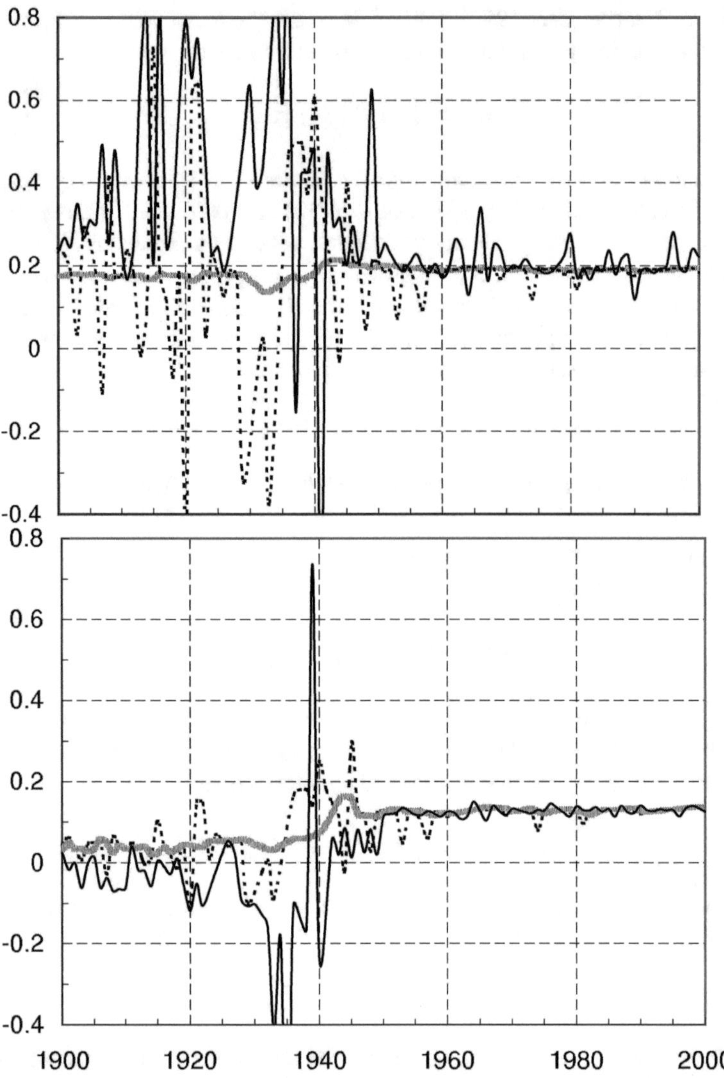


**Fig. 7.5** Marginal productivities of labour and substitutive work. Solid curves show direct estimate productivities of labour (top) and substitutive work (bottom) according to Eq. (7.28). Dashed curves represent results of calculation according to Eq. (7.22) at known values of capital stock productivities

marginal productivities are pulsating functions of time, and a maximum of one of the marginal productivities corresponds to a minimum of the other and vice versa. Averaged values of the quantities for the years 1950–2000 are estimated as

$$\beta \frac{L}{K} = (0.211 \pm 0.048) \text{ year}^{-1},$$
$$\gamma \frac{P}{K} = (0.117 \pm 0.012) \text{ year}^{-1}. \tag{7.32}$$

Note that negative and very large values of marginal productivities are omitted as connected with erroneous values of production factors.

Alternatively, the marginal productivities $\beta$ and $\gamma$ can be estimated from relations (7.22), assuming that capital productivity $\xi$ and values of the technological index $\alpha$ are known. The dashed lines in Fig. 7.5 show the alternative estimates of marginal productivities at empirical values of $\xi$ and empirical values of the technological index $\alpha$ calculated in Sect. 7.1.2. For the years 1950–2000 the alternative mean values of the marginal productivities and their standard deviations can be estimated as

$$\beta \frac{L}{K} = (0.192 \pm 0.025) \text{ year}^{-1},$$
$$\gamma \frac{P}{K} = (0.129 \pm 0.018) \text{ year}^{-1} \tag{7.33}$$

These values should be compared with the above estimates (formulae 7.32) of the same quantities.

The dotted lines in Fig. 7.5 represents the results of calculation of marginal productivities according to formulae (7.22) at empirical values of $Y/K$ and empirical values of the technological index $\alpha$ calculated in Sect. 7.1.2.

### 7.2.3 *What Is the Productivity of Capital?*

The results of estimating the marginal productivities confirm the validity relation (7.22). Thus, indeed, the marginal productivity of capital stock can be considered as the 'sum' of the marginal productivities of labour and substitutive work, and no other factors need to be included in the production function. Although one needs production equipment (capital stock) to attract extra amount of external energy to substitute labour, work (labour services) can be replaced only by work (capital services), not by capital stock. Productivity of capital stock is, in fact, productivity of labour and energy, and the main result of technological progress is the substitution of human efforts by the work of external energy sources by means of different sophisticated appliances. The production system of society is a mechanism which attracts a huge amount of energy to transform matter into things that are useful for human beings.

According to expression (6.18), the productivity of capital changes, if the technological coefficients and/or the production factors change. The former causes fast pulsation, while the latter provokes slow trends of the capital productivity. This quantity apparently depends on the definition of production capital $K$. Let us recall that the notion of production capital has to be refined by excluding some commodities from production investment. The share of core production capital in the total production investment remains unknown, but more importantly, the growth rate of the core production capital can differ from the growth rate estimated on available statistical data.

## 7.3 Productivity of Labour

The scientific and technical progress could be reduced to processes of introduction of innovations, that is successive replacement of instruments, materials, designs, adaptations and other objects with more perfect ones, from some point of view. Among all processes of replacement, the outstanding role is played by the processes of replacement of workers' efforts by machine work with the assistance of forces of nature. Substitution of efforts with machine work is a process of replacement which strongly influences the labour productivity, which is understood as the ratio of the value of output, measured in value units of constant purchasing capacity, to the expenditures of labour. According to (6.12), the productivity of labour can be written as

$$A = \frac{Y}{L} = \frac{Y_0}{L_0}\left(\frac{L_0}{L}\frac{P}{P_0}\right)^{\alpha}.$$

One should be assured that, at the definition of labour productivity, the output is measured in value units of constant purchasing capacity, as one speaks, to represent a 'physical' measure of output.

The values of labour productivity in the U.S. is depicted on Fig. 7.6. The quantity depends on the ratio of substitutive work to workers' efforts $P/L$; the increase in labour productivity cannot be understood without considering the phenomenon accompanying the progress of production—the attraction of natural energy sources (animals, wind, water, coal, oil and others) for performing work that replaces human efforts in production. However, the growth of labour productivity is connected also with modification of methods of utilisation of labour and energy, which is described by changing of the technological index $\alpha$; the essential transform for the U.S. is observed after year 1940.

The growth of labour productivity can be interpreted also on the base of equation for the growth rate, which follows Eq. (6.21) written on the assumption, that characteristics of the equipment do not change after its installation, that is, $\nu' = 0$ and $\eta' = 0$,

$$\frac{1}{A}\frac{\mathrm{d}A}{\mathrm{d}t} = \frac{(1-\overline{\lambda})(\nu+\mu)}{\overline{\lambda}} + \frac{1}{\xi}\frac{\mathrm{d}\xi}{\mathrm{d}t}. \tag{7.34}$$

In this equation, besides the known factor of amortisation $\mu$, there is a non-dimensional quantity $\overline{\lambda}$, which characterises the technology introduced into the production. The labour requirement $\overline{\lambda}$ appears to be the most important quantity to determine the growth rate of labour productivity. This quantity is a measure of the substitution of labour by energy. If $\overline{\lambda} = 1$, variations in technology do not occur, labour productivity is constant, and all incrementing of a product is connected only with an increase in human efforts. Human efforts are, certainly, the main motive

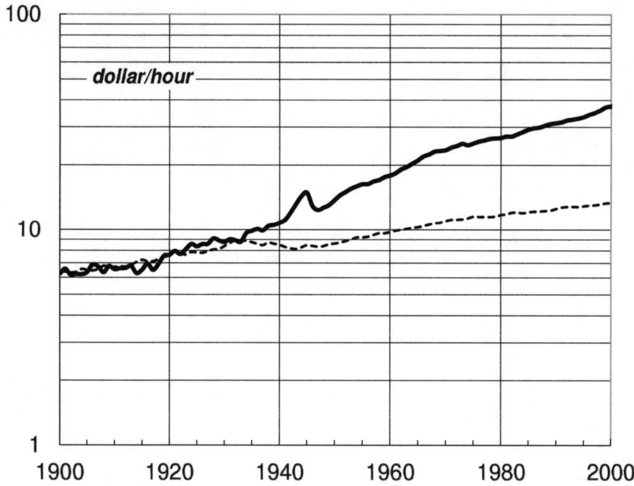


**Fig. 7.6** Labour productivity in the USA. The solid line represents actual values of labour productivity to the USA, measured in dollars of 1996 per one hour of worker. The changing of productivity has two components; the dashed line represents variation of the productivity due to the variation of $P/L$ at constant public resources ($\alpha = const$)

power, but, under the condition $\overline{\lambda} < 1$, the workers' efforts are partially replaced with the work of the machines movable by outer energy sources, and the labour productivity increases. This is a general description of the influence of scientific and technological progress, which occurs naturally in a picture of progress of mankind.

## References

1. Pokrovski, V.N.: Energy in the theory of production. Energy **28**, 769–788 (2003)
2. Pokrovski, V.N.: Productive energy in the US economy. Energy **32**, 816–822 (2007)
3. Morgenstern, O.: On the Accuracy of Economic Observation, 2 edn, completely revised. Princeton University Press, Princeton (1973)
4. Scott, M.F.G.: A New View of Tconomic Growth. Clarendon Press, Oxford (1989)
5. Blanchard, O.J., Fisher, S.: Lectures on Macroeconomics. MIT Press, Gambridge (1989)

# Chapter 8
# Social Production in Russia

**Abstract** The social production in Russia in years 1961–2060 is considered in terms of the theory formulated in the previous chapters, which provides the consecutive description of production functioning and the way to analyse the consisted trajectories of the future development. Scripts of the future take into account the technological level and possibilities of the usage of labour and outer energy, which distinguishes our method of the analysis of trajectories of development from the traditional methods used, for example, by the Ministry of Economic Development of the Russian Federation. The analysis of development of the national economy of Russia since 1961 and the script of 'conservative' future of Russia since 2016 are considered as an illustrative example.

## 8.1 Principles of Consistent Analysis

The considered period includes the unique events of destruction of economic system of the Soviet Union and creation of a market economy of the Russian Federation, which are described in detail by modern researchers (see, for example, [1, 2]). We can illustrate this dramatic history and complete the known analysis with the technological details, omitted in the existing investigations. We hope that history of Russian development and our analysis and methods of evaluation of scripts could be interesting for researchers of various economies.

The fundamental characteristic of production systems is the gross domestic product (GDP) that is determined by expression (6.16) through production factors: the set of production equipment (measured in its value) $K$, workers' efforts $L$ and substitutive work $P$. The theory describes capacity of the production equipment to operate with the usage of workers and energy: the effect of substitution of human efforts with work of some outer energy sources is taking into account. The production factors $K$, $L$ and $P$ are governed by simple balance equations (5.6)–(5.13) that contain technological characteristics of the used production equipment (quality of production capital). One of the characteristics, that is, the labour requirement—a quantity showing the amount of labour that is required for introduction of unit (by value) of production assets in action—appears to be the most significant. The method



V. N. Pokrovskii, *Econodynamics*, New Economic Windows,
https://doi.org/10.1007/978-3-319-72074-6_8

of the description allows to distinguish the influence of factors of production and the technological modification (structural and/or technological variations) of the system of production.

The development of the production system is determined eventually by possibilities to attract the production factors (see Sect. 5.4.1), but, to reveal the rates of potential growth of factors, a special research is required, and, to avoid it, it is possible to use for calculation of output $Y$ the simpler scheme described in Sect. 6.4.3, which assumes knowledge of four quantities: productivity of a fixed capital $\xi(t)$, the growth rate of expenditures of labour $\nu(t)$, the coefficient of depreciation $\mu(t)$ and the non-dimensional technological coefficient $\overline{\lambda}(t)$. On the basis of this method, it is possible to operate in following way: we suggest a hypothetical trajectory of the GDP to be set, assuming factor of depreciation, as well as the growth rates of expenditures of labour and investments known, and evaluate whether this trajectory can be realised at the existing assessments of efficiency of technology and the availability of power resources. We use the last method in this chapter for an appraisal of scripts of development of production system of Russia.

To identify the model of development of the national economy, we apply to empirical data for Russia that is collected in Appendix C in the form of the time series for some quantities, such as, first of all, the GDP, expenditures of labour, investments and value of the basic production assets. These time series are used to analyse and describe the economic development of the production system of Russia before the year 2016. To design a script of development after 2015, we suggest that expenditures of labour remain constant, but, nevertheless, we assume that the total expenses for maintenance of labour grow with the growth rate of GDP, so that the payment for expenditures of labour (the sum of wages and salaries) grows with the same rate. Such script can be realised only with substitution of expenditures of labour by work of the production equipment, which determines $\overline{\lambda} < 1$; considering stable development, we assume the value of coefficient of labour requirement to be constant. The listed assumptions determine a unique trajectory of development with the permanent growth rate of the GDP that is equal to 2%, so that the script of development is close to conservative script of the Ministry of Economic Development of the Russian Federation till the year 2030 [3].

## 8.2 Gross Domestic Product

The GDP represents an estimate of results of efforts of all members of the society that are spent for maintenance and development of the population and social structures. This quantity includes an assessment of tangible (buildings, equipment, food, etc.) and intangible (services, fundamental research, management, etc.) results of activity of the society. The intangible products, so as well as tangible products are necessary for the society to exist. Following Adam Smith and Karl Marx interpretation, it was considered in the USSR that work on creation of the intangible products does not produce value in economic sense, and the assessment of non-material products in the republics of the Soviet Union was not included in the national income, so that there was a tendency to neglect the sectors of non-material products. It is accepted now officially (see Appendix A) that estimates of non-material products are to be included in the balance, which is needed for more accurate and appropriating to the objective (see the discussion in Chap. 2) analysis of the activity of the society. Estimates of production of non-material products, which were required for restoration of the amount of the GDP for the period before capitalism restoration (1961–1990), were executed by Ponomarenko [4].

The estimated values of the GDP for the period till the year 2015 are collected in the table of Appendix C and shown in Fig. 8.1. To describe the development of the national economy in years after 2015, one can assume any dependence of the GDP in time, but not each trajectory is consistent with the characteristics of existing technology and availability of production factors. The tentative estimations show that the fast growth of GDP can appear incompatible with possibilities of the existing technologies and the availability of production factors, and, eventually, the growth rate of the GDP is taken as 2%, which provides a stable growth with constant substitution of expenditures of labour by work of the production equipment. The preset rate of development appears to be optimum under the reasonable assessments of efficiency of technology and availability of power resources. The continuous line in Fig. 8.1 represents extension of dependence at the preset growth rate. In process of delivery of new data, the trajectory of the future development can be more definite.

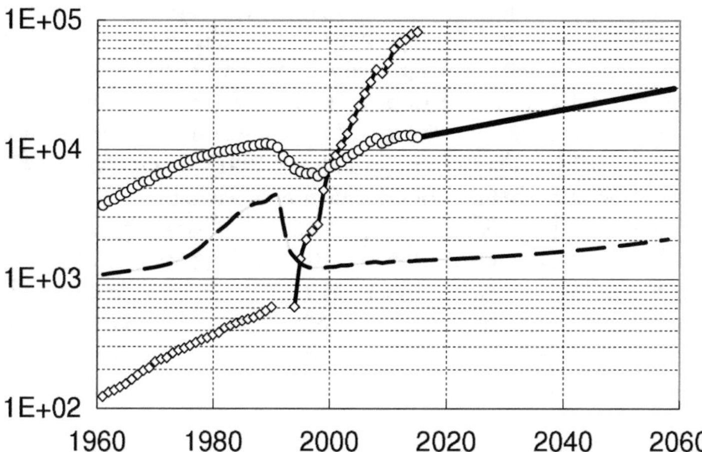


**Fig. 8.1** The GDP. The estimated values of the GDP are presented with rhombs (in rubles of current year, more abrupt and short curve) and empty circles (in rubles of 2000). The values for years 1961–1990 are given according to assessments by Ponomarenko [4], the values of the GDP from 1995 till 2015—according to Bessonov's analysis [5] and to the modern data of the Rosstat [6]. The extension of dependences of the GDP after 2015 corresponds to the 'conservative' script with the constant growth rate of 2%. All values are shown in billions ($10^9$) of rubles. The dashed median curve represents the created value in millions of power units at a choice of provisional scale of 50000 J

## 8.3 Population and Work Force

The entire work expenditures can be measured by a number of people working in a national economy, which is not an accurate measure of the efforts spent in production; the hours of work can increase or decrease at constant number of workers. The work expenditures could be estimated by working hours more accurately, though it is clear that the work can have various intensity, and working hours in various conditions do not represent equivalent work expenditures: even at the account of the working hours, it is necessary to introduce corrections for the quality of applied efforts as it is done, for example, in statistics of the United States. But it should be the subject matter of separate consideration.

The values of expenditures of labour in man-hours for Russia since 2005 can be found in issues of Rosstat [6]; however, for the previous years, it is possible to find only estimates of the number of workers. The assessment of the situation for these years is given by Voskoboynikov [7]: 'In official statistical publications for various years (The National economy of the USSR, The National economy of RSFSR) up to year 1990, one can find the data about mid-annual number of workers, state employees and collective farmers inclusively. In the Russian statistical year-book there are the data about mid-annual number of employed people from year 1980, while the existing values of mid-annual number of the employed in 1980, 1985 and 1990 mismatch to the mid-annual number of workers, state employees and collective farmers for the same years. Thus, simple association of the numbers of employment for the periods before and after 1990 has a certain error'. Voskoboynikov [7] offers, for the assessment of expenditures of labour up to 1990 inclusive, to use a mid-annual labour workforce, and then mid-annual number of employed people. As a compromised version, we suppose that till 1991, about half of population participates in production activity that corresponds to the official estimates of the number of workers for years 1980, 1985, 1990 and 1991 [8]. For the next years (after 1991), we use the data of Rosstat on number of employees in the national economy of Russia.

At known number of working people, the number of working hours can be estimated. By Rosstat assessments for 1980–1990 [8], an employee worked on average 230 days in a year, 7–8 h per day or about 35 h per week, for all considered period; however to harmonise the final results with official assessments of expenditures of labour after 2005, it is necessary to increase week performance for everyone till 40 working hours. Assessments of labour expenses in hours for a year are shown in the table of Appendix C.

The future numbers of workers are connected with the general demographic situation, which is influenced by the economic situation also. For forthcoming years, we assume the expenditures of labour to be constant. We also assume that the total expenses for maintenance of the labour force grow with the growth rate of the GDP, so that the payment for unit of expenditures of labour (wages and salaries) grows with the same rate.

## 8.4 Investments and the Basic Production Assets

Let us remind that investments are understood as estimation of value of equipment (the car, vehicles, the tool and stock, including expenses for installation of equipment for place of its constant operation, check and test of quality of installation) and constructions, established for production activity. The production investments $I$ determine the value of the basic production assets or production capital $K$ according to Eq. (5.6), that is,

$$\frac{\mathrm{d}K}{\mathrm{d}t} = I - \mu K. \tag{8.1}$$

Dynamics of the production equipment is characterised by the three quantities; if time series for two of them are known, for example, values of investments $I$ and the capital $K$, it is possible to estimate values of the third quantity, in this case, coefficient of depreciation $\mu$.

The time series for the investment $I$ and capital $K$ (in the current and comparable prices) *in Appendix C* are based on the values for the national economy for 1961–1990, as estimated by Ponomarenko [4]. For the extension of the time series for some of the next years, we used the values of investments from publications of Rosstat [6].

The values of capital $K$ for the period before the year 2016 are calculated according to formula (8.1) at known investments and coefficient of depreciation $\mu = 0.04$. The specified value of coefficient of depreciation corresponds to the service life of the equipment that is about 25 years for the Russian economy (see Table 1 in Bessonov and Voskoboynikov's work [9], p. 216), though this quantity could be greater. In the USA, for example, in the last years of the twentieth century, the value of the coefficient of depreciation reached 0.07 (see Fig. 2.6 in the second chapter). The initial (in the year 1960) value of the basic production assets is accepted to be equal to about two-thirds of the official estimate of value of production funds in statistics of the USSR. The values of the fixed capital $K$ after the year 2000 are based on the estimates of Rosstat [6]. The knowledge of investments and capital for this period allows to evaluate the coefficient of depreciation $\mu$, which in recent years takes values close to 0.07. The values of the established in this way investment and the basic production assets are collected in the table of Appendix C and shown in Fig. 8.2.

When designing the scripts of development of economy, it is necessary to take into account that investments cannot be appointed in an arbitrary way. The amount of the introduced fixed capital (quantity of investments) is determined by a possibility of attraction of additional production factors (work and substitutive work) and by the quality of the existing and establishing fixed capital, which is formalised in Eq. (5.13). In the simplest case, when there is no possibility to involve additionally neither work nor productive energy, Eq. (5.13) provides the restrictions on possible investments

$$\frac{I}{K} \leq \frac{\mu}{\lambda}, \quad \frac{I}{K} \leq \frac{\mu}{\varepsilon}. \tag{8.2}$$

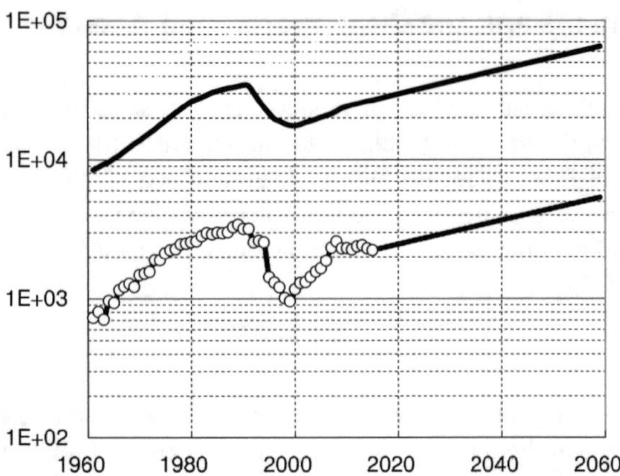


**Fig. 8.2** The investments and production capital. The values of investments for years 1961–1990 are represented by empty circles according to the estimates of Ponomarenko [4] and to the data of Rosstat [6] (from 1995 till 2015). The extension of dependences corresponds to the 'conservative' script with the constant growth rate equal to the growth rate of the GDP. The top curve represents values of the basic production assets, calculated according to formula (8.1). All quantities are shown in billions ($10^9$) rubles of 2000

This equation shows that, at constant quantity of the used production factors (expenditures of work $L$ and substitutive work $P$), the introduction of the new working equipment is possible only under condition of replacement of the old production assets, that is, at the coefficient of depreciation $\mu \neq 0$. The least available production factor limits the possible investments, and their quantity is determined also by the quality of the equipment to be installed, that is, by the values of labour requirement $\overline{\lambda}$ and energy requirement $\overline{\varepsilon}$. In case of the non-zero growth rates of the production factors, the restriction on possible investments is recorded in more complex form.

For the considered script of development of the national economy of Russia for the period after 2015, when, for simplicity, we consider that consumption of labour cannot be increased, $L = const$, by virtue of the first of Eq. (5.13), we find that investments are determined by the technological characteristics of the production system

$$\frac{I}{K} = \frac{\mu}{\overline{\lambda}}. \tag{8.3}$$

To provide the growth in the situation of the constancy of labour, the designed and introduced equipment should be based on the labour saving technologies that should be $\overline{\lambda} < 1$. The lesser the value of labour requirement (and accordingly, the greater use of substitutive work), the greater the investments and the greater the growth rate of GDP.

We assume that the values of investments grow with the growth rate of GDP. The values of the capital are calculated according to Eq. (8.1) with value of the coefficient of depreciation $\mu = 0.06$, which corresponds to the average value of the quantity after 2000, while the values of the labour requirement $\overline{\lambda}$ are determined by Eq. (8.3). The estimates of the labour requirement $\overline{\lambda}$ for the previous periods of development of Russia are given in Sect. 8.7.

## 8.5 Constituents of the GDP and Social Wealth

The GDP represents a set of products that have been created during a certain time (year, quarter) and used both for direct consumption and for production and non-production accumulation. As the set of products, the GDP can be divided, according to functionality of the products, into the three parts. First of all, it is necessary to exclude products (the foodstuffs, manufactured goods and services) straightforwardly consumed by the population over the considered period (year, quarter). Then, the products intended for maintenance and development of production or the investments in real form into the basic production capital (buildings, production equipment, networks of supply and so on) ought to be separated. If to estimate and exclude these parts of GDP (similar to that, as it has been done for the U.S., see Sect. 2.2.4), there remain huge (by value) quantity of products, which can be classified as the social accumulation or stocks. The last part of the GDP contains both tangible (buildings, the means of defence, parks, public constructions, etc.) and intangible (results of knowledge of laws of nature, works of art, codification of public laws and so forth) products. The intangible accumulation represents the results of human activity, vital for existence and progress of the society, and makes an essential part of all accumulation.

As an estimate of directly consumed products, which are necessary for the maintenance of the existence of population, can be taken, apparently, the quantity of the living standard multiplied by number of residents of the country (see Sect. 2.3.4). In initial years of the considered period for Russia, as a part of the USSR, the living standard is unknown, but for those years, the directly consumed product can be estimated as the sum of salaries (the total wage and salary fund) adding the expenses from public funds. The part of the salary fund was spent for accumulation, but this part assumingly was small, and we use the sum of salaries as an estimate of directly consumed product. In recent years (2000–2015), the living standard is estimated by Rosstat, and the directly consumed product can be calculated on the basis of official numbers. The future values of directly consumed product are determined both by development of the national economy and decisions of the operating bodies. We assume that the growth of directly consumed product occurs with the growth of the GDP. However, more accurate estimates of directly consumed product for Russia would be desirable.

The quantity that remains after a deduction from the GDP of investments, which, apparently, are estimated accurately (see the previous section), and the consumed product can be considered as the production and non-production accumulations. For more accurate estimates, it is necessary to consider, certainly, the balance of foreign trade, which is neglected here.

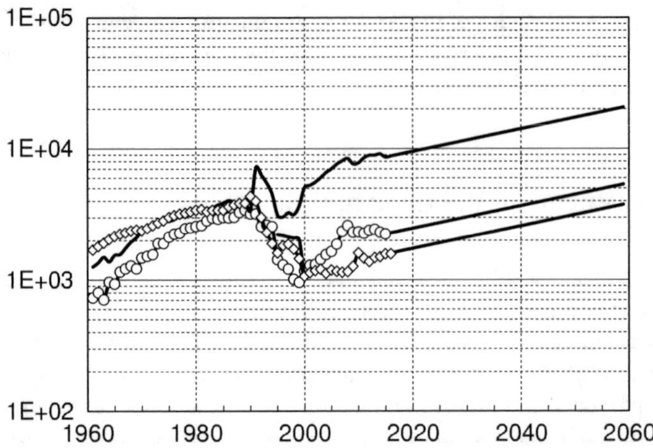


**Fig. 8.3** The components of the GDP. The investments are represented by open circles; the dependence is continued by a solid line. The quantity of straightforwardly consumed product is represented by rhombs and by the dotted curve. The top solid line represents the social accumulation, including a possible outflow of value abroad. All quantities are shown in billions ($10^9$) rubles of year 2000

The estimates of three specified components for Russia are shown in Fig. 8.3. The relationship between the parts of the GDP does not obey to any explicit law, but also is not, apparently, arbitrary one, being determined by the public discussions and the decisions of operating bodies. The empirical data for the U.S. (Chap. 2, Figs. 2.5, 2.6 and 2.7) and for Russia (Fig. 8.3) show that three parts are comparable and, one can think, are equally necessary for functioning the society. When designing the script of development, we should not allow any of the specified parts to prevail of others.

The production and non-production accumulation represent the contribution to the social wealth. To estimate the total value of the saved up social wealth $W$, it is possible to use Eq. (6.28), that is,

$$\frac{\mathrm{d}W}{\mathrm{d}t} = Y - C - \mu W, \tag{8.4}$$

where $C$ is the total direct consumption of products for a year.

## 8.6 Energy and Substitutive Work

The total consumption of primary energy carriers (for simplicity, one speaks about consumption[1]) $E$ can be divided, as it was explained in Sect. 2.5.2, in the two parts according to the functional role in production: a part of energy carriers (energy) is consumed directly by household and production enterprises for illumination, heating,

[1] See footnote 9 in Sect. 2.5.1.

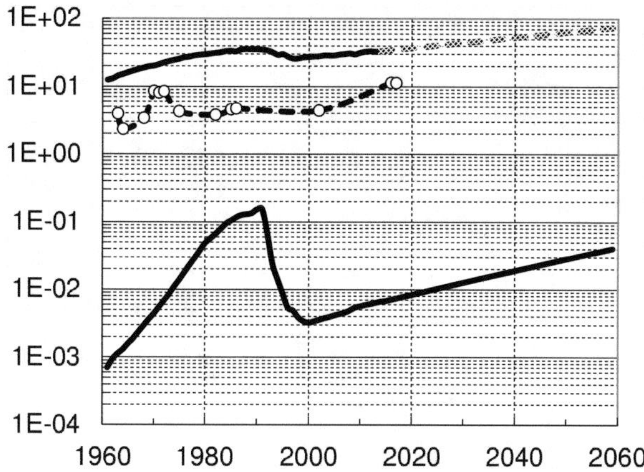


**Fig. 8.4** Energy in the national economy of Russia. The total primary consumption of energy carriers is presented by the top curve: the solid line shows the actual consumption till 2015, and the dotted line shows the possible consumption after 2015 according to the forecast of the Ministry for the Power Generating Industry of the Russian Federation [10]. The dashed line with the circles represents the primary productive energy consumed by the production equipment for performance of work of substitution; this quantity is calculated, as a part of primary energy, variations of which anticorrelate with variations of expenditures of labour. The bottom solid curve represents the true work of the production equipment on substitution of labour (substitutive work, the curve can be shifted upwards or downwards). All the quantities are estimated in quads (1 quad $= 10^{15}$ BTU $\approx 10^{18}$ J) for a year

chemical transformations and other aims, while another part—*primary substitutive work* $E_P$—is used to provide work of various appliances to substitute the efforts of people. The primary substitutive work $E_P$ is real work, and, unlike it, the other part of consumed energy $E - E_P$ is called by quasi-work (see Sect. 2.5.1). Such division has arisen in connection with the detailed investigation of the role of energy in production. For evaluation of the various parts of energy on the basis of empirical data, one can use the known methods (see Sects. 2.5.4 and 7.1.2), and the estimates for Russia are shown in Fig. 8.4 and in the table of Appendix C.

The values of the productive primary energy $E_P$ are estimated on the basis of the assumption that not only variations in true substitutive work $P$ but also variations in the productive primary energy $E_P$ anticorrelate with the variations in expenditures of labour $L$ (see Sect. 7.1.5). The substitutive work $P$ and the technological index $\alpha$ are calculated according to the known time series for the GDP $Y$, expenditures of labour $L$ and the production capital $K$, as explained in Sect. 7.1.2. One can see in the picture that the work $P$, actually substituting the worker efforts, essentially differs from the amount of primary productive energy $E_P$ and the total consumption of primary energy carriers $E$. The growth rate of substitutive work differs from the growth rate of primary energy; probably, this rate is closer to the growth rate of consumption of electricity, so as the electric motors are the main engines in production over the

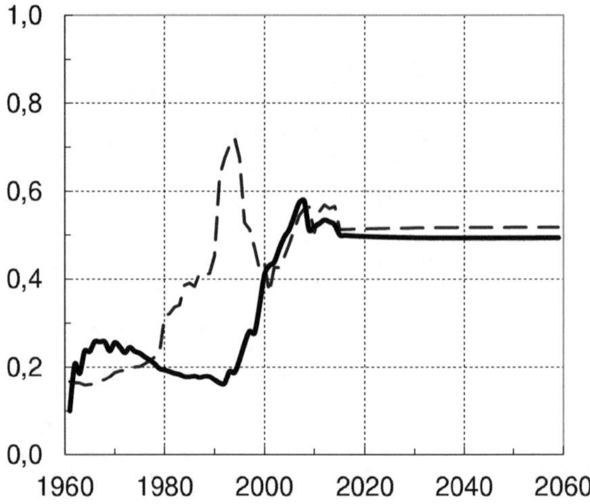


**Fig. 8.5** The technological index. The solid line represents values of $\alpha$, found according to Eqs. (7.2)–(7.4). The dashed line represents the values calculated for most favourable distribution of production factors (Eq. 8.5)

second half of the last century [11]. These results for production system of Russia can be compared with the findings for the U.S. (see Sect. 2.5.1).

The technological index $\alpha$ (see Fig. 8.5), which represents a combination of technological coefficients (see formula (5.20)), has been calculated simultaneously with true substitutive work. Also as for the substitutive work, the entire curve $\alpha$ can be shift upwards and downwards, so that it is important to know, at least, one absolute value of $\alpha$ at any point. Such a value can be found from the estimates of the total spending on the maintenance of work and the production equipment, as it has been demonstrated for the USA in Sect. 7.1.1, according to the formula

$$\alpha = \frac{pP}{cL + pP}. \tag{8.5}$$

To estimate the technological index $\alpha$, one needs to know the prices ($c$ and $p$) and quantities ($L$ and $P$) of production factors. The expenses on maintenance of labour force $cL$ are defined as value of the minimum quantity of products, which are necessary to support the labour force. This quantity can be estimated (see Sect. 2.2.4) through a threshold of poverty $c^*$ as $cL = c^*N$, where $N$ is the number of population. For Russia, such estimates can be executed for the period since 2000, so as from this year the minimal living standard was specified. For the Soviet era, instead of minimal living standard, it is possible to take wages (salaries) and number of workers. However, both in the first and in the second cases, the numbers are very approximate and are required to be corrected.

The results of the assessment according to formula (8.5) are shown in Fig. 8.5. One can notice that there is a correspondence of the assessments by the two methods in the beginning of 60th years—the period, which can be considered as rather stable period, and after the year 2000, but the distinction of the assessments in the 90th years has appeared; for these years, one cannot speak apparently about optimal allocation of production factors. The technological index $\alpha$ appears to be a rather conservative characteristic of production. The essential variations in the technological index are initiated by the extraordinary events, such as the influence of the World War II in years 1940–1945 on a national economy of the United States (see Fig. 7.1) or influence of the crisis of social system in Russia in years 1980–1990 (see Fig. 8.5).

## 8.7 Technological Characteristics of the Basic Production Assets

A convenient characteristic of introduced equipment is the amounts of production factors that are necessary for the equipment to be in operation, namely, the quantities of labour and substitutive work per unit (of value) of the introduced equipment, that is, the labour requirement $\overline{\lambda}$ and energy requirement $\overline{\varepsilon}$, which are discussed in Sect. 5.2.1. These quantities determine together (see formula (5.20)) the technological index $\alpha$ that is present in the production function as well.

Figure 8.6 shows values of the labour requirement and energy requirement, calculated by the method described in Sect. 7.1.3, according to the values of production factors and investments that was found earlier. The technological coefficients show the amounts of substitution of workers' efforts by work of outer forces in production processes. If the quantity $\overline{\lambda} < 1$ (and, accordingly, $\overline{\varepsilon} > 1$), expenditures of work (per unit of value of the production equipment to be installed) decrease, and substitutive work increases, which determines the tendency to an increase of labour productivity. One can note on the chart that average value of labour requirement $\overline{\lambda}$ is equal to 0.5 before the year 1980; it means that production equipment that had provided the reduction of expenditures of labour was introduced in that period. From the middle of the ninetieth years till 2008 $\overline{\lambda} > 1$, which shows the greater attraction of people to production. For the scenario of development in the period after 2015, the labour requirement is assumed to be constant and equal to 0.74. For comparison, we shall

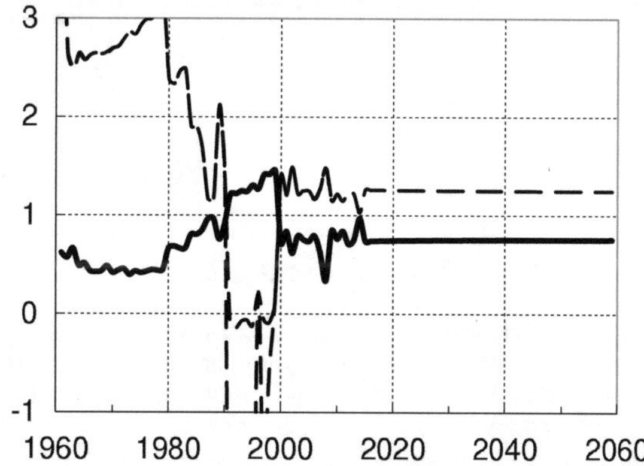


**Fig. 8.6** The technological characteristics. The solid curve represents the labour requirement, and the dotted one represents the energy requirement of the introduced production equipment in the national economy of Russia

remind (see Sect. 7.1.4) that the average value of the labour requirement in the second half of the twentieth century in the USA was nearby 0.8.

The total quantities of workers efforts and outer work can change arbitrarily. When the efficiency of use of primary energy by the equipment is improving, the total of consumption of energy carriers can decrease.

## 8.8 Marginal Productivities of Production Factors

The marginal productivities of production factors $\xi$, $\beta$ and $\gamma$ appears to be the fundamental characteristics of production system. To estimate their values, it is convenient to write, as it is done in Chap. 6, the production function in the form

$$dY - \Delta dt = \begin{cases} \xi\, dK \\ \beta\, dL + \gamma\, dP \end{cases}, \quad \Delta = Y \frac{1}{\xi} \frac{d\xi}{dt} = Y \ln \left( \frac{L_0}{L} \frac{P}{P_0} \right) \frac{d\alpha}{dt}. \tag{8.6}$$

To calculate the marginal productivities according to these relations, one needs to know the time series of output $Y$ and production factors $K$, $L$, as well as the calculated quantities of substitutive work $P$ and the technological index $\alpha$. The procedures of assessment of the marginal productivities have been described in Sect. 7.2 for the economy of the USA. The application of these procedures for the Russian Federation gives the values for the marginal productivities of the production factors shown in Fig. 8.7.

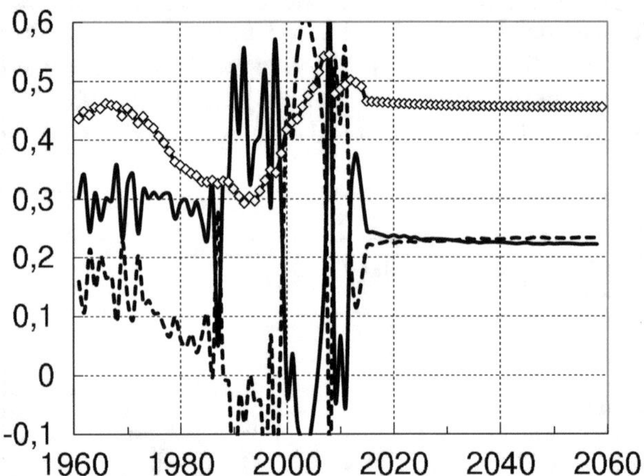


**Fig. 8.7** Productivity of production factors. The estimates of the marginal productivities, measured in units of year$^{-1}$, are presented by the curve with rhombs for the marginal capital productivity $\xi$, by the solid curve for the non-dimensional marginal labour productivity $\beta L/K$ and by the dotted curve for non-dimensional marginal productivity of substitutive work $\gamma P/K$

The productivity of the basic production assets (fixed capital), calculated as the ratio of output to value of the fixed capital

$$\xi = \frac{Y}{K}, \tag{8.7}$$

decreases, beginning since 1967, and falls by 1990 twice in comparison with the year 1970. The decrease in productivity of the fixed capital is equivalent to reduction of the total investments $I$ according to the rule

$$\frac{1}{\xi}\frac{\mathrm{d}\xi}{\mathrm{d}t} = -\frac{I}{K} + \mu, \tag{8.8}$$

so that the real increase of investments over these years did not influence on output. The reduction of capital productivity is one of the main characteristics of evolution of the national economy of Russia in the years 1970–1990.[2] It means, in simple words, that the depleted and out-of-date equipment in the national economy kept working: the newly introduced equipment did not replace outdated and worn-out things, or, in some cases, did not work at all. The rules of the game between the centre and the producing units encourage the accumulation of reserves of equipment, which were used also in interests of separate persons. From the middle of 60th years, the difficulties in management of the production system were revealing more and more strongly, at the same time informal (shadow) production relationship had been developing[3]: the bad management provided the permanent degradation of the production system.

The values of productivity of capital for the period after 2015 corresponds to the assumed values of GDP and production factors. The forecasted values of productivity of a fixed capital $\xi$ can be detailed by consideration of the production system as an ensemble of many cooperating sectors. Indeed, in multi-sector approximation (the input–output model), the technological index $\alpha$, for example, is connected with fundamental technological matrices and with the differences in growth of separate sectors (see Chap. 9, Eq. 9.21).

## 8.9 Decomposition of the Growth Rate of GDP

To separate influence of the growth rates of production factors from influence of variations in efficiency of the system, which are determined with some internal modifications that are formally connected with the quantities $\xi$ and $\alpha$, discussed

[2]The start of the process of reduction gave the reform of the year 1965, which had introduced profit as an assessment of activity of the managing subject; that had given a basis for degradation of the socialist system and to the subsequent establishment of capitalist relationship [12].

[3]Bogdanov [13] demonstrated that the developed economic criminality in the USSR existed from the end of 1940th years.

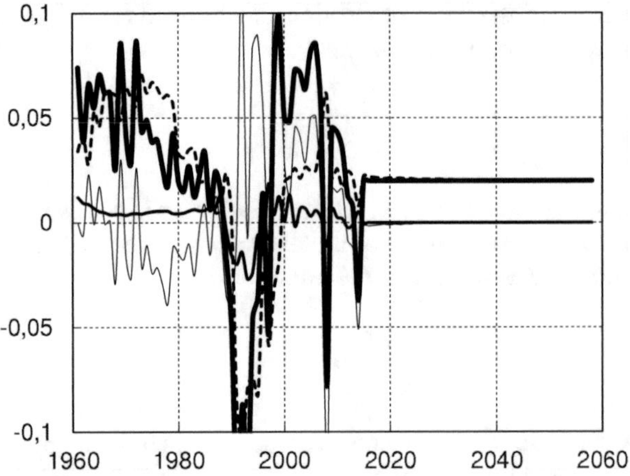


**Fig. 8.8** Decomposition of the growth rate of GDP. The growth rate of GDP (the thick solid line) is decomposed in three contributions: the expenditures of labour (the thin solid line), the substitutive work (the dashed line) and the changing of quality of production equipment (the structural contribution, the very thin solid line)

earlier, let us pass from relations (8.6) to expressions for the growth rate of output

$$\frac{1}{Y}\frac{\mathrm{d}Y}{\mathrm{d}t} = \begin{cases} \dfrac{1}{K}\dfrac{\mathrm{d}K}{\mathrm{d}t} + \dfrac{1}{\xi}\dfrac{\mathrm{d}\xi}{\mathrm{d}t}, \\ (1-\alpha)\dfrac{1}{L}\dfrac{\mathrm{d}L}{\mathrm{d}t} + \alpha\dfrac{1}{P}\dfrac{\mathrm{d}P}{\mathrm{d}t} + \ln\left(\dfrac{L_0}{L}\dfrac{P}{P_0}\right)\dfrac{\mathrm{d}\alpha}{\mathrm{d}t} \end{cases} \tag{8.9}$$

The first terms of Eq. (8.9) represent the contribution to the growth rate of GDP due to variations in the production factors $K$, $L$, $P$ at constant 'structure' of production system. The last terms are connected with structural and/or technological variations in production system, that is, with evolution of system of production straightforwardly.

To fulfil the actual decomposition of the growth rate of GDP, we need in the time series of the output $Y$, fixed capital $K$ and estimates of workers' efforts $L$. The decomposition of the growth rate of GDP is unambiguous and does not require any other information: no any arbitrary parameters are included in the theory. Decomposition of the growth rate of output for Russia, according to Eq. (8.9), is shown in Fig. 8.8. Let us notice that the quantity

$$\frac{1}{\xi}\frac{\mathrm{d}\xi}{\mathrm{d}t} = \ln\left(\frac{L_0}{L}\frac{P}{P_0}\right)\frac{\mathrm{d}\alpha}{\mathrm{d}t} \tag{8.10}$$

appears to be similar to the total factor productivity (the residual of Solow) in the neoclassical theories of economic growth, where two production factors $L$ and $K$ are considered.

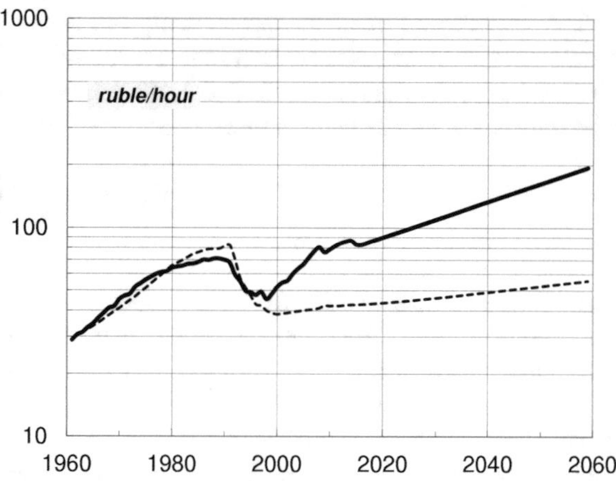


**Fig. 8.9** Labour productivity. The solid line represents actual values of labour productivity in Russia measured in rubles of year 2000 per working hour. The changes of productivity have two components; the dashed line represents the variations in productivity connected only with changes of the ratio $P/L$ at constant efficiency of the usage of the social resources ($\alpha = const$)

## 8.10 Productivity of Labour

The labour productivity, determined as the ratio of value of output to expenditures of labour, is a measure of efficiency of the social production (see Sect. 7.3). The increase in labour productivity cannot be understood without taking into account the phenomenon accompanying progress of production—attraction of outer energy sources (domestic animals, wind, water, coal, oil and other) for performance of economic works instead of efforts of humans. As it has been shown in the previous chapters, the substitution effect describes observable dynamics of production and determines expression (in monetary units of constant purchasing capacity) for the labour productivity

$$A = \frac{Y}{L} = \frac{Y_0}{L_0}\left(\frac{L_0}{L}\frac{P}{P_0}\right)^{\alpha}. \tag{8.11}$$

The values of labour productivity in the Russian Federation are presented in Fig. 8.9. The labour productivity increases till year 1989 and reveals sharp fall over the blurred period. Then, one can observe the growth of labour productivity after 1998, and this growth can, according to our assumptions, proceed after 2015 with the growth rate nearby 0.02, which is presented in Fig. 8.9. The values of labour productivity in Russia remain to be less than in any developed capitalist country.[4]

[4]Let us, for example, compare it with labour productivity in the USA. According to Rosstat [6], [year 2011], in the Russian Federation in 2010, it was expended 148977 million man-hours and made the output (GDP) by value of 44939153 million rubles, so that one worker produced about 302 rubles in current (2010) prices per an hour in the national economy of Russia. For comparison, in the USA in 2010, it was expended 247228 million man-hours and output (GDP) of 14526500 million dollars was made. One can easily see that it was produced 58.76 dollars per one man-hour or, accepting, that one dollar is equal to 30 rubles, 1763 rubles. Thus, average labour productivity in the USA exceeded this quantity in Russia about six times in 2010.

The growth rate of labour productivity is connected (see Sect. 6.4.2) with the characteristics of used technology: the productivity of a fixed capital $\xi$, the non-dimensional technological coefficient $\overline{\lambda}$, the coefficient of depreciation $\mu$, as well as with the growth rate of expenditures of labour $\nu$

$$\frac{1}{A}\frac{\mathrm{d}A}{\mathrm{d}t} = \frac{(1-\overline{\lambda})(\nu+\mu)}{\overline{\lambda}} + \frac{1}{\xi}\frac{\mathrm{d}\xi}{\mathrm{d}t}. \tag{8.12}$$

One of the quantities, namely, the technological coefficient $\overline{\lambda}$—labour requirement—plays a crucial role; this quantity controls the mechanism of substitution and can be considered as a measure of substitution. If $\overline{\lambda} = 1$, labour productivity changes only in connection with variations in the marginal productivity of the capital $\xi$. If the quantity $\overline{\lambda} < 1$, the consumption (per unit of a fixed capital) of work decreases, and consumption of services of the capital increases, which determines an increase of labour productivity. The situation is opposite, if the quantity $\overline{\lambda} > 1$. One can see that the main motive power of progress is humans' efforts, but at the same time, the technological innovations, if $\overline{\lambda} < 1$, allow one to involve additional energy into production processes by means of various complex adaptations. It is a general description of the mechanism of scientific and technological progress, which is naturally included into the description of process of production of value. The desire to boost the labour productivity requires the greater and greater attraction of substitutive work and, together with it, consumption of a lot of energy carriers.

## 8.11 Final Remarks

In this chapter, the first attempt to show the possibility of the new method of analysis and forecast of development of a national economy is presented. The method takes into account the energy spend in a national economy and includes the description of technological quality of the fixed capital. The method allows to analyse scripts of economic growth and to separate the available programmes of development.

As an illustrative example, the script of development with the constant value of expenditures of labour was considered; the requirement of a constancy and the 'conservatism' defines the unique value of the growth rate of GDP equal to 2%. The greater growth rates are possible when the labour saving equipment is introduced. Certainly, it is impossible to expect that the actual trajectory of the progress of production after 2015 will follow this straight line. For the design of more adequate scripts of development, it is necessary to have expert assessments of the future technology (the technological coefficients $\overline{\lambda}$ and $\overline{\varepsilon}$) and assessments of availability of production factors (humans' work and substitutive work).

Apparently, the most essential result of the analysis for Russia is the indication of a possibility of development, when additional volumes of energy are attracted. However, available energy, alongside with work, can become a factor limiting the economic growth. Certainly, the situation is not fatal; investments into sectors of production of energy and into development of the energy-saving technologies will assist overcoming the crisis. This result has thermodynamic accuracy, that is, the conclusion does not depend on details of the description.

In this chapter, we use an approximation, in which all national economy is considered as one sector. For more detailed analysis and forecasting of development, it is necessary to also address the models on the basis of the theory of input–output balance, general principles of which are developed by Vasily Vasilevich Leontiev (see a description and discussion in Chaps. 4 and 9). For an assessment of the situation, it is necessary to address, first of all, the elementary three-sector model described in Sect. 2.1.

## References


1. Khanin G.I.: Ekonomicheskaya istoriya Rossii v noveyshee vremya (Economic History of Russia in the XX Century), vols. 1–3. State Technical Univerity, Novosibirsk (2008–2014)
2. Katasonov, V.Yu.: Ekonomika Stalina (The Stalin's Economy). Institut russkoy tsivilizatsii (Institute of the Russian Civilization), Moscow (2014)
3. Minekonomrazvitiya, R.F.: Prognoz dolgosrochnogo sotsial'no-ekono-micheskogo razvitiya Rossiyskoy Federatsii na period do 2030 goda (Scenarios of Long-term Social and Economic Progress of the Russian Federation for the Period Till 2030), Moscow (2013)
4. Ponomarenko, A.N.: Retrospektivnye natsional'nye scheta Rossii: 1961–1990 (The Retrospective National Accounts of Russia: 1961–1990). Finansy i statistika, Moscow (2002)
5. Bessonov, V.A.: Problemy postroeniya proizvodstvennykh funktsiy v rossiyskoy perekhodnoy ekonomike (Problems of formulation of the production function for the Russian transitive economy). In: Bessonov, V.A., Tsukhlo, S.V. (eds.) Analiz dinamiki rossiyskoy perekhodnoy ekonomiki (The Analysis of Dynamics of the Russian Transitive Economy), pp. 5–89. Institut ekonomiki perekhodnogo perioda, Moscow (2002)
6. Rosstat: Rossiyskiy statisticheskiy ezhegodnik. Ctatisticheskiy sbornik (The Russian Statistical Year-Book). Rosstat, Moscow (2011 and the last years). http://www.gks.ru/. Assessed 15 July 2017
7. Voskoboynikov I.B.: Otsenka sovokupnoy faktornoy proizvoditelnosti rossiyskoy ekonomiki v period 1961–2001 gg. s uchetom korrektirovki dinamiki osnovnykh fondov (Assessment of cumulative factors productivity of the Russian economy during 1961–2001 in view of updating dynamics of a fixed capital). Pre-print WP2 of SU HSE, Moscow (2003)
8. Rosstat: Narodnoe khozyaystvo Rossiyskoy Federatsii: Statisticheskiy ezhegodnik (The National Economy of the Russian Federation: Statistical Year-Book). Respublikanskiy informatsionno-izdatel'skiy tsentr (Republican information-publishing centre), Moscow (1992)
9. Bessonov, V.A., Voskoboynikov, I.B.: O dinamike osnovnykh fondov i investitsiy v rossiyskoy perekhodnoy ekonomike (On dynamics of the fixed capital and investments in the Russian transitive economy). Ekonomicheskiy zhurnal VSHE (Economic Journal of HSE) **10**(2), 193–228 (2006)
10. Minenergo, R.F.: Energeticheskaya strategiya Rossii na period do 2030 goda. Utverzhdena rasporyazheniem Pravitel'stva Rossiyskoy Federatsii ot 13 noyabrya 2009 g. 1715-r (The energy strategy of Russia for the period till 2030. Approved by the Government of the Russian Federation from November 13, 2009, No 1715-.). Moscow (2009)
11. Ayres, R.U., Ayres, L.W., Pokrovski, V.N.: On the efficiency of electricity usage since 1900. Energy **30**, 1092–1145 (2005)

12. Antonov, M.F.: Kapitalizmu v Rossii ne byvat'! (Never to Be Capitalism in Russia!). Yauza, Eksmo, - Moscow (2005). http://m-antonov.chat.ru/capital/index.htm
13. Bogdanov, S.V.: Khozyaystvenno-korystnaya prestupnost' v SSSR 1945–1990 gg.: faktory vosproizvodstva, osnovnye pokazateli, osobennosti gosudarstvennogo protivodeystviya (Economic-Mercenary Criminality in the USSR 1945–1990: the Factors of Reproduction, the Basic Parameters, Peculiarities of the State Counteraction). Dissertatsiya ... doktora istoricheskikh nauk: 07.00.02 (The Dissertation... Doctors of History: 07.00.02), Kursk (2010)

# Chapter 9
# Dynamics of Production in Many-Sector Approach

**Abstract** In this chapter, we are returning to the many-sector model of production system, consisting of $n$ production sectors, each of them creating its own product. The sectors are interacting with each other, as was described in Chap. 4, but to develop a description of the time behaviour of the system, one has to formulate the elementary laws of evolution for each sector. One can assume that production of value and dynamics of production factors in every sector are described in the same way as for the entire economy, according to the rules set up in Chaps. 5 and 6. The schematization of the production system allows us to formulate the simplest evolution theory including only three production factors in every sector. This allows one to draw a picture of economic growth, taking into account the specific features of each sector. As an illustration, development of the three-sector production system is considered.

## 9.1 Dynamics of the Production Factors in Sectors

### 9.1.1 *Dynamics of Technological Coefficients*

Similar to the entire production system, the sector can be thought of as a collection of production equipment $K^i$, which is activated by labour $L^i$ and substitutive work $P^i$. One can assume that all speculations of Chap. 5 can be reproduced for a separate sector, so that Eqs. (5.6) and (5.13) for the production factors can be rewritten in a simplified form as

$$\frac{dK^i}{dt} = I^i - \mu K^i, \quad \frac{dL^i}{dt} = \lambda^i I^i - \mu L^i, \quad \frac{dP^i}{dt} = \varepsilon^i I^i - \mu P^i. \tag{9.1}$$

The technological coefficients $\lambda^i$ and $\varepsilon^i$ $(i = 1, 2, \ldots, n)$ characterise quality of the production equipment introduced in sector $i$. The coefficient of depreciation $\mu$ has to be given as a parameter of the problem.



V. N. Pokrovskii, *Econodynamics*, New Economic Windows,
https://doi.org/10.1007/978-3-319-72074-6_9

It is also convenient to introduce the non-dimensional technological variables for every sector,

$$\overline{\lambda}^i = \lambda^i K^i / L^i, \qquad \overline{\varepsilon}^i = \varepsilon^i K^i / P^I$$

and assume that the speculations of Sects. 5.3 and 5.4 can be repeated for each sector so that one supposes the technological variables are determined by equations similar to Eqs. (5.30) and (5.31), that is,

$$\frac{d\overline{\lambda}^i}{dt} = -\frac{1}{\tau^i}\left(\overline{\lambda}^i - \frac{\tilde{\nu}^i + \mu}{\tilde{\delta}^i + \mu}\right), \quad \frac{d\overline{\varepsilon}^i}{dt} = -\frac{1}{\tau^i}\left(\overline{\varepsilon}^i - \frac{\tilde{\eta}^i + \mu}{\tilde{\delta}^i + \mu}\right). \tag{9.2}$$

The time of crossover from one technological situation to another $\tau^i$ can be different for different sectors. It is determined by internal processes of replacement of technology within the sector. The symbols for the rates of potential growth of capital, labour and substitutive work in every sector, $\tilde{\delta}^i$, $\tilde{\nu}^i$ and $\tilde{\eta}^i$ $(i = 1, 2, \ldots, n)$, are introduced here. These quantities apparently relate to the rates of potential growth of the production factors for the entire production system and are assumed to be given as functions of time.

It is also convenient to introduce the sector technological index

$$\alpha^i = \frac{1 - \overline{\lambda}^i}{\overline{\varepsilon}^i - \overline{\lambda}^i}$$

and write an equation for it

$$\frac{d\alpha^i}{dt} = \frac{\tilde{\delta}^i - \tilde{\nu}^i - \alpha^i\,(\tilde{\eta}^i - \tilde{\nu}^i)}{\tau^i(\overline{\varepsilon}^i - \overline{\lambda}^i)(\tilde{\delta}^i + \mu)}. \tag{9.3}$$

One can see that, if the potential rate of capital growth $\tilde{\delta}^i(t)$ is determined by equation

$$\tilde{\delta}^i = \tilde{\nu}^i + \alpha^i(\tilde{\eta}^i - \tilde{\nu}^i),$$

the technological index appears to be an integral of evolution and, therefore, can be considered as a very important characteristic of the production system.

### 9.1.2 Investment

To determine the sector investment $I^i$ in Eq. (9.1), one has to take into account internal and external restrictions on the development of the system. The internal restrictions, imposed by the technological structure of the production system and by the necessary private consumption, determine the potential growth rate of capital stock $\tilde{\delta}^i$, which is connected with the potential investments in sector $i$,

$$\tilde{I}^i = (\tilde{\delta}^i + \mu)\, K^i.$$

It is assumed that there is a method, for example, the method described in Sect. 4.3.3, to calculate the sector potential investments and, consequently, according to the above relation, the rate of potential growth of capital stock.

The external restrictions are imposed by the availability of labour and substitutive work, which is assumed to be described by their rates of potential growth for every sector $\tilde{\nu}^i(t)$ and $\tilde{\eta}^i(t)$. These exogenous characteristics are given as functions of time.

Therefore, one can rewrite relation (5.27) for each sector,

$$I^i = \chi^i\, K^i + \min\left\{\tilde{I}^i,\; (\tilde{\nu}^i + \mu)L^i/\lambda^i,\; (\tilde{\eta}^i + \mu)P^i/\varepsilon^i\right\}, \tag{9.4}$$

where $\chi^i$ is a centrally planning intervention in sector $i$. In the many-sector model, interventions $\chi^i$ can be understood as coefficients of reallocation of investments among different sectors, so that

$$\sum_{i=1}^{n} \chi^i\, K^i = 0.$$

By relation (9.4), three modes of the development of each sector are determined. The first choice in relation (9.4) corresponds to the internal restrictions. The second choice in (9.4) is valid in the case of abundance of substitutive work and raw materials and a lack of labour. In this case, labour is used completely and there is a possibility of attracting extra substitutive work. The latter is a reason for technological changes. Internal processes lead to decrease of labour and increase of substitutive work in production processes. The last choice in Eq. (9.4) is valid in the case of a lack of substitutive work and an abundance of labour and raw materials. In one-sector approach, one has three modes of development, and, approaching the production system with the many-sector model, one has three modes for every sector and, consequently, many possible modes of development of the production system.

## 9.2 Sector Production of Value

### *9.2.1 Sector Production Functions*

As was described in Chap. 2, the output of a sector $i$ is characterised by the three quantities: gross output $X^i = X_i$, final output $Y_i$ and sector production of value $Z^i$. These quantities are connected with each other by Eqs. (4.6) and (4.7), which can be written in the form

$$X_i = \frac{1}{1-a^i} Z^i, \quad Y_j = \sum_{i=1}^{n} \frac{\delta^i_j - a^i_j}{1-a^i} Z^i, \quad a^i = \sum_{i=1}^{n} a^i_j, \tag{9.5}$$

where $a^i_j$ is a component of the matrix of intermediate production consumption (the input–output matrix) introduced by relation (4.2).

Our immediate task is to determine the output vectors as functions of production factors. From relation (4.12) and the above relations, gross output $X^i = X_i$, final output $Y_i$ and production of value $Z^i$ in each sector ($i = 1, 2, \ldots, n$) are connected with the vector of amount of the sector production equipment or sector capital stock $K^i$ by the relations

$$X_i = \frac{1}{b^i} K^i, \quad b^i = \sum_{j=1}^{n} b^i_j, \tag{9.6}$$

$$Y_j = \sum_{i=1}^{n} \xi^i_j K^i, \quad \xi^i_j = \frac{\delta^i_j - a^i_j}{b^i} = \xi^i \frac{\delta^i_j - a^i_j}{1-a^i}, \tag{9.7}$$

$$Z^i = \xi^i K^i, \quad \xi^i = \sum_{j=1}^{n} \xi^i_j = \frac{1-a^i}{b^i}, \quad i, j = 1, 2, \ldots n, \tag{9.8}$$

where coefficients $a^i_j$ comprise a matrix of intermediate production consumption (input–output matrix), and coefficients $b^i_j$ comprise a matrix of fixed capital (capital–output matrix). The components of the fundamental technological matrices $a^i_j$ and $b^i_j$ are combined to form the components $\xi^i_j$ of a matrix of capital productivities. The quantities $\xi^i_j$ show how an increase in production equipment in sector $i$ affects the final output in sector $j$. One can see that, to calculate the final product, the all components of the matrix of capital productivity are needed.

Formulae (9.6)–(9.8) connect the vector characteristics of output $X_i$, $Y_j$ and $Z^i$ with the amount of sector capital $K^i$. On the other hand, production of value is connected with the sector consumption of production factors: labour $L^i$ and substitutive work $P^i$. In the general case, the sector productivities can depend on the production factors consumed in all the other sectors. It is possible to expect that quantities of marginal productivities are determined by a mutual market of production factors. In the simplest case, one can consider production of value $Z^i$ in a sector to be determined by consumption of production factors in the same sector. In this case, we refer to the procedure, which was used in Sect. 6.3, to determine

$$Z^i = Z^i_0 \frac{L^i}{L^i_0} \left( \frac{L^i_0}{L^i} \frac{P^i}{P^i_0} \right)^{\alpha^i}, \quad \alpha^i = \frac{1-\overline{\lambda}^i}{\overline{\varepsilon}^i - \overline{\lambda}^i}. \tag{9.9}$$

The constant $Z^i_0$ is controlled by initial values of variables. Then, the gross output $X_i$ and the final output $Y_i$ can be easily found with the help of relations (9.5).

### 9.2.2 Marginal Productivities and Technological Change

Equations (9.8) and (9.9) appears to be the two complementary expressions for production of value in sector $i$,

$$Z^i = \begin{cases} \xi^i K^i, & \xi^i = \dfrac{1-a^i}{b^i}, \\ Z_0^i \dfrac{L^i}{L_0^i}\left(\dfrac{L_0^i}{L^i}\dfrac{P^i}{P_0^i}\right)^{\alpha^i}, & \alpha^i = \dfrac{1-\overline{\lambda}^i}{\overline{\varepsilon}^i - \overline{\lambda}^i} \end{cases}, \quad i = 1, 2, \ldots n,$$

so that, comparing them, one can obtain

$$\xi^i = Z_0^i \frac{L^i}{L_0^i K^i}\left(\frac{L_0^i}{L^i}\frac{P^i}{P_0^i}\right)^{\alpha^i}, \quad \alpha^i = \frac{1-\overline{\lambda}^i}{\overline{\varepsilon}^i - \overline{\lambda}^i}. \tag{9.10}$$

To separate the effects of production factors and technological change, one can consider the differential of production of value

$$dZ^i - \Delta^i dt = \begin{cases} \xi^i \, dK^i \\ \beta_i \, dL^i + \gamma_i \, dP^i \end{cases}, \quad i = 1, 2, \ldots n, \tag{9.11}$$

where the capital marginal productivity $\xi^i$ is defined above and, from the above definitions,

$$\beta_i = \xi^i (1-\alpha^i) \frac{K^i}{L^i}, \quad \gamma_i = \xi^i \alpha^i \frac{K^i}{P^i}, \tag{9.12}$$

$$\Delta^i = -\frac{K^i}{b^i}\frac{da^i}{dt} - \frac{(1-a^i)K^i}{b^i}\frac{db^i}{dt} = Z^i \ln\left(\frac{L_0^i}{L^i}\frac{P^i}{P_0^i}\right)\frac{d\alpha^i}{dt} = Z^i \frac{1}{\xi^i}\frac{d\xi^i}{dt}. \tag{9.13}$$

The quantity $\Delta^i$ is connected with changes of components of the technological matrices $\mathsf{A}$ and $\mathsf{B}$ and can be called the technological change within the sector labelled $i$.

Let us note that relations (9.11) are valid for the case, when all prices of products do not depend on time. In the opposite case, we ought to use a new quantity $\hat{Z}^i$, namely, production of value measured in the current money unit, for which we have

$$\frac{d\hat{Z}^i}{dt} = p_i \left(\frac{dZ^i}{dt}\right)_{p_i} + \hat{Z}^i \frac{d \ln p_i}{dt}, \tag{9.14}$$

where $\left(\dfrac{dZ^i}{dt}\right)_{p_i}$ is the derivation of production of value in sector $i$ at constant prices given by formula (9.11). The indexes of prices of products $p_i$ must be considered to be new variables. One needs extra equations to include them for consideration in this case. Some equations will be discussed in Chap. 10. Further, we restrict ourselves to the case when all prices are constant.

## 9.3 Rules of Aggregation and Structural Shift

### 9.3.1 Production Factors

To return to the one-sector description of the production system, we can refer to natural definitions (which were discussed in Sect. 2.3.2, Eq. 2.24 and in Sect. 5.1, Eq. 5.1) of production factors as sums of corresponding sector quantities,

$$K = \sum_{i=1}^{n} K^i, \quad L = \sum_{i=1}^{n} L^i, \quad P = \sum_{i=1}^{n} P^i.$$

One can sum Eq. (9.1) for the dynamics of sector production factors to obtain Eqs. (5.6) and (5.13) for the dynamics of the entire amounts of production factors. The procedure defines the technological coefficients for the production system as a whole,

$$\lambda = \sum_{j=1}^{n} \frac{I^j}{I} \lambda^j, \qquad \varepsilon = \sum_{j=1}^{n} \frac{I^j}{I} \varepsilon^j, \tag{9.15}$$

where $I^j$ is the gross investment in sector $j$, and $I = \sum_{j=1}^{n} I^j$ is the gross investment in the entire economy.

It is convenient to introduce the symbols for the growth rates of sector production factors: capital, labour and substitutive work, $\delta^i$, $\nu^i$, $\eta^i$, respectively, and to define the growth rates of production factors for the entire system by the relations

$$\delta = \frac{1}{K} \sum_{i=1}^{n} \delta^i K^i, \quad \nu = \frac{1}{L} \sum_{i=1}^{n} \nu^i L^i, \quad \eta = \frac{1}{P} \sum_{i=1}^{n} \eta^i P^i. \tag{9.16}$$

The introduced averaged growth rates depend, generally speaking, on time, even if the sector growth rates do not.

In the introduced symbols, Eqs. (9.1)–(9.4) can be rewritten as

$$I^i = (\delta^i + \mu) K^i, \qquad \lambda^i = \frac{\nu^i + \mu}{\delta^i + \mu} \frac{L^i}{K^i}, \qquad \varepsilon^i = \frac{\eta^i + \mu}{\delta^i + \mu} \frac{P^i}{K^i}, \tag{9.17}$$

and one can easily see that formulae (9.16) and (9.17) allow us to return to expressions (written in Sect. 5.2.2) for investment and the technological coefficients of the production system as a whole,

$$I = (\delta + \mu)K, \quad \lambda = \frac{\nu + \mu}{\delta + \mu}\frac{L}{K}, \quad \varepsilon = \frac{\eta + \mu}{\delta + \mu}\frac{P}{K}.$$

### 9.3.2 Production of Value

The final output of the entire system has to be calculated according to any of the two equations

$$Y = \begin{cases} \sum_{j=1}^{n} Y_j \\ \sum_{i=1}^{n} Z^i \end{cases}$$

while the sector production of value $Z^i$ is defined by expressions (9.8) and (9.9). One can use Eq. (9.8) to obtain the final output and its derivative

$$Y = \sum_{i=1}^{n} \xi^i K^i, \quad \frac{dY}{dt} = \sum_{i=1}^{n} Z^i \left(\delta^i + \frac{1}{\xi^i}\frac{d\xi^i}{dt}\right). \tag{9.18}$$

One can separate the growth rate of the total capital stock $\delta$ and use Eq. (9.13) for the technological change $\Delta^i$ to obtain an expression for the growth rate of output in the form

$$\frac{1}{Y}\frac{dY}{dt} = \delta + \frac{1}{Y}\sum_{i=1}^{n}\left[\Delta^i + Z^i\,(\delta^i - \delta)\right]. \tag{9.19}$$

The deviations of the growth rate of final output from the growth rate of capital $\delta$ are connected with the sector technological changes and the diversity of sector development. Comparison of this equation with Eqs. (6.17) and (6.19) allows us to determine an expression for the total technological change

$$\Delta = \sum_{i=1}^{n}\left[\Delta^i + Z^i\,(\delta^i - \delta)\right]. \tag{9.20}$$

It allows us to calculate the technological index according to the equation

$$\frac{d\alpha}{dt} = \frac{1}{Y}\ln^{-1}\left(\frac{L_0}{L}\frac{P}{P_0}\right)\sum_{i=1}^{n}\left[\Delta^i + Z^i\,(\delta^i - \delta)\right]. \tag{9.21}$$

Let us note that the growth rate of the sector final output $\delta_j$ is, generally speaking, different from the growth rate of the production of value in sector $\delta^i$. The quantities are connected, due to Eq. (9.2), by the relation

$$\delta_j = \sum_{i=1}^{n} \delta^i \frac{\delta^i_j - a^i_j}{1 - a^i} \frac{Z^i}{Y_j}. \tag{9.22}$$

## 9.4 Equations of Evolution

### 9.4.1 Basic Relations

The preceding results allow one to write a closed set of equations for the dynamics of the production system, which is assumed to consist of $n$ sectors and is characterised by the parameters listed below. The equations can be written in different equivalent forms (see Sect. 6.4), by making use of the different sets of assumptions and variables. Here, we refer to the approach considered in Sect. 6.4.3 and assume that every sector (labelled $i,\ i = 1, 2, \ldots, n$) is described by variables:

$Z^i$ – production of value,
$K^i$ – value of production equipment,
$L^i$ – consumption of labour,
$P^i$ – consumption of substitutive work *and*
$\alpha^i$ – technological index.

Additionally, the gross and final outputs can be calculated according to Eq. (9.5), that is,

$$X_i = \frac{1}{1 - a^i} Z^i, \quad Y_j = \sum_{i=1}^{n} \frac{\xi^i_j}{\xi^i} Z^i, \quad \xi^i = \sum_{i=1}^{n} \xi^i_j.$$

It is convenient to list the fundamental characteristics of production system, which are needed for the description:

$\xi^i_j$ — the components of the matrix of capital marginal productivity. The components have to be given as functions of time or can be calculated through the components of the fundamental technological matrixes: input–output matrix $a^i_j$ and capital–output matrix $b^i_j$ as

$$\xi^i_j = \frac{\delta^i_j - a^i_j}{b^i}, \quad b^i = \sum_{l=1}^{n} b^i_l. \tag{9.23}$$

Components of the matrixes $a^i_j$ and $b^i_j$ ought to be estimated empirically from definitions (4.2) and (4.12).

$s_j$ — share of investment product in final output of sector $j$. One can assume that investment sectors are separated, so that for them $s_j = 1$, whereas for others $s_j = 0$.

$\mu^i$ — coefficient of depreciation of production equipment. As a simplification, it can be accepted that it has the same value for all products in all situations and is constant.

$\nu^i$ — the rate of growth of labour in every sector should be given as a function of time.

$\overline{\lambda}^i$ — the dimensionless technological coefficient should be given as a function of time. If the quantity $\overline{\lambda}^i < 1$, the consumption (for unit of capital stock) of labour decreases and the consumption of substitutive work (work of production equipment) increases. The situation is opposite if the quantity $\overline{\lambda}^i > 1$.

One can see that these quantities determine the applied technology of the production system, and we can say that the technology in an input–output model is given if we know these parameters.

We refer to the results of Sect. 6.4 and this chapter to collect the set of equations for the listed variables:

$$
\begin{aligned}
&Z^i = \xi^i K^i, \\
&\frac{dK^i}{dt} = \delta^i K^i, \quad \delta^i = \frac{\nu^i + (1 - \overline{\lambda}^i)\mu^i}{\overline{\lambda}^i}, \quad \mu^i = \mu^i(t), \quad \overline{\lambda}^i = \overline{\lambda}^i(t), \\
&\frac{dL^i}{dt} = \nu^i L^i, \quad \nu^i = \nu^i(t), \\
&\frac{dP^i}{dt} = \eta^i P^i, \quad \eta^i = \frac{\delta^i - (1 - \alpha^i)\nu^i}{\alpha^i}, \quad \alpha^i = \frac{\ln\left(\dfrac{Z^i}{Z^i_0}\dfrac{L^i_0}{L^i}\right)}{\ln\left(\dfrac{L^i_0}{L^i}\dfrac{P^i}{P^i_0}\right)}
\end{aligned}
\tag{9.24}
$$

It is assumed that there are no other restrictions.

## 9.4.2 About Solution of the System of Equations

Relations (9.24) comprise a set of evolutionary equations, so the initial values of all the variables, that is,

$$Z^i(0), \quad K^i(0), \quad L^i(0), \quad P^i(0), \quad \alpha^i(0) \tag{9.25}$$

should be given, while the initial value of capital stock must correspond to the initial values of output and marginal productivity $K(0) = Y(0)/\xi(0)$. The initial values of labour $L(0) = L_0$ and substitutive work $P(0) = P_0$ can be chosen independently.

The problem—in this case, one has a Cauchy problem—can be solved by numerical methods. The equations determine a trajectory of evolution of the production system. The final consumption and the storage of intermediate products appear to be the consequence of evolution of the system.

Note that the above system of equations is valid for the case when all prices or, in other words, money units to measure value, are considered to be constant during the time. The system can be reformulated for the case when prices change. In this case, one has to introduce the new variables for the final output measured in the current money unit $\hat{Z}^i$ instead of variables $Z^i$ (using Eq. 9.14). Some equations for the indexes of prices of the product $p_i$, $i = 1, 2, \ldots, n$, that must be considered to be new variables have to be added as well. Some principles of the design of equations for the price indexes will be discussed in Chap. 10.

One supposes that the equations reflect some universal features of production systems. In our theory, economic growth is coupled with technological changes and growth of consumed energy. Understanding of the energy–economy coupling in production systems was considered crucial for the design of proper models to generate scenarios of development, and much effort and money have been spent to create some simulation models of energy–economy systems [1]. Such models provide us with many details of internal processes but require much input information (many empirical parameters). In contrast, phenomenological models, such as the one above, deal with aggregate variables and look simpler. They can be helpful in generating reliable scenarios of the evolution of global and national economies in macroeconomic terms for the government use.

## 9.5 Dynamics of the Three-Sector System

As a simple example, we consider dynamics of the production system consisting of the three 'pure' (according to Leontiev) sectors that were described in Sect. 2.1. The scheme of the production processes is presented in Sect. 2.2.2 and in Table 2.2. The first sector produces the means for production and supplies all sectors with material resources; this sector corresponds to the first Marx's production division. The second sector is connected with the governmental projects and creates also principles of the organisation (science, research and development, codification, works of art), necessary for the organisation of production in all three sectors. The third sector produces goods and services for current personal and social consumption, the product of the sector is completely consumed; this sector corresponds to the second Marx's production division that creates consumer goods. We apply our reasoning to dynamics of the production system of Russia in the twenty-first century.

### 9.5.1 Balance Equations

The production is organised for creation of products for final consumption, but many other products are necessary to approach this target. The part of the gross output of each of three described sectors is distributed, generally speaking, over all sectors for intermediate production consumption, so that Eq. (4.6) in this simple case can be recorded in the following form:

$$\begin{aligned} X_1 &= a_1^1 X_1 + a_1^2 X_2 + a_1^3 X_3 + Y_1 \\ X_2 &= a_2^1 X_1 + a_2^2 X_2 + a_2^3 X_3 + Y_2 \\ X_3 &= Y_3. \end{aligned} \tag{9.26}$$

The final outputs of the sectors $Y_1$, $Y_2$ and $Y_3$ are used for direct consumption and accumulation. The final product of the first sector $Y_1$ represents total investments of the production equipment $Y_1 = I^1 + I^2 + I^3$ to all sectors. Having accumulated, the investments form the basic production funds (the production equipment in a broad sense); distribution of a fixed capital over the sectors, according to general expressions (4.12), is considered to be proportional to the gross outputs of the sectors

$$\begin{aligned} K_1 &= b_1^1 X_1 + b_1^2 X_2 + b_1^3 X_3, \\ K_2 &= 0, \\ K_3 &= 0. \end{aligned} \tag{9.27}$$

The second sector creates an intermediate non-material product, in particular, investments in 'human capital' and technical progress; a part of the product, $Y_2$, is stored. The final product of the third sector is completely consumed with norm of consumption for a worker $c$. On the assumption that the norm of consumption is equal for workers of all three sectors, it is possible to record $Y_3 = c(L^1 + L^2 + L^3)$.

The value, created in the sectors, according to the general Eq. (2.13) can be recorded as

$$\begin{aligned} Z^1 &= X_1 - a_1^1 X_1 - a_2^1 X_2 = cL^1 + A^1 + M^1, \\ Z^2 &= X_2 - a_1^2 X_1 - a_2^2 X_2 = cL^2 + A^2 + M^2, \\ Z^3 &= X_3 - a_1^3 X_1 - a_2^3 X_2 = cL^3 + A^3 + M^3, \end{aligned} \tag{9.28}$$

where $A^1$, $A^2$, $A^3$ designate value of a part of production assets of the sector used at creation of the final output (amortisation), and $M^1$, $M^2$, $M^3$ are the surplus products in three sectors, accordingly.

By consideration of Eqs. (9.26) and (9.28) it is possible to establish property of model: though value of the product created in the sector $Y_j$ is not equal, generally speaking, to the value created in the sector $Z_j$ (see the discussion in Sect. 2.2.2), the sum of values of all products is equal to production of value in all sectors

$$Y = \sum_{j=1}^{3} Y_j = \sum_{j=1}^{3} Z_j = Z.$$

### 9.5.2 Equations of Evolution

Our main goal consists in determining of the time dependence of the components of the gross domestic product (GDP), the dynamic of which is obeyed to Eq. (9.7), that is,

$$Y_j = \sum_{i=1}^{3} \xi^i_j \, K^i, \quad \xi^i_j = \frac{\delta^i_j - a^i_j}{b^i}, \quad j = 1, 2, 3. \tag{9.29}$$

The basic production assets $K^j$, $j = 1, 2, 3$ can be found according to the first of Eq. (9.1), that is,

$$\frac{dK^i}{dt} = I^i - \mu K^i, \quad i = 1, 2, 3. \tag{9.30}$$

The special choice of three sectors allows us to determine the total investments into all sectors as the final product of the first sector, so that it is possible to write

$$Y_1 = I^1 + I^2 + I^2 = \xi^1_1 \, K^1 + \xi^2_1 \, K^2 + \xi^3_1 \, K^3. \tag{9.31}$$

Now, establishing fractions of investments into each separate sector $x_1$, $x_2$, $x_3$, $(x_1 + x_2 + x_3 = 1)$, we can formulate the system of equations for the component of a vector of a fixed production capital

$$\begin{aligned}
\frac{\mathrm{d}K^1}{\mathrm{d}t} &= (x_1\xi^1_1 - \mu^1)\, K^1 + x_1\xi^2_1 \, K^2 + x_1\xi^3_1 \, K^3, \\
\frac{\mathrm{d}K^2}{\mathrm{d}t} &= x_2\xi^1_1 \, K^1 + (x_2\xi^2_1 - \mu^2)\, K^2 + x_2\xi^3_1 \, K^3, \\
\frac{\mathrm{d}K^3}{\mathrm{d}t} &= x_3\xi^1_1 \, K^1 + x_3\xi^2_1 \, K^2 + (x_3\xi^3_1 - \mu^3)\, K^3.
\end{aligned} \tag{9.32}$$

The system of Eqs. (9.29) and (9.32) defines components of output, if the time-dependent quantities are given: the components of the tensor of productivity $\xi^i_j$, the fractions of investments into the three sectors $x_1$, $x_2$, $x_3$, and the sector coefficients of depreciation of the fixed capital $\mu^i$, $i = 1, 2, 3$. We can note that all characteristics of the production system and its variations are connected directly with actions of people; we can tell that people define *the programme of development*. This is a particular solution of the problem discussed in Sect. 6.4. As it was already noticed, the problem of evolution of production system can be formulated in the different approximations.

### 9.5.3 *Identification of the Initial State*

Considering evolution of production system of Russia, the year 2000 is accepted as an initial point in time; an estimate of the initial situation of the system is made according to the publications of Rosstat [2, 3] and the results of analysis in the previous chapter.

#### 9.5.3.1 Fundamental Matrixes

The production relationship among various sectors, that is, the structure of production system, is described by the fundamental technological matrixes $\mathsf{A}$ and $\mathsf{B}$, which, in the considered case, by virtue of definition of sectors (see balance Eqs. 9.26 and 9.27), have a simple form

$$\mathsf{A} = \begin{Vmatrix} a_1^1 & a_1^2 & a_1^3 \\ a_2^1 & a_2^2 & a_2^3 \\ 0 & 0 & 0 \end{Vmatrix}, \quad \mathsf{B} = \begin{Vmatrix} b_1^1 & b_1^2 & b_1^3 \\ 0 & 0 & 0 \\ 0 & 0 & 0 \end{Vmatrix}. \tag{9.33}$$

The all components of matrixes are non-negative. Let us notice that matrixes are degenerate, and it not exception, but a rule, so as there are sectors, product of which is completely consumed (sector 3), as well as sectors, that do not create the investment products (sectors 2 and 3). Matrixes $\mathsf{A}$ and $\mathsf{B}$ are individual characteristics of the system, and the assessment of values of the components of matrixes is carried out on the basis of results of studying of actual production system.

To assess the components of the matrix $\mathsf{A}$, we address available tables of *input–output* [3] for 2003 and suggest that numbers are suitable for description of the situation in 2000. The final table of direct expenditure for 22 sectors is reproduced in Appendix A. To pass to the three-sector model, apparently, we should judge, first of all, what part of a product of each of 22 sectors is used for the direct consumption, and what parts for the accumulation, and then to combine the numbers in columns and lines of the table in appropriate way. Unfortunately, such procedure appears rather tedious, and the result appears to be dependent from the experience and qualification of an expert, who is doing the assessments of the sector products, to say nothing of reliability of statistical estimates. Nevertheless, attempt to estimate components of a matrix $\mathsf{A}$ has been made by this method that has led to the following result:

$$\mathsf{A} = \begin{Vmatrix} 0.31 & 0.15 & 0.28 \\ 0.08 & 0.12 & 0.33 \\ 0 & 0 & 0 \end{Vmatrix}. \tag{9.34}$$

The matrix $\mathsf{B}$ is closely connected with the matrix of marginal productivity of capital (see formula 9.29), and, while assessing the components of the matrix $\mathsf{B}$, one can require, that after calculation the components of the matrix of marginal productivity of capital, one would have a known value of aggregate productivity, which, as it is found in the previous chapter, matters 0.0417 in the year 2000. On the additional assumption that the sector productivities of capital have the identical values, the requirement determines the matrix

$$\mathsf{B} = \begin{Vmatrix} 1.45 & 1.75 & 0.95 \\ 0 & 0 & 0 \\ 0 & 0 & 0 \end{Vmatrix} \text{ year}. \tag{9.35}$$

The values of the components of matrix $\mathsf{B}$ can be estimated more accurately by consideration of sector distribution of a fixed capital over the sectors that can be found in Table 12.39 of the Statistical Year Book [2].

The matrixes $\mathsf{A}$ and $\mathsf{B}$ describe the organisation of the developing production system, and the components of the matrixes change in due course; the above-specified values are attributed to 2000. By virtue of the above-discussed circumstances, the accuracy of the assessments of the components of matrixes $\mathsf{A}$ and $\mathsf{B}$ is not great, so the found numbers should be considered only as an illustrative example.

The components of the matrixes $\mathsf{A}$ and $\mathsf{B}$ combine, according to the rule (9.29), to create the fundamental matrix of capital marginal productivity (the matrix of capital productivity), and the specified values of the components (9.34) and (9.35) determine the matrix

$$\Xi = \begin{Vmatrix} 0.48 & -0.086 & -0.29 \\ -0.055 & 0.50 & -0.35 \\ 0 & 0 & 1.052 \end{Vmatrix} \text{ year}^{-1}. \tag{9.36}$$

The remarks concerning the matrixes $\mathsf{A}$ and $\mathsf{B}$ can be done about the matrix $\Xi$ as well.

#### 9.5.3.2 Initial Values of Variables

According to Rosstat [2, 4], the value of the Gross Domestic Product $Y$ in year 2000 is equal to $7320 \cdot 10^9$ rubles. We can refer to the results of the previous chapter (Sect. 8.5) to establish that in the initial year, the components of final output had the values

$$\begin{aligned} Y_1(2000) &= 1167 \cdot 10^9 \text{ rubles (2000)/year}, \\ Y_2(2000) &= 1069 \cdot 10^9 \text{ rubles (2000)/year}, \\ Y_3(2000) &= 5084 \cdot 10^9 \text{ rubles (2000)/year}. \end{aligned} \tag{9.37}$$

These values, according to formulae (9.5), define (at known values of the components of the technological matrixes $\mathsf{A}$ and $\mathsf{B}$) the initial values of production of value in the sectors

$$
\begin{aligned}
Z^1(2000) &= 2759 \cdot 10^9 \text{ rubles (2000)/year}, \\
Z^2(2000) &= 2578 \cdot 10^9 \text{ rubles (2000)/year}, \\
Z^3(2000) &= 1982 \cdot 10^9 \text{ rubles (2000)/year}.
\end{aligned}
\tag{9.38}
$$

The value of the basic production assets (in a broad sense, in view of domestic property and other material accumulation) $K$, by the assessments of Rosstat ([2], Table 12.35; [4], Table 11.21) is equal in 2000 to about $21000 \cdot 10^9$ rubles of 2000 (values are recalculate to rubles of 2000); however, due to the further re-estimation, Rosstat has reduced (see Sect. 8.4 of the previous chapter) the figure to $17500 \cdot 10^9$ rubles of year 2000. The relationship between fixed capital and output of the system (9.7), at a known matrix of capital productivity (9.36), allows to find the quantities and distribution of fixed capital production over sectors

$$
\begin{aligned}
K^1(2000) &= 6558 \cdot 10^9 \text{ roubles (2000)}, \\
K^2(2000) &= 6181 \cdot 10^9 \text{ roubles (2000)}, \\
K^3(2000) &= 4830 \cdot 10^9 \text{ roubles (2000)}.
\end{aligned}
\tag{9.39}
$$

The sum of the sector production assets is equal to $17569 \cdot 10^9$ rubles of 2000 and practically coincides with the last (2015) assessment of Rosstat, which confirms the assessment of the components of the matrix of a capital productivity (9.36).

The amounts of expenditures of labour and substitutive work in the national economy of Russia in 2000 can be found in Appendix C

$$
\begin{aligned}
L(2000) &= 118.8 \cdot 10^9 \text{ man} \cdot \text{h/year}, \\
P(2000) &= 3.2 \cdot 10^{12} \text{ J/year}.
\end{aligned}
\tag{9.40}
$$

Each of these quantities can be allocated into three parts over the sectors, but their absolute values do not enter into the further discussion and, consequently, we do not demonstrate the results.

The specified quantities are the initial values of the basic variables of the system. Besides, we shall need in some other quantities that will be discussed later in the formulation of the programme of development.

### 9.5.4 Evolution of the System at Given Programme

Considering the empirical dependences of the GDP and its components Fig. 9.1, we separate two periods of development. The average growth rate of GDP in the first period (2000–2007) is equal to 0.066, the coefficient of depreciation is equal to 0.043, the productivity of basic production assets increases from 0.417 up to 0.542; the development occurs at reduction of expenditures of labour with the rate of 0.015 and the increase of substitutive work with the growth rate 0.053. In the second

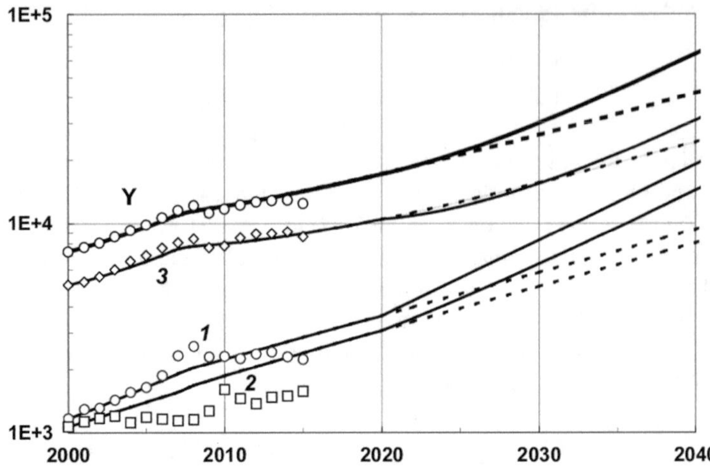


**Fig. 9.1** Evolution of the Russian economy. The trajectories of sector outputs (numerated curves) are calculated according to Eqs. (9.29) and (9.32). The top thick solid lines represent the total output of the system. The beginnings of the curves (till 2010) reproduce empirical values of variables (the points with symbols). The dashed lines present the trajectories of development in a non-varying situation. The solid lines represent outputs at variation of investment policy in 2020. All assessments are in billions ($10^9$) of rubles 2000

period (2008–2015), the average growth rate of GDP decreases to the value 0.006, the coefficient of depreciation increases up to 0.064 and the productivity of the basic production assets falls to the value 0.5 by the end of the period; progress occurs at growth of substitutive work with the growth rate 0.047 and practically constant expenditures of labour.

For the description of the actual trajectory of development and its hypothetical extension, we use Eqs. (9.29) and (9.32) with the known initial values of variables. The values of the sector coefficient of depreciation of fixed capital $\mu^i$, $i = 1, 2, 3$ are considered to be known (see above) and equal in all sectors. The matrix of productivity $\xi^i_j$ is estimated by comparison with productivity of the basic production assets. The increase in productivity in the first period of development (till the year 2008) is connected with the reduction of values of the components of the matrix capital–output $b^i_j$ and with corresponding increase of the components of the matrix $\xi^i_j$ till 2008; in the following years, the values of the components retained constant.

The distribution of investments over the sectors, that is, $x_1$, $x_2$, $x_3$, is being chosen in such a way that to achieve the best conformity of empirical assessments to theoretical dependences of both the GDP and its components. This requirement for an initial period of development determines the fractions of investments into various sectors as

$$x_1 = 0.41, \quad x_2 = 0.38, \quad x_3 = 0.21.$$

We consider two versions of development of the situation shown in Fig. 9.1. The dashed lines represent components of the GDP in the case, if neither structural shifts defined by variations of the components of the fundamental matrixes; no investment policy fixed by values of the quantities $x_1$, $x_2$, $x_3$ occurs after the year 2008. The trajectories of development are the exponent curve with the asymptotic growth rate 0.044. We shall notice that for the maintenance of such development, the growth rate of substitutive work should exceed the growth rate of the GDP. The solid lines represent trajectories of development in the case, when the investment policy changes in year 2020: it is supposed the increase in investments in the first sector that creates the means of production. The asymptotic value of the growth rate of GDP in this case is 0.086; this script, from the point of view of availability of substitutive work, looks less realistic. The trajectories calculated in this way are trajectories of potential growth. For more accurate description, it is necessary to consider possibilities of attraction of work and substitutive work, considering a more complex problem, formulated, for example, in the first sections of this chapter.

The described method allows to demonstrate the influence of structural variations, investment policy, policy of replacement of the equipment and other prospective possible influences on the system. Considering the problem, it was repeatedly marked the assumptions that were made at the assessment of the quantities needed for calculation, in particular, at the assessment of the components of matrixes (9.34), so that the described picture of development should be considered not as any forecast, but as a demonstration of the method and rules of designing of scripts of development.

## 9.6 Technological Coefficients and Technological Matrices

To characterize the technology of the production system, two pairs of quantities were introduced earlier: the fundamental technological matrices $\mathsf{A}$ and $\mathsf{B}$ in Chap. 4 and technological coefficients $\lambda$ and $\varepsilon$ in Sect. 5.2. We can guess that these pairs of quantities are connected with each other, or at least we can try to establish some relations in the frame of the many-sector economic model.

One can assume that the production equipment, produced in sector $i$, is characterized by the primary technological coefficients $\lambda_i$ and $\varepsilon_i$. The basic production equipment to be installed in the sector $j$ is a mixture of products with different technological coefficients,

$$\lambda^j = \sum_{i=1}^{n} \frac{I_i^j}{I^j}\lambda_i, \qquad \varepsilon^j = \sum_{i=1}^{n} \frac{I_i^j}{I^j}\varepsilon_i, \qquad I^j = \sum_{i=1}^{n} I_i^j, \tag{9.41}$$

where $I_i^j$ is the gross investment of product $i$ in sector $j$.

These last relations can be written in another form. First of all, we transform the first of Eq. (9.41) by summing up as follows:

$$\sum_{i=1}^{n} \lambda^i I^i = \sum_{l=1}^{n} \lambda_l I_l.$$

Then, we use formulae (4.18), assuming that coefficients $\overline{b}_l^i$ are constant, for investment $I_l$, which gives

$$\sum_{i=1}^{n} \lambda^i I^i = \sum_{i,l=1}^{n} \overline{b}_l^i \lambda_l I^i$$

and, from the arbitrary values of the sector investments $I^i$, we have for the technological coefficients

$$\lambda^j = \sum_{i=1}^{n} \overline{b}_i^j \lambda_i, \qquad \varepsilon^j = \sum_{i=1}^{n} \overline{b}_i^j \varepsilon_i. \tag{9.42}$$

It is clear that the primary technological coefficients $\lambda_i$ and $\varepsilon_i$ depend only on the amount of products used in production. Therefore,

$$\begin{aligned} \lambda_i &= \lambda_i(X_1^i, X_2^i, \ldots, X_n^i), \\ \varepsilon_i &= \varepsilon_i(X_1^i, X_2^i, \ldots, X_n^i). \end{aligned}$$

We can assume that the technological coefficients do not depend on a scale of production, so we can consider the technological coefficients to be uniform functions of the zeroth power, that is, their arguments have to be combined in ratios. From Eq. (4.2), the technological coefficients can be written as a function of components of technological matrix $\mathsf{A}$,

$$\begin{aligned} \lambda_i &= \lambda_i(a_1^i, a_2^i, \ldots, a_n^i), \\ \varepsilon_i &= \varepsilon_i(a_1^i, a_2^i, \ldots, a_n^i), \end{aligned} \tag{9.43}$$

where $i$ is the label of the sectors which create the production equipment.

Thus, one can write for the sector technological coefficients

$$\begin{aligned} \lambda^j &= \sum_{i=1}^{n} \overline{b}_i^j \lambda_i(a_1^i, a_2^i, \ldots, a_n^i), \\ \varepsilon^j &= \sum_{i=1}^{n} \overline{b}_i^j \varepsilon_i(a_1^i, a_2^i, \ldots, a_n^i). \end{aligned} \tag{9.44}$$

The technological coefficients are determined as functions of the components of the fundamental technological matrices $\mathsf{A}$ and $\mathsf{B}$, which are considered to depend on time (see Sect. 4.1). Further on, dependence (9.44) ought to be approximated from empirical data, and one can see that possible approximation depends on our choice of the sectors. The number of sectors we should take into account must be rather large to describe the technological changes properly.

We can illustrate relations (9.44) by our simple example of the three-sector production system, in which only one sector produces production equipment with technological coefficients $\lambda_1$ and $\varepsilon_1$. In this simple case, all basic equipment in all sectors consist of the product of the first sector. Therefore,

$$\lambda^1 = \lambda^2 = \lambda^3 = \lambda_1, \quad \varepsilon^1 = \varepsilon^2 = \varepsilon^3 = \varepsilon_1.$$

According to expression (9.43), the technological coefficients can be written as a function of components of technological matrix $\mathsf{A}$

$$\lambda_1 = \lambda_1(a_1^1, a_2^1), \qquad \varepsilon_1 = \varepsilon_2(a_1^1, a_2^1).$$

The values of the technological coefficients are determined by technological research and the level of development of science. Therefore, we assume that an increase of products of science, research and development (the products of sector 2) ensures technological progress, so that dependence can be approximated in a simple way by the relations

$$\lambda_1 \sim \left(\frac{a_2^1}{a_1^1}\right)^{-v}, \qquad \varepsilon_1 \sim \left(\frac{a_2^1}{a_1^1}\right)^{-u}.$$

These relations determine a decrease in the values of coefficients $\lambda_1$ and $\varepsilon_1$ at positive values of indexes $v$ and $u$.

## References

1. Wene, C.-O.: Energy-economy analysis: linking the macroeconomic and systems engineering approaches. Energy **21**, 809–824 (1996)
2. Rosstat: Rossiyskiy statisticheskiy ezhegodnik 2003 (The Russian Statistical Year-Book 2003). Goskomstat Rossii, Moscow (2003)
3. Rosstat: Sistema tablits Zatraty - Vypusk Rossii za 2003 god (The System of the Inrput-Output Tables for Russia in 2003). Rosstat, Moscow (2006)
4. Rosstat: Rossiyskiy statisticheskiy ezhegodnik 2008 (The Russian Statistical Year-Book 2008). Rosstat, Moscow (2008)

# Chapter 10
# Mechanism of Social Estimation of Value

**Abstract** In previous chapters, the notion of price was used as an empirical estimate of value of a product. The price is not an intrinsic characteristic of the product as a thing, but it emerges as a result of a bilateral assessment: *a producer* estimates efforts and expenses necessary to create a thing and *a consumer* estimates usefulness of that thing for him. The price emerges as a result of an agreement between the producer and consumer, and it thus appears connected with features of behaviour of economic agents. However, this does not mean that price is a subjective quantity; the price of a product exceeds expenses (cost) of manufacture for an amount, which the consumer can pay willingly, so that the attribution of value of a set of products to the production factors is not unreasonable. The relationship between producers and consumers in a process of exchange of products is a market of products. The theory of prices is a theory of the market. In this chapter, the theory of prices is considered for simple schemes that can be described in macroeconomic terms.

## 10.1 The System of Production and Consumption

On the basis of economic activity of human beings, one can detect explicit or implicit desires of people to satisfy their vital needs. In a society, each member needs the services of other persons, which he receives in exchange for his services to other members of the society. Thus, it is possible to assert that the essential content of economic activities of a person is an exchange of services. It is the main thing, which unites human beings in society. 'Every man thus lives by exchanging or becomes in some measure a merchant, and the society itself grows to be what is properly a commercial society' ([1], Chap. 4). The exchange of services has the form of exchange of products.



V. N. Pokrovskii, *Econodynamics*, New Economic Windows,
https://doi.org/10.1007/978-3-319-72074-6_10

### 10.1.1 Behaviour of Economic Agents

*An economic agent (actor)* as a carrier of will and a source of decisions on the operation of exchange appears in the theory. Strictly speaking, each active member of the society is an economic agent (actor). Moreover, each economic agent is a consumer and a producer simultaneously. As a consumer, he uses a set of products $\check{x}_1, \check{x}_2, \ldots, \check{x}_n$. As a producer, the individual spends some efforts of various types $\hat{e}_1, \hat{e}_2, \ldots, \hat{e}_m$ to create products. It is assumed that every agent knows what is good and what is bad for him, in other words, he has a certain *system of values*, which can be different for different agents, though some values can be mutual ones. Every economic agent tries to improve his situation, however beautiful it may be.

It is possible to assume that each individual as an economic agent, tries to spend less effort to get more products. To describe formally the behaviour of an economic agent and the aspiration to improve one's situation, it is introduced traditionally, following Walras [2] and many others, the characteristic of the economic agent, namely, the function of utility,

$$u(\hat{\mathsf{e}}, \check{\mathsf{x}}) = u(\hat{e}_1, \hat{e}_2, \ldots, \hat{e}_m,\ \check{x}_1, \check{x}_2, \ldots, \check{x}_n). \tag{10.1}$$

It is a decreasing function in relation to coordinates $\hat{e}_1, \hat{e}_2, \ldots, \hat{e}_m$ and an increasing function in relation to coordinates $\check{x}_1, \check{x}_2, \ldots, \check{x}_n$. It is more convincing to start with a description of the preferences of the agent to formally determine the agent's subjective utility function $u(\hat{\mathsf{e}}, \check{\mathsf{x}})$. We will return to this in Sect. 10.2.

As a producer, the economic agent tries to reduce his efforts and, as a consumer, he tries to increase the number of products that he can obtain for the efforts; thus, speaking formally, the agent aspires to reach the greatest value of the function of utility under some restrictions. For a single economic agent, the restriction can be recorded in the form of an inequality

$$\sum_{i=1}^{n} p_i \check{x}_i \le \sum_{j=1}^{m} w_j \hat{e}_j, \tag{10.2}$$

where $p_i$ is the price of a product with index $i$, and $w_j$ is a money estimate of a unit of effort of type $j$. Restriction (10.2) means that the agent cannot spend more than he or she gets for his/her efforts. Some other restrictions can be taken into account as well.

In modern societies, a single person can be considered as a consumer, but products are usually created by enterprises, which unite the efforts of several (many) individuals and should be considered as economic agents themselves. The behaviour of an enterprise is determined by tendency to get the greatest profit at given prices $p_i$ and $w_j$; speaking formally, the aim of the enterprise is to maximise the objectivity function

$$\pi(\check{\mathsf{e}}, \hat{\mathsf{x}}) = \sum_{i=1}^{n} p_i \hat{x}_i - \sum_{j=1}^{m} w_j \check{e}_j \geq 0. \tag{10.3}$$

Here we suppose that among a set of products $\hat{x}_1, \hat{x}_2, \ldots, \hat{x}_n$ there can be quantities both with a positive sign (output) and with a negative one (input). The enterprise aims to increase the output (at the same time to reduce expenses) $\hat{x}_1, \hat{x}_2, \ldots, \hat{x}_n$ and to decrease the consumed effort $\check{e}_1, \check{e}_2, \ldots, \check{e}_m$.

### 10.1.2 Elementary Economic System

The theory of the market was originally developed by Walras [2] for a system consisting of an assembly of individual consumers and production units, cooperating only through the prices of products and services. A simple heuristic model of the society allowed Walras to consider the mechanism of estimation of value.

To develop an appropriate theory, one has to consider all the economic subjects simultaneously, assuming that economic agents depend on each other through the exchange of efforts and products. According to Walras [2], the economic system can be imagined as consisting of many agents, say, of $s$ consumers and $r$ producers. Each consumer ($\alpha = 1, 2, \ldots, s$) is characterised by an offer of efforts $\hat{\mathsf{e}}^{\alpha}$ and a demand of products $\check{\mathsf{x}}^{\alpha}$, the variables being the arguments of the function of utility of type (10.1). Each producer ($\gamma = 1, 2, \ldots, r$) is characterised by a demand of efforts $\check{\mathsf{e}}^{\gamma}$ and a supply of products $\hat{\mathsf{x}}^{\gamma}$, the variables being the arguments of the function of profit of a type (10.3). All economic agents are independently running businesses according to their rules, but, nevertheless, they depend on each other through the exchange of efforts and products.

To calculate the amounts of efforts and products, it is necessary to find maximum points of $s$ functions of utility of type (10.1) and $r$ functions of profit of type (10.3), considering restrictions which follow from the balance of efforts, products and profits. Walras [2] recorded a system of algebraic equations for all variables in which, naturally, the prices of products and wages, $p_1, p_2, \ldots, p_n$ and $w_1, w_2, \ldots, w_m$ also are appearing. Later, Wald [3], McKenzie [5] and Arrow and Debreu [4] showed that the system of equations has a non-negative solution, which confirms the consistency of a considered model. The proof has required powerful and esoteric mathematical tools, fascinating a few generations of mathematical economists who have written thousands of papers.[1] Thus, it was shown that the proposed mechanism of exchanges can be taken as a basis for explaining the mechanism of human estimation of prices and, consequently, of the value of any set of products.

---

[1] Mark Blaug ([6], p. 17) paradoxically evaluates that 'The result of all this is that we now understand almost less of how actual markets work than did Adam Smith or even Léon Walras'.

The theory of Walras and his followers, though valid only for the simplified system and for equilibrium situations,[2] has a unique importance. The theory determined the principles of description of economic agents, which, in general form, are universal, they are applicable to any economic system, whether it be capitalism or socialism. These formal principles are laid in the foundation of the theory of exchange, or the theory of the market. At the same time, this theory is a theory of prices.

The area of applicability of the Walras' theory is severely restricted, which has been noted by many researchers [7–9]. To approach the description of the real production–consumption system, one has to take into account the real hierarchical structure of economic systems that can be very complex. Some works have demonstrated the diversity of architecture of economic systems and the variety of behaviour of economic agents (e.g. [9, 10]). Examples of construction of the theory for some simple cases are considered by Weidlich [11]. Nevertheless, there are ample opportunities for research.

## 10.2 Subjective Utility Function

The modern introduction of the utility function [12, 13] starts with a formalisation of human preferences. Introduced in this way, a subjective utility function is a characteristic of an economic agent rather than a characteristic of a fixed set of products, which is an argument of the utility function and can be represented as a vector,

$$\mathsf{x} = \begin{Vmatrix} x_1 \\ x_2 \\ \cdot \\ x_n \end{Vmatrix}. \tag{10.4}$$

The first candidate for the function which can characterise the given amount of products in their relation to a human's needs is the value of these products. However, one has to decline this candidate: the value of a set of products does not appear to be a function of amounts of products at all (see Sect. 11.2). Instead of the value function, one uses the utility function, which we introduce here, following the classics [12, 13], in the following way.

To compare two sets of products, that is, vectors $\mathsf{x}^1$ and $\mathsf{x}^2$, one introduces relations between the vectors. A human as a consumer can estimate if he prefers a set $\mathsf{x}^1$ to

[2]Economic equilibrium assumes that all macroeconomic variables which define the economic system, are constant, aside from fluctuations. From the point of view of thermodynamics, it is a steady-state situation, where the processes of production and consumption of products are occurring. Economic equilibrium is an idealisation of reality, which has been emphasised many times. However, this concept appears to be a very useful idealisation, just like thermodynamic equilibrium or steady-state situations in physics.

a set $\mathsf{x}^2$, or, on the contrary, a set $\mathsf{x}^2$ to a set $\mathsf{x}^1$, or if he cannot distinguish between two sets, correspondingly:

$$\mathsf{x}^1 \succ \mathsf{x}^2, \quad \mathsf{x}^2 \succ \mathsf{x}^1, \quad \mathsf{x}^1 \sim \mathsf{x}^2.$$

It was shown [12] that a monotonically increasing function can be defined on the space of dimension $n$ in such a way, that

$$\begin{aligned} \mathsf{x}^1 \succ \mathsf{x}^2 &\Rightarrow u(\mathsf{x}^1) > u(\mathsf{x}^2), \\ \mathsf{x}^1 \sim \mathsf{x}^2 &\Rightarrow u(\mathsf{x}^1) = u(\mathsf{x}^2). \end{aligned} \tag{10.5}$$

The arrow $\Rightarrow$ shows that the right-hand side relation follows the left-hand side one. The properties of the utility function $u(\mathsf{x})$ follow the simple assumptions.

If the amount of at least one single product in the set $\mathsf{x}^1$ is greater than the amount of the same product in the set $\mathsf{x}^2$, then

$$\mathsf{x}^1 \succ \mathsf{x}^2 \Rightarrow u(\mathsf{x}^1) > u(\mathsf{x}^2).$$

This means that all partial derivatives of the utility function are positive

$$\frac{\partial u}{\partial x_j} > 0. \tag{10.6}$$

Then, it is assumed that a mixture of two sets $\mathsf{x}^1$ and $\mathsf{x}^2$ is preferred to any of the sets; consequently,

$$\begin{aligned} &u(\lambda \mathsf{x}^1 + (1-\lambda)\mathsf{x}^2) > u(\mathsf{x}^1), \\ &u(\lambda \mathsf{x}^1 + (1-\lambda)\mathsf{x}^2) > u(\mathsf{x}^2), \quad 0 < \lambda < 1. \end{aligned}$$

The property is followed by the relation

$$u(\lambda \mathsf{x}^1 + (1-\lambda)\mathsf{x}^2) > \lambda u(\mathsf{x}^1) + (1-\lambda) u(\mathsf{x}^2).$$

This means that the utility function $u$ is strictly convex and a matrix of second partial derivatives,

$$u_{ij} = \frac{\partial^2 u}{\partial x_i \partial x_j} \tag{10.7}$$

that is, the Hesse matrix is negatively determined.

The function $u(x_1, x_2, \ldots, x_n)$ with described properties is called a utility function, or more precisely, a subjective utility function. The properties of the utility function are determined by postulates which are the reflection of empirical evidence. Note that any monotonically increasing transformation of variables of the utility function determines a new utility function with the same properties. These functions should be considered identical.

## 10.3 Demand Functions

One can consider a separate consumer who has to decide which products are the most necessary ones for him. The consumer has some income $M$ that is obtained in exchange for his efforts $e_i$, while label $i$ ($i = 1, 2, \ldots, m$) enumerates types of efforts. This income has a money form and increases when the agent's efforts increase, so in the simplest case, one can write

$$M = \sum_{i=1}^{m} w_i \; e_i, \tag{10.8}$$

where $w_i$ is a money estimate of a unit of effort of type $i$.

The money is spent to acquire products according to the consumer's preferences. One can assume that the main thing is to get money; if it is obtained, there is no difficulty to acquire anything. This assertion can be formalised in the theory [2] as

$$p_1x_1 + p_2x_2 + \cdots + p_nx_n \leq M, \tag{10.9}$$

where $p_i$ is the price of product, and $x_i$ is the quantity of acquiring products. We suppose here that there are no other restrictions.

However, the assumption about the unrestricted market distribution of products is not always valid. In certain societies, there is another mechanism for the distribution of products. In centrally planned societies, the existence of money does not mean that a person can buy any products; it is necessary to have a special right to buy products. This right is reached by efforts to achieve of a certain social rank, so the main aim of a person's activity is to increase in social rank [14]. It is not surprising that 'scientific norms of consumption' were elaborated in such societies.

Then, one can assume that the consumer is characterised by a subjective utility function,

$$u(x_1, x_2, \ldots, x_n),$$

where $x_1$, $x_2$,...,$x_n$ are amounts of products.

On the assumption that the consumer chooses a set of products that provide the biggest value to the utility function, one can describe the consumer's behaviour as an effort to maximise the utility function

$$\max u(x_1, x_2, \ldots, x_n)$$

with restriction (10.9). The amount of money $M$ for acquiring the desirable set of products is fixed, as well as the prices $p_i$ of all products. Note that budget restriction (10.9) can be written in the form of an equality and the problem of the choice of products can be solved as a problem of searching for a conditional maximum

$$\max u(x_1, x_2, \ldots, x_n), \quad \sum_{i=1}^{n} p_i x_i = M, \quad \mathsf{x} \geq 0. \tag{10.10}$$

There are no difficulties in adding other restrictions, if any.

To solve the problem, one starts with the Lagrange function

$$\mathcal{L}(\mathsf{x}, \lambda) = u(\mathsf{x}) - \lambda \left( \sum_{i=1}^{n} p_i x_i - M \right),$$

where $\lambda$ is a Lagrange multiplier. One should equate partial derivatives of the Lagrange function to zero to obtain a set of equations for the unknown quantities,

$$\frac{\partial u}{\partial x_j} - \lambda p_j = 0, \tag{10.11}$$

$$M - \sum_{i=1}^{n} p_i x_i = 0. \tag{10.12}$$

The first of the equations determines the ratio of product prices as a ratio of the marginal utilities of the products, that is, as a ratio of the first derivatives of the utility function,

$$\frac{\partial u}{\partial x_i} : \frac{\partial u}{\partial x_j} = \frac{p_i}{p_j}, \qquad i, j = 1, 2, \ldots, n. \tag{10.13}$$

These relations were written by Marshall ([15], Mathematical Appendix II).

Solutions of Eqs. (10.11) and (10.12) can be represented as functions of the quantities $x_i$ and $\lambda$ depending on the parameters of the problem,

$$x_i = x_i(\mathsf{p}, M), \quad \lambda = \lambda(\mathsf{p}, M). \tag{10.14}$$

A change of scale of value does not change the problem, so one can define a demand function as a uniform function of its arguments, that is,

$$x_i = x_i\left(\frac{\mathsf{p}}{M}\right). \tag{10.15}$$

The effects of prices and money on demand can be investigated without knowledge of the explicit form of the utility function [16, 17]. If the utility function is given, the demand functions can be found easily. For example, for the utility function

$$u = x_1^{\alpha_1} x_2^{\alpha_2} \ldots x_n^{\alpha_n}, \quad \sum_{i=1}^{n} \alpha_i < 1$$

a solution to the problem (10.10) is

$$x_i(\mathsf{p}, M) = \frac{\alpha_i M}{\alpha p_i}, \quad \lambda(\mathsf{p}, M) = \alpha \frac{u}{M}, \quad \alpha = \sum_{i=1}^{n} \alpha_i. \tag{10.16}$$

The speculations in this section determine the demand functions, which themselves can be set in the foundation of the theory. In fact, one should assume special properties of the utility function in order for the demand functions (10.15) to describe the empirical facts. However, the demand function can be determined from empirical data independently. The demand functions are usually decreasing functions of prices, though there are some exceptions to this rule.

## 10.4 Welfare Function

The utility function considered in Sect. 10.2 is based on the consumer's subjective preferences and can be called *a subjective utility function*. It can be introduced for each consumer, who tries to choose a situation to maximise this function. To investigate links between these utility functions and the objective utility function, which will be considered in the next chapter, one has to introduce a subjective utility function for the whole community.

Consider a society consisting of $s$ independent consumers, who together hold some amount of products

$$Q_1, Q_2, \ldots, Q_n,$$

while the consumer $\alpha$ owns parts of the products

$$x_1^\alpha, x_2^\alpha, \ldots, x_n^\alpha, \quad \alpha = 1, 2, \ldots, s,$$

where

$$\sum_{\alpha=1}^{s} x_i^\alpha = Q_i, \quad i = 1, 2, \ldots, n. \tag{10.17}$$

One assumes that a consumer's utility function

$$u^\alpha(\mathsf{x}^1, \mathsf{x}^2, \ldots, \mathsf{x}^s), \quad \alpha = 1, 2, \ldots, s \tag{10.18}$$

depends both on variables with label $\alpha$ and on all other variables, while

$$\frac{\partial u^\alpha}{\partial x_i^\alpha} > 0, \quad \alpha = 1, 2, \ldots, s, \quad i = 1, 2, \ldots, n,$$

$$\frac{\partial u^\alpha}{\partial x_i^\nu} \leq 0, \quad \alpha \neq \nu, \quad \alpha, \nu = 1, 2, \ldots, s, \quad i = 1, 2, \ldots, n.$$

Every consumer maximises his utility function, as was described in the previous section, but now we should consider maximisation of $s$ functions simultaneously. There is apparently no single point, which gives maximum values for all functions simultaneously. However, one can exclude all points where values of the utility functions can be enlarged simultaneously. The remaining points make up what is called the Pareto-optimal set. No consumer in the Pareto-optimal point can improve his/her welfare without diminishing someone else's welfare. The problem of distribution of products among the members of the society was posed and investigated by Pareto [18].

To find a Pareto-optimal set, one can construct the welfare function for the entire community,

$$U(\mathsf{x}^1, \mathsf{x}^2, \ldots, \mathsf{x}^s) = \sum_{\nu=1}^{s} \alpha_\nu \, u^\nu(\mathsf{x}^1, \mathsf{x}^2, \ldots, \mathsf{x}^s), \quad \alpha_\nu \geq 0, \quad \sum_{\nu=1}^{s} \alpha_\nu = 1. \quad (10.19)$$

A point of maximum of function (10.19) at any values of multipliers $\alpha_\nu$ belongs to the Pareto-optimal set.

To find Pareto-optimal points, we can use the Lagrange method for solving the problem of maximisation of function (10.19) at restrictions (10.17) and we find, that the Pareto-optimal points obeyed to the set of equations

$$\sum_{\nu=1}^{s} \alpha_\nu \frac{\partial u^\nu}{\partial x_i^\mu} - p_i = 0, \quad \mu = 1, 2, \ldots, s; \quad i = 1, 2, \ldots, n, \quad (10.20)$$

where $p_i$ are Lagrange multipliers of the problem which are functions of the multipliers $\alpha_\nu$.

One can reasonably assume that the consumer's utility function depends mainly on his own choice, so that relation (10.20) can be rewritten in a simpler form,

$$\alpha_\nu \frac{\partial u^\nu}{\partial x_i^\nu} - p_i = 0, \quad \nu = 1, 2, \ldots, s; \quad i = 1, 2, \ldots, n.$$

This gives the relation for every consumer,

$$\frac{\partial u^\nu}{\partial x_i^\nu} : \frac{\partial u^\nu}{\partial x_j^\nu} = \frac{p_i}{p_j}, \quad \nu = 1, 2, \ldots, s; \quad i, j = 1, 2, \ldots, n. \quad (10.21)$$

The equation for each consumer has the same form as Eq. (10.13), and one can assert that the set of Lagrange multipliers of the problem is a set of prices, which are identical for all the participants.

Welfare function (10.19) is a function of amounts of products distributed among the members of the society. In line with this function, one can consider a function of the entire amounts of products,

$$U(Q_1, Q_2, \ldots, Q_n), \quad Q_i = \sum_{\alpha=1}^{s} x_i^{\alpha}, \quad i = 1, 2, \ldots, n.$$

To relate this function to function (10.19), we assume that all consumers are in identical situations, that is, the amounts of products are distributed equally among all consumers, which means that the quantity $\mathsf{x}^{\nu}$ in function (10.19) does not depend on the index $\nu$ and can be replaced by the quantity $\mathsf{Q}/s$. Then, for every consumer,

$$\alpha_{\nu}\frac{\partial u^{\nu}}{\partial x_j^{\nu}} = \frac{\partial U}{\partial Q_j}, \quad \nu = 1, 2, \ldots, s; \quad j = 1, 2, \ldots, n.$$

One can see that the following relation is valid:

$$\frac{\partial U}{\partial Q_i} : \frac{\partial U}{\partial Q_j} = \frac{p_i}{p_j}, \quad i, j = 1, 2, \ldots, n. \tag{10.22}$$

One can compare the properties of the welfare function as a function of the entire amounts of products with the properties of the objective utility function introduced in the next chapter (Sect. 11.2). The relation (10.22) is exactly relation (11.7) in which $U$ is an objective utility function. Therefore, one can suppose that, as characteristics of a set of products, the objective utility function and the welfare function are indistinguishable. Although the objective utility function does not present the value of a set of products, nevertheless, one expects that the function is connected with the estimations of value.

## 10.5 The Simplest Markets

Not only a separate individual, but a group of people, or even a society as a whole can be considered to be an economic agent. Further, in this section, we shall consider very simple cases, which will allow us to demonstrate equations for prices. To consider special examples in more detail, we assume that the economy consists of $n$ producers—each sector is a separate producer, which outputs a single product—and only one consumer, the society as a whole. Every one of the $n + 1$ economic agents has its own aims, plans its activity and can make decisions.

The aim of a sector is to obtain a bigger amount of production of value $Z^j$. To imitate the behaviour of the sector, we solve a maximisation problem, that is, we find the amount of gross output which determines the greatest value of $Z^j$. We refer to the results of Sect. 4.2.2 to say that the gross output should be planned to be as big as possible at the restriction given by the production factors, so a supply function for each product can be determined as a function of prices,

$$Y_j^{\mathsf{supply}}(\mathsf{p}) = Y_j(p_1, p_2, \ldots, p_n), \quad j = 1, 2, \ldots, n. \tag{10.23}$$

The aim of the entire society is to obtain the greatest usefulness from the products, that is, to find quantities of the output which would maximise a utility function at restricted amount of money $M$. In this way, one can determine the demand functions,

$$Y_j^{\text{demand}}(\mathsf{p},\ M) = Y_j(p_1, p_2, \ldots, p_n,\ M), \quad j = 1, 2, \ldots, n. \tag{10.24}$$

### 10.5.1 Free-Price Market

The problem of simultaneous maximisation of $n+1$ objective functions can be reduced to the problem of simultaneous consideration of the demand and supply functions for each product. One can consider each product separately and introduce the excess demand function

$$Z_j(\mathsf{p},\ M) = Y_j^{\text{demand}}(\mathsf{p},\ M) - Y_j^{\text{supply}}(\mathsf{p}), \quad j = 1, 2, \ldots, n, \tag{10.25}$$

where in contrast to what will be considered in the next section, one assumes that there is only one price for each product on the market.

In the situations which we call *economic equilibrium*,[3] demand is equal to supply, so that

$$Z_j(\mathsf{p},\ M) = 0, \quad j = 1, 2, \ldots, n. \tag{10.26}$$

This system of equations defines *equilibrium prices*, whereby it is assumed that the demand and supply functions have dependencies of prices such that a stable solution of system (10.26) exists. One can consider the demand and supply functions of a chosen product as functions of its own price only. Situations in the market, as have usually been considered [19], are depicted in the plot of Fig. 10.1, which shows demand and supply curves.

Empirical observations assure us that excess demand for a product provokes an increase in the price and vice versa. At small deviations of the demand from the supply, one can write a simple rule for the growth rate of price,

$$\frac{dp_i}{dt} = k_i\ Z_i\ (\mathsf{p},\ M), \quad k_i \geq 0. \tag{10.27}$$

---

[3] Economic equilibrium assumes that all macroeconomic variables which define an economic system are constant, aside from fluctuations. It recalls the definition of equilibrium in thermodynamics: all thermodynamic variables are constant on average, though there is movement of constituents particles of the thermodynamic system. Similarly, the material constituent of an economic system at equilibrium is not in thermodynamic equilibrium: there are processes of production and consumption of products. Economic equilibrium is an idealisation of reality, which has been stressed many times [7]; nevertheless, it is a very useful idealisation, like that of thermodynamic equilibrium in physics.

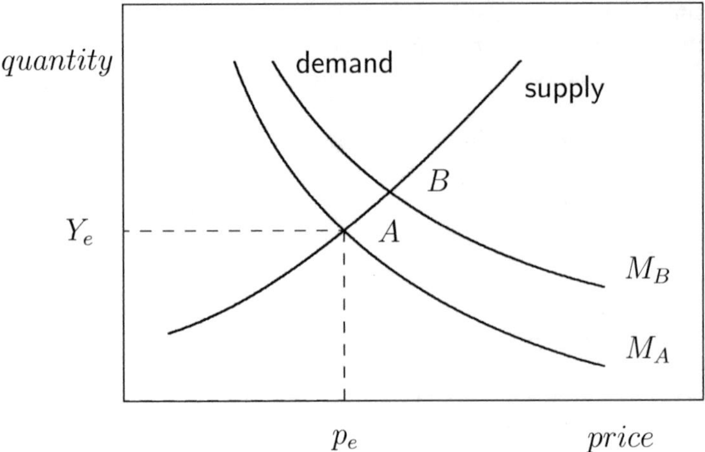


**Fig. 10.1** Situations on a free-price market of single product. The intersection of the demand and supply curves determines an equilibrium point $(p_e, \ Y_e)$. At $M_B > M_A$, the demand increases and a new equilibrium price and a new equilibrium quantity, which are greater than the previous ones, appear

However, the trajectory of the approach to the equilibrium point can differ from a straight line and can resemble a cobweb, as has been described many times [19, 20].

The conditions for equilibrium prices can be written in the form of the Walras law,

$$\sum_{i=1}^{n} p_i \ Z_i \ (\mathsf{p}, \ M) = 0. \tag{10.28}$$

The last relation defines equilibrium prices $p_j \geq 0$, if the relation is valid at $Z_i(\mathsf{p}) \leq 0$. The existence of a system of prices is connected with the behaviour of the excess demand functions $Z_i(\mathsf{p})$. So, as we rely on the empirical observation that equilibrium prices exist, the problem is in defining a class of functions which can describe demand and supply. The problem has been studied in many details [19].

The essential component of a real market is money which is considered as a special artifact. The amount of money $M$ should correspond to the value of the commodities in the sense discussed in Chap. 3. Relation (10.26) is valid for equilibrium situations and is an analogy of relation (3.25) in Chap. 3. In non-equilibrium situations, there is an excess of money, which should be equated to the excess demand,

$$\Delta M = \sum_{i=1}^{n} p_i \ Z_i \ (\mathsf{p}, \ M). \tag{10.29}$$

If money is taken as a separate product, the situation with the demand excess can be considered as an equilibrium, but it is convenient to refer to such situations as non-equilibrium ones. The name is justified, as the system tends to an equilibrium state when relation (10.26) is valid.

On the plot of Fig. 10.1, a point $A$ of intersection of curves determines the equilibrium price and quantity of the product. An excess or shortage of money causes a displacement of the demand curve up ($M_B > M_A$) or down ($M_B < M_A$). A new point $B$ of equilibrium appears, whereby the price and quantity of the product take new values.

Thus, the money excess provokes changes of prices, which can be determined with the help of relations (10.27) and (10.29). One should know the demand and supply functions to determine these changes. However, when we restrict ourselves to investigating a system which includes only one sector which needs money, the situation is simple. We can assume that, in the case of the discussed three-sector model, money is only needed for products of the third sector. Therefore, Eqs. (10.27) and (10.29) can be rewritten as

$$\frac{dM}{dt} = p_3 \, Z_3 \, (p_1, p_2, p_3, M),$$
$$\frac{dp_3}{dt} = k_3 \, Z_3 \, (p_1, p_2, p_3, M).$$

One can see that these equations are followed by the simple relation

$$\frac{dp_3^2}{dt} = 2 \, k_3 \, \frac{dM}{dt}. \tag{10.30}$$

The growth rate of the squared price of the commodities for immediate consumption is equal to the amount of emitted money.

### 10.5.2 Fixed-Price Market

In cases when there is a monopoly on the production of all commodities, fixed prices of the commodities can be set. This is a case of a *centrally planned economy*, designed in countries where the state is the only owner of the whole production system. Analysis shows [14] that the real owner of the production is a *nomenclature class* which governs the economy on the behalf of the state.

As in the previous section, we consider a market where $n$ producing sectors and only one consumer are participating. We assume that the economic agents are characterised by demand and supply functions, but, in contrast to the assumption in the previous section, we suppose that each product has, generally speaking, two prices: a wholesale price $p'$ for the producer and a retail price $p''$ for the consumer. The existence of the two sets of prices is, according to Polterovich ([21], p. 185), a phenomenon of centrally planned economy'. To describe a situation on the market, instead of function (10.25), one should introduce an excess demand function which, in contrast to the function in the previous section, depends on the two sets of prices,

$$Z_j\ (\mathsf{p}', \mathsf{p}'', M) = Y_j^{\mathsf{demand}}(\mathsf{p}'',\ M) - Y_j^{\mathsf{supply}}(\mathsf{p}'), \quad j = 1, 2, \ldots, n. \tag{10.31}$$

Instead of (10.26), equilibrium situations are defined by the relations

$$Z_j\ (\mathsf{p}', \mathsf{p}'', M) = 0, \quad j = 1, 2, \ldots, n. \tag{10.32}$$

The profound meaning of the two-price-set market for the owner, the state, is the possibility to obtain income that is equal to the quantity

$$\sum_{i=1}^{n} (p_i'' - p_i')\ Y_i. \tag{10.33}$$

The state declares (and has the means to persuade its citizens) that the social interests are the most important ones, so that, in addition to the optimisation criteria for each participant of the market, considered in the previous section, the situation in the market is determined by the requirement of maximisation of quantity (10.33) as well.

Though the state controls the prices, the fixed prices $p_i'$ and $p_i''$ cannot be set quite arbitrarily. One can see that system (10.32) can determine a set of equilibrium retail prices if a set of the wholesale prices is given, and vice versa. Aside from this, the enterprises of the production system keep some independence and interest in obtaining profit. This is provoked by the requirements of finance for self-support of each enterprise. Therefore, the wholesale prices are fixed at such levels that enterprises (at least, some) have to obtain some moderate profit.

The equilibrium quantity of the product, which is being sold and bought, can be less or greater than the coordinate of the point of intersection of the demand and supply curves (the true equilibrium value which could be reached in the free-price market), as is shown on the plots of Fig. 10.2. It is a well-known fact [21] that there are commodities of both types in a centrally planned economy. Initial equilibrium situations for different products are depicted with lines $A'' - A'$ and $A' - A''$. In one case, at $p'' > p'$, the state gets a profit, which is called the turnover tax; in the other case, at $p' > p''$, the state is urged to subsidise the poor sectors.

The excess of money provokes crossover to a new equilibrium situation (line $B'' - B'$ in Fig. 10.2 (top) and $B' - B''$ in Fig. 10.2 (bottom)) which has to be determined by the state, as it controls both production and prices. To begin with, the state usually announces some changes in wholesale prices. This announcement is accompanied by statements that the decision does not concern retail prices. Despite the statements, after some time, the retail prices go up.

The rules of crossover can be understood on the basis of relation (10.32) and the requirement to get the greatest value of criterion (10.33). One can see that the greater the quantity of the profit product and the difference between the retail and wholesale prices $p'' - p'$ are, the greater the profit, which is favourable for the state. However,

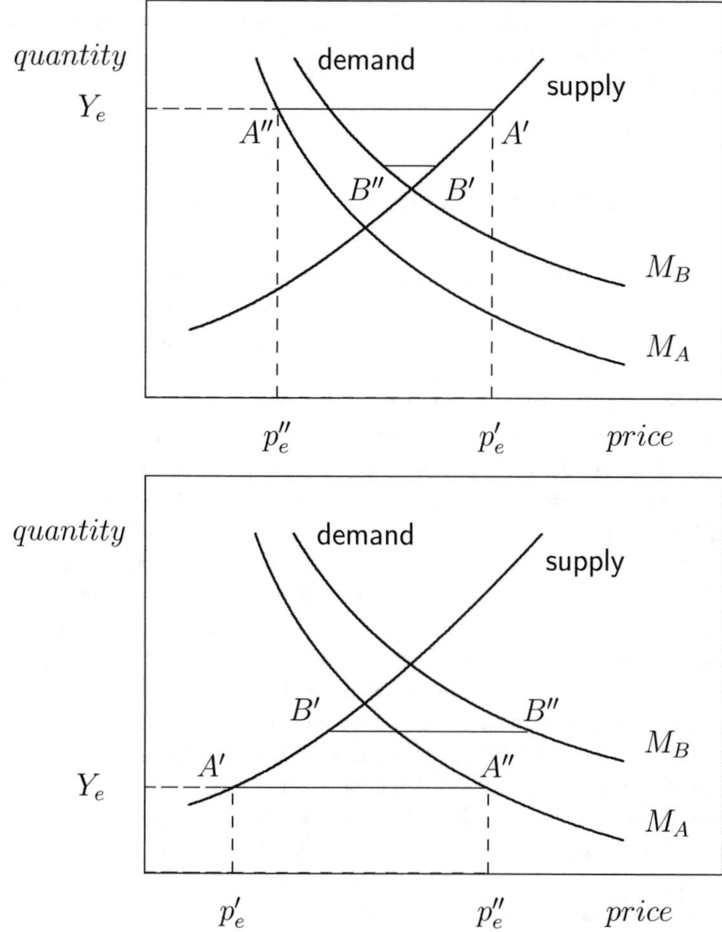


**Fig. 10.2** Situations on a fixed-price market of single product. Top: Subsidised product. The wholesale price is greater than the retail price ($p'_e > p''_e$), and the government is urged to pay subsidies $(p'_e - p''_e)\ Y_e$ to the sector. Bottom: Profit product. The wholesale price is less than the retail price ($p'_e < p''_e$), and the government obtains a profit $(p''_e - p'_e)\ Y_e$

as one can see from Fig. 10.2 (bottom), these requirements are conflicting for the profit products, and it is difficult to forecast the resulting equilibrium situation in this case.

For the subsidised products, in contrast, the lower both the quantity of the output and the difference between the prices are, the better for the state. These are not conflicting requirements, so one expects that the output would not be greater in the new equilibrium situation (Fig. 10.2, top). Therefore, it is favourable for the state interest to have a low level of production and consumption of the subsidised products, in which set almost all food products except vodka and, perhaps, a few other things were included in the USSR. However, there was apparently a 'natural' lower level of consumption; the production system has to provide for the survival and reproduction of the labour force.

## 10.6 The Problem of Coordination of Interests

Apparently, various patterns of the organisation of production and the mechanisms of management can be imagined, whereby it is possible to separate the limiting cases about which one speaks in terms of a centrally planned economy and a market economy. The Walras model describes the economic system as an assembly of individual consumers, independent in their decisions, and the production associations, cooperating at free prices only through an exchange of products and services (see Sects. 10.1 and 10.5.1). This model is considered as the most idealised model: interests of producers and consumers are taken into account completely, whereas the possible interests and demands of the society as a whole are entirely ignored. Another idealised model also, schematised in Sect. 10.5.2, describes an economic system, as the unite enterprise; it is supposed, that the central government on behalf of the entire society and in the common interests organises the production, appoints the prices and decides, what part of the social product should be directed to the expansion of the production and what part could be used for direct consumption. Though some general principles of behaviour of the producers and consumers appear to be universal, the results of the theory appear to depend on assumptions about the architecture of the system, which can be rather complicated and cannot be universal.

Apparently, none of the models is actually realised in this extreme formulation; the real picture for all countries always appears to be an intermediate version between the two extreme cases. Even in the post-war Soviet Union, that was known to be rigid central-planned system, it was declared that all means of production belonged to the society, the personal property of some things and the possibility of usage of the land plots allowed persons to be engaged in productive activity, which could not be controlled by the state. There were also cooperative enterprises (the artels) and not advertised enterprise activity in those years, For the adequate description of a national economy of Soviet Union it is necessary to consider, apparently, the activity existed outside of the centralised control.

On the other hand, when a private property on the means of production flourishes, the government is compelled to keep the decisive influence on the enterprises of those industries, which matter for the country as a whole, but are ignorant by private proprietors. It can be power, transport, communication, protection against epidemics and acts of nature, roads, mail, education, information service, a security, social insurance, care of old people and invalids as well as cares of protection of borders and conservation of an inhabitancy. The state establishes tools to take a part of the income of the enterprises and provides order and protection against external enemies, stability of the national currency and social protection.

Modern economic systems are characterised by a complex hierarchical organisation: national state monitors the general situation and organises national projects, creating the general conditions for ability to people to live. The sovereign government (state) is the highest in hierarchy body, influencing on production and distribution of products. At the same time numerous economic subjects, both producers and consumers, take part in the organisation and functioning of processes of manufacture. Both the government and individual subjects, proceeding from their interests and reasons, make efforts to defend the position. There is a problem of the optimum organisation of production and distribution, when the interests of both the entire society and the different associations of producers and consumers are taken into account. Apparently, various schemes of the organisation of production and distribution are possible.

## References


1. Smith, A.: An Inquiry into the Nature and Couses of the Wealth of Nations. Clarendon Press, Oxford (1976) (In two volumes)
2. Walras, L.: Elements d'economie politique pure ou theorie de la richesse sociale. Corbaz, Lausanne (1874)
3. Wald, A.: On some system of equations of mathematical economics. Econometrica **19**, 368–403 (1951)
4. McKenzie, L.: On the existence of general equilibrium for a competitive market. Econometrica **27**, 54–71 (1959)
5. Arrow, K.J., Debreu, G.: Existence of an equilibrium for a competitive economy. Econometrica **22**, 265–290 (1954)
6. Blaug, M.: Disturbing currents in modern economics. Challenge **41**, 11–34 (1998)
7. Kornai, J.: Anti-Equilibrium: On Economic Systems Theory and the Tasks of Research, 2nd edn. North-Holland Publishing, Amsterdam (1975)
8. Soros, G.: The Crisis of Global Capitalism: Open Society Endangered. Public Affairs, New York (1998)
9. Baumol, W.J.: Economic Theory and Operations Analysis, 4th edn. Prentice-Hall, London (1977)
10. Akerlof, G.A.: The market for "lemons": quality uncertainty and the market mechanism. Q. J. Econ. **84**(3), 488–500 (1970)
11. Weidlich, W.: Sociodynamics. A Systematic Approach to Mathematical Modelling in the Social Sciences. Harwood Academic Publishers, Amsterdam (2000)
12. Von Neumann, J., Morgenstern, O.: Theory of Games and Economic Behaviour, 3rd edn. Princeton University Press, Princeton (1953)
13. Bridges, D.S., Mehta, G.B.: Representation of Preference Orderings. Springer, Berlin (1995)
14. Voslenskii, M.S.: Nomenklatura. The Soviet Ruling Class. Doubleday, Garden City (1984)
15. Marshall, A.: Principle of Economics, 8th edn. Macmillan, London (1920)
16. Slutsky, E.: Sulla teoria del bilancio del consumatore. Giornale degli economisti **60**, 19–23 (1915)
17. Hicks, J.R.: Value and Capital, 2nd edn. Clarendon Press, Oxford (1946)
18. Pareto, V.: Manuel d'economie politique. Girard et Briere, Paris (1909)
19. Intriligator, M.D.: Mathematical Optimisation and Economic Theory. Prentice-Hall, Englewood Cliffs (1971)
20. Allen, R.G.D.: Mathematical Economics, 2nd edn. Macmillan, London (1963)
21. Polterovich, V.M.: Ekonomicheskoje ravnovesije i khozjaistvennyi mekanizm (Equilibrium and economic mechanism). Nauka, Moscow (1990)

# Chapter 11
# Value from a Physicist's Point of View

**Abstract** As a specific concept of economics, value does not need to be reduced to any scientific concepts, but as far as a production process can be considered as a process of transformation of 'wild' forms of matter into forms useful for humans (dwellings, food, clothes, machinery and so on), one can look for analogies in thermodynamics. Thermodynamic laws are quite general and are applicable to any system, no matter how big and complicated it is, while they do not require a knowledge of the structure of the system in all details. The phenomenon of production can be considered from a general point of view, assuming all our environment to be *a thermodynamic system*, which is in a far-from-equilibrium state. For the matter to be transformed into shapes of different commodities (complexity), one needs in the creative work of production equipment. The process of production can be regarded as process of materialisation of information, whereby the cost of materialisation is the work of the production system. To maintain complexity in a thermodynamic system, fluxes of matter and energy must flow through the system.

## 11.1 Energy Principle of Evolution

### *11.1.1 Thermodynamics of the Earth*

One can consider the upper layers of the Earth as *a thermodynamic system* in a non-equilibrium state. Due to Prigogine [1, 2], we know that stable dissipative structures can exist in such states. All biological organisms and all artificial things on the Earth can be considered dissipative structures which exist due to fluxes of energy [2, 3]. There are also dissipative structures of larger scale: convection flows in the atmosphere and oceans, ecological systems, socio-economic systems, systems of knowledge and others. The human population itself is an example of large-scale dissipative structure. However, we only begin to recognise and describe a large-scale dissipative structure in the upper layers of the Earth.



V. N. Pokrovskii, *Econodynamics*, New Economic Windows,
https://doi.org/10.1007/978-3-319-72074-6_11

In the most rough approximation, the Earth, as a thermodynamic system, can be characterised by internal energy $E$, temperature $T$ and entropy $S$. These quantities are connected with each other by the first law of thermodynamics, which, assuming that the work of the external gravitational forces, leading to deformation of the Earth's form, can be neglected in comparison with other terms, is recorded in the form of

$$dE = T\,dS. \tag{11.1}$$

The values of the internal energy $E$ and entropy $S$ of the Earth, as *an open thermodynamic system*, are influenced by external fluxes and internal processes, so that a variation of the entropy of the Earth is defined by the formula

$$T\,dS = \Delta Q - \sum_{i} \Xi_i\,\Delta\xi_i + \sum_{j=1}^{K} \mu_j\,\Delta N_j. \tag{11.2}$$

This fundamental equation[1] shows that various influences on the system lead to the variation of one universal quantity—entropy—and the variation can be connected both with the fluxes of heat $\Delta Q$ and substances $\Delta N_j$ through the borders of the system, and with variations of the internal structure of the system. It is assumed that the internal complexity of the system is described with a set of some internal variables $\xi_1,\ \xi_2,\ \ldots$.

The non-equilibrium state of the Earth is supported by the energy fluxes: the Earth receives the fluxes of radiant energy from the Sun and re-radiates heat, and a number of transformations take place with energy on its way through the system (Fig. 11.1). On the modern assessments [8, 9], from the total incoming radiation energy that is $5.5 \cdot 10^{24}$ J per year, about a third is reflected by clouds and the surface of the Earth, and another part (about $3 \cdot 10^{24}$ J per year) is absorbed by the atmosphere and the surface of the Earth. Assessments show that photosynthesis is the basic mechanism of absorbing the solar energy. The flux of the external radiation energy has been changing the thermodynamic characteristics of the Earth during the time of its evolution.

In the current epoch, the energy influx from the Sun is balanced by a radiation outflow [8], so the Earth can be considered to be in a steady state, which implies that the thermodynamic characteristics of the Earth do not change. Though the entropy of the stationary Earth is constant, there is internal production of entropy in the system, as in any non-equilibrium thermodynamic system. The production of entropy is connected with the variation of the internal variables,

$$d_i S = -\frac{1}{T} \sum_{j} \Xi_j\,d\xi_j, \quad d_i S \geq 0, \tag{11.3}$$

[1] One can note that Eq. (11.2) represents a generalisation of the equation for variation of entropy, established by Prigogine ([5], Eq. (3.52)) for systems with chemical reactions (see also [6]). The generalisation of the theory can be found in work [7].

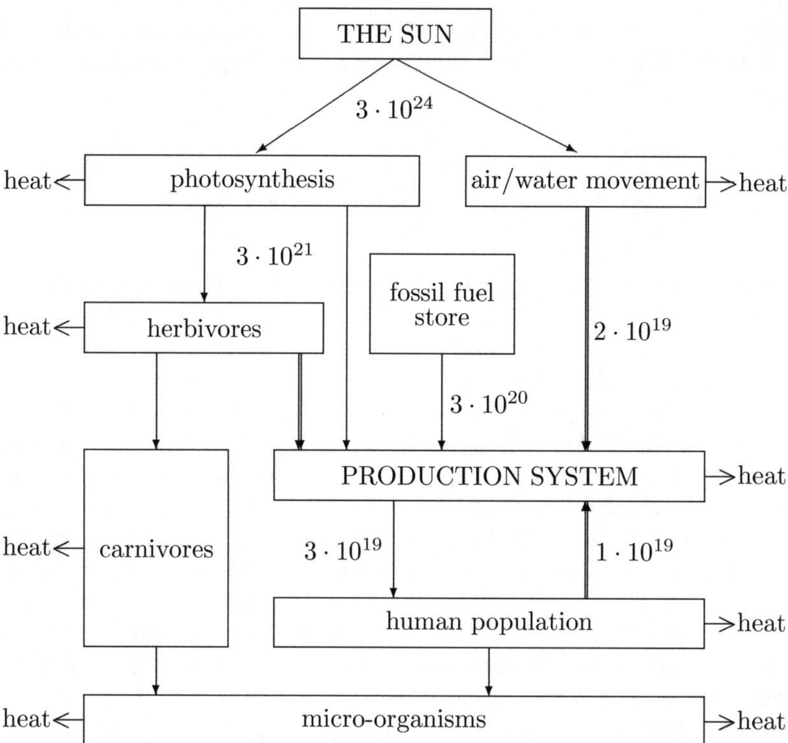


**Fig. 11.1** The energy flows in terrestrial systems. The main fluxes of chemical energy (single lines) and mechanical energy (double lines) are shown. Estimates of fluxes are in joules per year, according to [4]

To estimate this quantity in the case when the Earth is considered to be in a steady state, we note that the internal production of entropy is compensated by a change of entropy, due to the outgoing fluxes of energy,

$$d_i S = -d_e S = -\frac{1}{T}\left(\Delta Q + \sum_{j=1}^{K} \mu_j \, \Delta N_j\right). \tag{11.4}$$

The Earth receives a solar flux of high-energy photons (appropriating temperature about 6000 K with chemical potential $\mu = 0$) and radiates heat at a much lower temperature about 300 K. These fluxes of radiation energy contribute to a change of entropy $d_e S$ that has been estimated by many researchers [9, 10] and appears to be negative $d_e S < 0$. From Eq. (11.4), this evidences the internal production of entropy, which can be considered a measure of complexity of the Earth. To create and maintain the special complexity (far-from-equilibrium objects or dissipative structures), as in any thermodynamic system [1–3], there is a need for energy fluxes moving through the system.

### 11.1.2 Human Population and Fluxes of Energy

The life of every biological population is based on energy fluxes which come to the population through food and organisms of species. One can refer to those fluxes as *biologically organised fluxes of energy*. For the human population, for example, the biologically organised flux of energy is about $4 \cdot 10^9$ J/year · man $= 4 \cdot 10^6$ Btu/year · *man* in all the centuries of human existence (see the discussion in Sect. 2.4.1).[2]

In line with the biologically organised fluxes of energy, the human population has *socially organised fluxes of energy*, when the *production system* of society plays the role of a mechanism attracting energy from out-of-body sources. Prime sources of energy (the remains of the former biospheres: wood, coal and oil; direct and indirect solar energy in the form of wind, water and tides; and energy of fission and fusion of nuclei) are used via different appliances to transform matter of the natural environment into things of the artificial environment that are useful for the human. The ways in which energy has been utilised by humans have been considered in the previous chapters in some detail.

The socially organised consumption of energy per capita from traditional sources (mainly firewood and charcoal, animal dung, and agricultural wastes) and commercial sources (oil, coal, gas, hydroelectricity, nuclear power and so on) for the world and for some countries is shown in Fig. 11.2. This quantity has reached the amount of biologically organised flux of energy during the agricultural era. In the middle of the nineteenth century, the amounts of the two fluxes were approximately equal. Nowadays, in developed countries, the socially organised flux exceeds by 50–100 times the biologically organised flux of energy. It is much bigger now for developed countries. For the U.S. economy, for example, the consumption of primary energy is approximately $4 \times 10^8$ Btu $\approx 4 \times 10^{11}$ J per person per year in the year 2000. The entire population of the world need more than $3 \times 10^{21}$ J per year to live as the U.S. citizens live now. This is only one thousand times less than the amount of energy received by the Earth from the Sun!

### 11.1.3 Principle of Evolution

According to the *energy principle of evolution*, those populations and their associations (ecosystems) which can utilise the greater amount of energy from their environment have an advantage for survival [11–13]. One can state, taking into account

[2]British thermal unit (Btu) = 252 cal = 1053.36 J $\approx 10^3$ J.

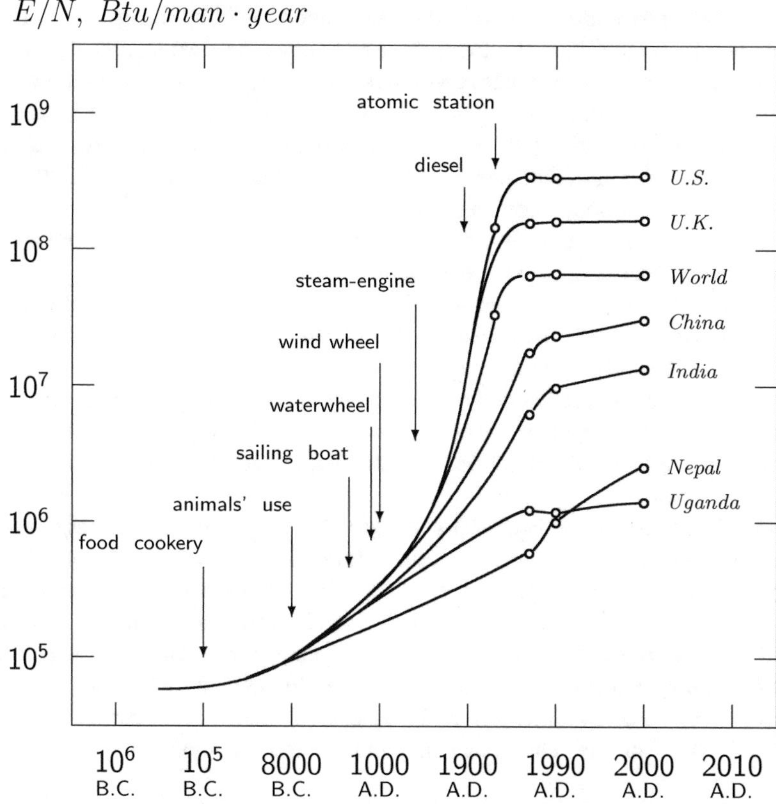


**Fig. 11.2** Socially organised flux of energy per capita. An increase in consumption of energy is connected with the invention of more and more sophisticated devices for energy utilisation (e.g. wind, running water, coal and oil.). The pictured data does not include the work of animals, which should be added to the fluxes. These corrections are essential for Uganda and Nepal, but can be neglected for other countries. Values for points are taken from Energy Statistics Yearbook (2003 and previous issues)

the existence of the two fluxes of energy, that the energy principle of evolution is also valid for the human population. Indeed, as was argued in the previous chapters, the principle, according to which the production system is developing, can be stated as a principle of the maximal captivation of available resources: the production system tries to swallow up all available production factors. In fact, this principle of progress is a principle of evolution, stated by Lotka [11], Pechurkin [12] and Odum ([13], p. 20): the trajectory of evolution of a system is defined by trends of the system to use the greatest quantity of available energy.

This formal statement is a description of the total of joint actions of many entrepreneurs trying to obtain the largest profit. The real path of evolution of the production system is determined by the availability of labour and energy. The development of the production system is eventually determined by the growth of labour supply and by possibilities of attracting an extra amount of external energy, so that the set of evolution equations of the production system contains two important quantities: potential growth rates of labour and substitutive work, $\tilde{\nu}(t)$ and $\tilde{\eta}(t)$. These quantities have to be given as exogenous functions of time, but, as we argued in Sects. 2.4.1 and 2.5.5, in fact, they are endogenous characteristics in the problem of evolution of human population on the Earth.

Plenty of energy is used by the human population through the improvements of technology. Managing the huge amount of energy allows the human population to survive in every climate zone of the Earth and expand itself in great measure. Moreover, one can see from the history of mankind that the nations which controlled the available energy had an advantage over other nations. One can refer to the classic example: the industrial revolution and prosperity of Britain began from the invention of the steam engine proper, which allowed utilisation of energy stored in coal in great amounts. The history of nations can be rewritten as the history of the struggle for fluxes of energy.

In the beginning of the twenty-first century, the main primary sources of energy are oil, gas and coal—the remains of the former biospheres. It is assumed that the energy needs will increase; however, the peak of the fossil era has passed. Fossil fuel consumption will grow more slowly than total primary energy needs. The future development of mankind is connected with available energy. Only an abundance of available energy would ensure the prosperous development of the human population. Can we then get energy directly from the Sun, which gives to the Earth $3 \cdot 10^{24}$ J per year or $3 \cdot 10^{21}$ Btu per year, or can we find new sources of energy?

## 11.2 Thermodynamic Interpretation of Value

One can note that all the artificial things around us have special forms and are adjusted for use in special aims. This means that there is some *complexity* in the environment, a complexity that is created by man and for man. A human being is encircled by artificial products that can be sorted and counted, so one considers the amounts of quantities in natural units of measurement, to be given

$$Q_1, Q_2, \ldots, Q_n.$$

All these objects: buildings, machines, vehicles, sanitation, clothes, home appliances and so on make up the national wealth, which can be characterised from different points of view. However, a general characteristic of artificial objects appears to be value. As a specific concept of economy, value should not be reduced to any other known concepts, but, so as processes of production can be considered as processes of transformation of 'wild' forms of substances into forms useful to people (mainly, without variation of internal energy), it is possible to look for analogies in thermodynamics.

### 11.2.1 Value of a Stock of Products

From a conventional point of view, a stock of products is characterised by its value, and an empirical estimate of *the value of each product* exists. The value of a unit of the product is its *price*. It is assumed that the prices of all products are given as

$$p_i, \quad i = 1, 2, \ldots, n.$$

This allows one to estimate the increase in value of a stock of products,

$$dW = \sum_{j=1}^{n} p_j \, dQ_j. \tag{11.5}$$

It is necessary to take into account that the price of a product is not an intrinsic characteristic of the product. As was already noted earlier in Sect. 2.2, the price depends on the quantities of all products which are in existence at the moment. As a rule, the price decreases if the quantity of the product increases, though the situation can be more complicated. One can observe that there are coupled sets of products, such that an increase in the quantity of one product in a couple is followed by an increase (in the case of a couple of complementary products) or a decrease (in the case of a couple of substituting products) of the price of the other product of the couple. Therefore, one ought to consider the price of a product to be a function of quantities of, generally speaking, all products,

$$p_i = p_i(Q_1, Q_2, \ldots, Q_n).$$

The dependence of prices $p_i$ on the amounts of products is arbitrary, so that one can hardly expect that form (11.5) is a total differential of some function $W(Q_1, Q_2, \ldots, Q_n)$. One cannot say that $W$ is a characteristic of the set of the products which is independent of the history of their creation. In other words, value is not a function of a state of the system. However, a function of a state, which is closely related to value, can be introduced.

### 11.2.2 Objective Utility Function

Using some assumptions, a function of a state of a system, which is called the utility function, can be introduced on the basis of relation (11.5). Indeed, the linear form (11.5) can be multiplied by a certain function, called the integration factor,

$$\phi = \phi(Q_1, Q_2, \ldots, Q_n),$$

so that, instead of form (11.5), one has a total differential of a new function,

$$dU = \sum_{j=1}^{n} \phi(Q_1, Q_2, \ldots, Q_n)\, p_j(Q_1, Q_2, \ldots, Q_n)\, dQ_j. \tag{11.6}$$

Requirements on the integrating multiplier are connected with properties of prices as functions of products, so existence of the function $U$ depends on the properties of prices. The integrating multiplier can be taken to be positive, so the linear form (11.6) defines a monotonically increasing function of each variable. It resembles the behaviour of value. One can also expect other properties of function $U$ to relate to the properties of value. In particular, as one could expect for the partial derivatives of the value function (if it exists), the ratio of the partial derivatives of the function $U$ is equal to the ratio of the prices

$$\frac{\partial U}{\partial Q_i} : \frac{\partial U}{\partial Q_j} = \frac{p_i}{p_j}. \tag{11.7}$$

One can see that the important properties of the artificial environment can be characterised by a function $U$. The introduced function $U$ is called *the (objective) utility function*, taking into account that the properties of function $U$ coincide with those of the conventional *utility function* which is introduced as *a subjective* utility function, connected with the sensation of preference of one aggregate of products against another (see Sect. 9.2). As a matter of fact, these functions have to be considered identical, because the utility function is defined with accuracy to monotonic transformation of its variables. The above transformation of value to utility reminds us the transformation of heat to entropy. In other terms, an analogy between the theory of utility and the theory of heat was discussed by von Neumann and Morgenstern (see item 3.2.1 of their work [14]).

The understanding of the fact that utility function $U$ can replace the non-existing value function in theoretical considerations was achieved in the second half of the nineteenth century and is considered as a revolution in economic theory [15], completed by the work of prominent investigators.

### 11.2.3 Thermodynamics of Production

To understand what changes from the thermodynamic point of view occur on the Earth at productions, it is convenient to consider two subsystems. We shall separate *a subsystem of artificial things* which includes a human population and its close environment: buildings, clothes, cars, sewer networks and all other subject matters created by the person. We believe that natural characteristics of an artificial environment are quantities of products that can be listed and measured: $Q_1, Q_2, \ldots, Q_n$. We considered the remaining part *as a subsystem of a habitat of dwelling (the environment)*, containing all natural formations that is not touched by the hand of a human. The artificial environment is supported and developed by the production system that performs work, that is, eventually after many transformations, work of energy that the Earth receives from the Sun. To run the production, resources of the natural environment are used: the habitat decreases, the artificial environment increases; the set of numbers $Q_1, Q_2, \ldots, Q_n$ defines, thus, 'the volume' of the subsystem of artificial things and the borders between two considered subsystems.

Each of the subsystems is an open thermodynamic system, exchanging heat and substances with each other and with the outer space. In the further consideration, it is used as the simplified schematization: the natural environment subsystem receives energy from the Sun and passes it to the other subsystem in chemical form. For maintenance and development of the artificial environment, there is a production system that fulfils work, which is, after many transformations, the work of energy from the Sun. In conformity with the general laws of thermodynamics, it is possible to formulate thermodynamic relations for the considered subsystems. Designating with a prime symbol the characteristic quantities related to the subsystem of the human population and artificial things, one records

$$\mathrm{d}E' = T\,\mathrm{d}S' - \mathrm{d}A, \quad \mathrm{d}A = \frac{1}{T}\sum_j a_j\,\mathrm{d}Q_j,$$
$$\mathrm{d}S' = -\frac{Q'}{T} - \frac{1}{T}\sum_j \Xi'_j\,d\xi'_j + \frac{G}{T}, \tag{11.8}$$

where $T$ is the temperature, $\mathrm{d}A$ is the work of system of human population and artificial environment on the shift of borders with a natural environment, $a_j$ is the work on creation of unit of a product $j$, $Q'$ is the flux of heat from the considered subsystem in the environment, and $G$ is the flux of chemical energy from the natural environment subsystem to the considered subsystem. Work $\mathrm{d}A$ is fulfilled by people and external power sources by means of the production equipment and energy sources. This work changes both the natural and the artificial environment.

The subsystem of the natural environment does not fulfil any work, so that for it

$$\mathrm{d}E'' = T\,\mathrm{d}S'', \quad \mathrm{d}S'' = \frac{Q' - Q}{T} - \frac{1}{T}\sum_j \Xi''_j\,d\xi''_j - \frac{G}{T}, \tag{11.9}$$

where $Q$ is a flux of heat disseminated by the Earth, $Q'$ is a flux of heat from the subsystem of artificial things and $G$ is a flux of chemical energy from the subsystem of the natural environment to the subsystem of the artificial environment.

The sum of parities (11.8) and (11.9) gives naturally the description of the thermodynamics of the Earth, disposed in Sect. 10.1.1, in particular, relation (11.1), whereby one gets an expression for the increment of entropy,

$$\mathrm{d}E = T\,\mathrm{d}S, \quad dS = -\frac{Q}{T} - \frac{1}{T}\sum_j \Xi_j\,d\xi_j - \frac{1}{T}\sum_j a_j\,\mathrm{d}Q_j. \tag{11.10}$$

The last two terms in the formula for the increment of entropy have an identical form; consequently, this formula shows that quantities of products $Q_1, Q_2, \ldots, Q_n$ can be introduced into the list of internal variables (parameters of internal complexity) of the thermodynamic system of the Earth. Like any internal variables, these variables, left to themselves, disappear, which leads to additional dissipation of energy. The creation of artificial things is connected with the reduction of entropy of the Earth,

$$\mathrm{d}S = -\frac{1}{T}\sum_j a_j\,\mathrm{d}Q_j, \tag{11.11}$$

where $a_j$ is the work for creation of a unit of product $j$. It is the work of the production system and is, eventually after many conversions, the work of the energy flux which the Earth receives from the Sun.

The process of creation of artificial things (the production process) can be interpreted as a process of creation complexity, or negative entropy, of the system. Negative entropy $-S$ is a natural measure of complexity of a non-equilibrium state of the system, in this case, the quantities and complexities of the artificial environment. The total work on the environment, done by humans and by mechanisms which use external sources of energy, is work as it is understood by thermodynamics, so the above relation is the known thermodynamic relation between entropy $S$ and work $A$,

$$-\,dS = \frac{1}{T}\,dA. \tag{11.12}$$

This simple scheme becomes complicated: simultaneously with useful products, the production system also creates useless and harmful products (waste and pollution), while all real processes are irreversible. The Earth is not an isolated system and heat is radiated into space. The production of useful things also stimulates processes of mixing, dispersion and diffusion, so that matter necessary for production will become progressively unavailable [16]. In other words, the chemical elements become increasingly mixed together and thus more and more difficult to separate from each other. Substances necessary for manufacture all become more and more inaccessible. But given the availability of energy, the materials could be recovered from waste, like an ore pile [17]. The Earth is not waiting for a diffusion death: despite some processes of degradation of matter, the essence of production processes is the creation of useful complexity in the environment.

### *11.2.4 Do Negative Entropy and Utility Function Coincide?*

Relation (11.12) shows that negative entropy $-S$ can be a natural characteristic of complexity of the artificial environment from the thermodynamic point of view, while quantities of products $Q_1, Q_2, \ldots, Q_n$ can be considered as internal (complexity) parameters. Note that an increment of entropy includes both useful complexity and useless, sometimes harmful, but inevitable consequences of production. The other characteristic of a set of products is the utility function $U$, defined by equation (11.6). In contrast to negative entropy, the utility function conventionally describes only useful complexity, though estimates of damage of the environment are included in prices of the products.

One has two functions: $U$ and $-S$ as characteristics of a set of commodities.[3] Either function is a monotonically increasing function of all variables; the functions are identical, if one neglects all by-products: useless and harmful complexity connected with production. One can choose the function $U$ to be a characteristic function that describes useful man-made complexity in the system and interprets this function as negative entropy of the system of useful artificial things.

Due to Eqs. (11.6) and (11.11), one might think that value is an analogy to negative entropy which can be considered as a measure of complexity in the system—in our case, the complexity which is useful for human beings. The correspondence is not accurate, because entropy is a function of state in contrast to value, which is not a function of state. The increase in value $dW$ is close, but might be different from a change of negative entropy $dS$. From relation (11.12), as far as the change of internal energy of the system can be neglected and the conditions considered to be isothermal, one has proportionality between increase in value of the set of products and total work done on the system, that is,

$$dW \sim dA. \tag{11.13}$$

The relation is valid, if the value is measured with absolute energy units. Otherwise, when money units of constant purchase power are used, as usually, for estimates of value, the relation between completed work and produced value (see Eq. 11.15) is not universal, in contrast to relation (11.12).

The proportionality between increase in value $dW$ and total work $dA$ was investigated with the analysis of empirical data for the U.S. economy [19, 20]. The authors consider their results as confirmation of what is called the embodied energy theory of value. However, it is not energy but complexity which is left in matter after the work has been done. This complexity exists and can be, in principle, estimated in another way. Therefore, the attempts to calculate embodied energy in artificial things remind us of the attempts to calculate the amount of phlogiston in a body. From all points of view, it is better to regard value as something which is very close to negative entropy of our close environment. Fluxes of products should be considered as fluxes of negative entropy, not as fluxes of energy.

[3]The existence of the utility function is justified by the fact that there is a preference relation on the set of products. Similar to that, the existence of entropy is justified by an acceptability relation on the space of thermodynamic variables. The similarity between the utility representation problem in economics and the entropy representation problem in thermodynamics was demonstrated by Candeal et al. [18]. Astonishingly, it seems to be not just a formal analogy: the two functions appear to be different estimates of a set of products.

## 11.3 Energy Content of a Monetary Unit

In Marx's theory of value, it is postulated that expenditure of labour is an absolute measure of value, a source of all created wealth (products). When one accounts the effect of substitution of labour with true work of the production equipment, one could expect that the total amount of work, including properly accounted labour work and work of production equipment, could be an absolute measure of value [21]. We shall test this statement, following the work by Beaudreau and Pokrovskii [22].

The total work on the production of value in a unit of time is the sum of the work of substitution $P$, measured in power units, and the work of humans $h\,L$, where $h \approx 4,18 \cdot 10^5$ J/h is an estimate of the work of one person per hour (see Sect. 2.4.1). The total can be recorded as

$$A = P + h\,L. \tag{11.14}$$

This work fulfils 'useful' changes in our environment (in the form of useful consumer goods and services), which can be estimated by production of value $Y$ (in money units, for year, for example), written, from relations (6.12) and (6.13), as

$$Y = \beta\,L + \gamma\,P = \gamma\left(P + \frac{\beta}{\gamma}\,L\right). \tag{11.15}$$

If value is estimated by monetary units of constant purchasing capacity, marginal productivities $\beta$ and $\gamma$ depend on production factors. Under the choice of monetary unit with constant 'energy content', the marginal productivities appear to be constant.

Expressions (11.14) and (11.15) allow us to determine the true work, necessary to complete a thing or service, which costs one monetary unit or, in other words, 'the energy content' of the monetary unit

$$\frac{A}{Y} = \frac{1}{\gamma}\,\frac{P + h\,L}{P + (\beta/\gamma)\,L} = \frac{1}{Y}\,(P + h\,L). \tag{11.16}$$

The assessments of 'energy content' of a monetary unit for the economic systems of the United States and Russian Federations are shown in Fig. 11.3. The 'energy content' of the dollar is $(1 - 2) \times 10^5$ J per dollar of 1996; its mean value in the last years of the century (1960–2000) is $1.4 \times 10^5$ J. The 'energy content' of the 2000 ruble is less: the mean value for the same years (1960–2000) is $0.1 \times 10^5$ J. Pulsations of this quantity (see Fig. 11.3) can be connected with natural variations in the contribution of work (in terms of energy), contrary to our assumption of a constancy of the coefficient $h$. The 'energy content' of the dollar is 14 times that of the ruble, whereas the official exchange rate was about 30 roubles for dollar. The purchasing power parity of ruble is about twice less than the official rate of exchange, so that the calculated 'energy content' can be a measure of purchasing capacity of monetary units. Some differences can be also attributed to the difference of values

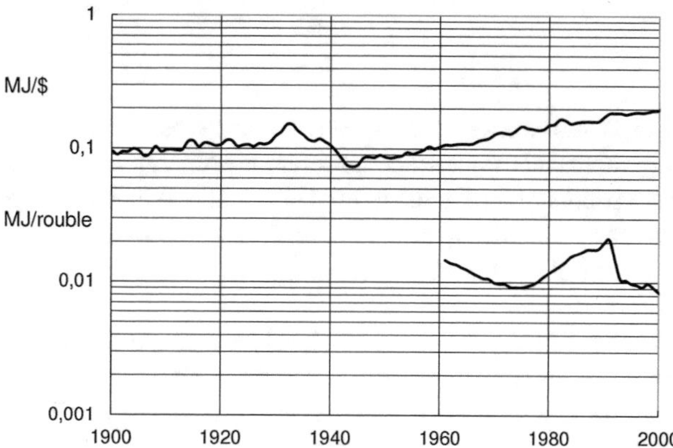


**Fig. 11.3** 'Energy content' of monetary units. Curves show the amount of work necessary for creation of a product with a value of one dollar of 1996 (top curve) and one ruble of 2000 (bottom curve) in different years. Adapted from [22] with some changes of values for Russia

for $h$ in the American and Russian economies. Lastly, it should be noted that the Russian data is, in general, less reliable than the U.S. data.

The values of the 'energy content' shown in Fig. 11.3 appear not to be constant in time: when the contribution of the substitutive work is dominating, which was the case in the U.S. in the second half of the last century (see Chap. 2, Fig. 2.9), the expression (11.16) for "the energy content' can be written as

$$\frac{A}{Y} \approx \frac{P}{Y}, \quad P \gg h\,L.$$

So as substitutive work increases faster than output, when the production grows, this relation shows that 'the energy content' is an increasing function of time. Under the exponential approximation, when the variables are given by Eqs. (2.30) and (2.36), 'the energy content' of a money unit grows as

$$\frac{A}{Y} \sim e^{(\eta-\delta)t}. \tag{11.17}$$

For the U.S. in the second half of the last century, for example, $\eta - \delta = 0.0272$. Indeed, one can see the increase of 'the energy content' on the plot of Fig. 11.3 for the U.S. economy after year 1950.

Here, it is necessary to recollect that output in expressions (11.15)–(11.17) is estimated by a special monetary measure with constant purchasing capacity at all times, but this unit of value apparently changes with time. An absolute measure of value is equivalent to some amount of energy. Thus, the expression for output can be written as

$$Y = \frac{A}{\epsilon_{\text{ref}}}, \tag{11.18}$$

where work $A$ is the real work of humans and machines, as determined by expression (11.14), and $\epsilon_{\text{ref}}$ represents some standard reference 'energy content' of a monetary unit. To measure production of value in the U.S., it is possible to accept the amount of $10^5$ J, but, to draw a distinctive curve in Fig. 2.1, we calculate time dependence of production of value (GDP of the U.S. economy) according to the formula (11.18) at $\epsilon_{\text{ref}} = 50000$ J.

Other approaches to the problem [13, 23, 24] are giving the assessments of the total consumption of energy (or exergy) for output, taking into account all 'previous' expenditures of energy needed for production 'from the very beginning'. Such estimates include all losses of energy during production. In this case, universality is lost: these estimates of 'energy content' depend on the efficiency of transformation of energy in processes of production. Note that our assessments of 'energy content' of a monetary unit naturally appear to be lower than 'the total exergy or emergy content' [13, 23], because substitutive work is a small part of the total consumed energy.

## 11.4 Thermodynamics of Production Cycle

The performance of thermic machines, which are designed to transform heat into mechanical work, is described by thermodynamics by means of thermodynamic cycles. Thermodynamics itself has emerged from studying this kind of thermodynamic cycle [6]. Investigating the processes of production, one deals with thermodynamic cycles of another kind, i.e. production cycles which are designed to transform the 'wild' forms of substances into 'useful' forms (dwellings, food, clothes, machinery and so on). We shall consider a simple example of production cycle and demonstrate what one needs in order to create a non-equilibrium state of matter artificially.

The main elements of a production process are production equipment, which is able to perform special operations, remaining (excluding wear and tear) unchanged after the cycle, and some substances, the forms of which (one can assume that the change of energy can be neglected) are being changed due to special work of the production equipment. A production cycle can be considered to be a sequence of elementary operations $j_1, j_2, \ldots$, while a set of elementary operations is given. The index $j_l$ is an index of an elementary operation which is fulfilled as number $l$ in the sequence of operations. The unique choice of indexes determines where, when and how forces are allowed to act to perform work which can be calculated as a sum of work at elementary operations,

$$\Delta A = A_{j_1} + A_{j_2} + \cdots$$

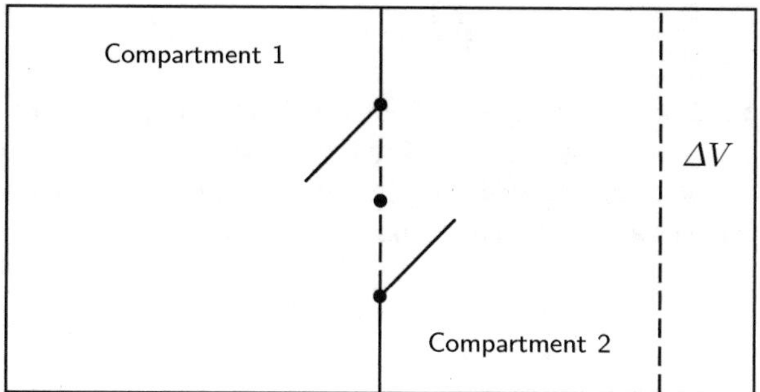


**Fig. 11.4** Scheme of the production container. The compartments contain a gas, the state of which can be affected by a sequence of the two production operations: A—the compartments can be connected or isolated and B—the volume of the second department can be diminished or restored to its previous volume

### *11.4.1 A Simple Production Process*

So as not to be too abstract, let us consider, as an example, a system of $2N$ particles (ideal gas) in a container consisting of two compartments of volume $V$ each, as shown in Fig. 11.4. There are some devices, which allow the compartments to be connected or isolated (let us call this operation A) and the volume of the second department to be diminished or restored to previous volume (operation B).

Let us assume that in an initial state each compartment has volume $V$ and the compartments are connected with each other, while the gas is in an equilibrium state, so that each compartment has on average $N$ particles. One can imagine that a deliberate sequence of operations can be applied to the system. We consider isothermal processes consisting of several elementary operations, while the operation is fulfilled in a reversible manner. After one has performed the sequence: B—decreasing of volume 2 in $\Delta V$, A—isolating of the compartments, B—increasing of volume 2 in $\Delta V$, and A—connecting the compartments, the configuration of the outer devices is initial, but the gas appears to be in a non-equilibrium state. The mean number of particles in each compartment can be found to be

$$N_1 = N(1 + \xi), \quad N_2 = N(1 - \xi), \quad \xi = \frac{(\Delta V/2V)}{1 - (\Delta V/2V)} \tag{11.19}$$

The entropy of the system can be directly estimated according to Boltzmann formulae applied to this case,

$$S = k \ln W, \quad W = \frac{(2N)!}{N_1!\, N_2!},$$

so that the change of entropy of the system can be calculated as

$$\Delta S = -kN\xi^2. \tag{11.20}$$

The terms of third order and higher are neglected here and further on.

The work $\Delta A$ which is needed to pass the system through the cycle can be calculated as work of/on the ideal gas in every four steps of the cycle. One finds eventually that external forces have to produce extra work during the cycle,

$$\Delta A = kTN\xi^2. \tag{11.21}$$

The internal energy of the system

$$E = 3NkT$$

as the internal energy of the ideal gas does not change in the process, so the first law of thermodynamics can be written, considering every step of the process to be reversible, in the form

$$T\,\Delta S = -\Delta A,$$

and the change of entropy of the system in the process can be estimated as

$$\Delta S = -kN\xi^2. \tag{11.22}$$

The considered particular non-equilibrium state cannot be created without work of external forces and without the deliberate choice of sequence of elementary operations. Somebody possesses certain sources of energy and has the aim to create a unique non-equilibrium form of matter. To achieve the goal, the creator sends the message in codes of elementary operations: BABA. No other messages can be helpful. The information content of the deliberate message can be estimated if one takes into account that this message is one in eight possibilities. So, the message carries the information entropy in the amount

$$\Delta I = -\log_2 \frac{1}{8} = 3. \tag{11.23}$$

The information content of the message can be considered to be materialised in non-equilibrium form (complexity) of matter. The cost of materialisation is the work of production system $\Delta A$.

### 11.4.2 Output of the Production Cycle

Returning to the general case, the inputs of information $\Delta I$ and work $\Delta A$ eventually determine a new organisation of matter which, due to production processes, acquires forms of different objects. This is a special complexity, which, from a thermodynamic point of view, can be characterised by negative entropy. The decrease in entropy of the entire environment $\Delta S$ due to processes of production can be connected with special work of the production system,

$$- \Delta S = \frac{1}{T} \Delta A. \tag{11.24}$$

This is the same relation as (11.12), and one can state that only properly organised work of the production system is needed to transform the natural environment into an artificial environment.

The artificial things, created in the process of production, can also be estimated by their value. The relations among entropy, value and work were discussed already in the previous sections (Sects. 11.2.4 and 11.3). Equation (11.15) determines value of output (in monetary units of constant purchasing power) via efforts of humans and work of machines. When output is measured in absolute energy units, the marginal productivities $\beta$ and $\gamma$ in Eq. (11.15) are equal and the equation turns into a simple relation

$$Y \sim \Delta A. \tag{11.25}$$

Useful work $\Delta A$ includes muscle work by animals or humans, work by electric motors, mobile engine power as delivered, for example, to the rear wheels of vehicles, but, in this context, excluding heat, delivered in a space of buildings or in processes.

## References

1. Prigogine, I.: From Being to Becoming. Time and Complexity in the Physical Sciences. Freeman & Company, New York (1980)
2. Nicolis, G., Prigogine, I.: Self-Organisation in Non-Equilibrium Systems: From Dissipative Structures to Order through Fluctuations. Wiley, New York (1977)
3. Morowitz, H.J.: Energy Flow in Biology: Biological Organisation as a Problem in Thermal Physics. Academic Press, New York (1968)
4. Newman, E.I.: Applied Ecology. Blackwell Scientific Publications, London (1993)
5. Prigogine, I.: Introduction to Thermodynamics of Irreversible Processes, 2 revised edn. Interscience Publishers (1961) (a division of Wiley, New York and London)
6. Kondepudi, D., Prigogine, I.: Modern Thermodynamics: From Heat Engines to Dissipative Structures. Wiley, Chichester (1999)
7. Pokrovskii, V.N.: A derivation of the main relations of non-equilibrium thermodynamics, vol. 2013, article ID 906136. Hindawi Publishing Corporation: ISRN Thermodynamics (2013). http://dx.doi.org/10.1155/2013/906136
8. Harries, J.E.: Physics of the Earth radiative energy balance. Contemp. Phys. **41**, 309–322 (2000)
9. Kleidon, A.: Non-equilibrium thermodynamics and maximum entropy production in the Earth system: applications and implications. Naturwissenschaften **96**, 653–677 (2009)
10. Rebane, K.K.: Energija, entropija, sreda obitanija (Energy, Entropy, Habitat). Znanije, Moscow (1985)
11. Lotka, A.J.: Elements of Physical Biology. Williams and Wilkins, Baltimore (1925)
12. Pechurkin, N.S.: Energeticheskije aspekty nadorganizmennykh sistem (Energy Aspects of Superorganism Systems). Nauka, Novosibirsk (1982)

13. Odum, H.T.: Environmental Accounting. Emergy and Environmental Decision Making. Wiley, New York (1996)
14. Von Neumann, J., Morgenstern, O.: Theory of Games and Economic Bhaviour, 3rd edn. Princeton University Press, Princeton (1953)
15. Blaug, M.: Economic Theory in Retrospect, 5th edn. Cambridge University Press, Cambridge (1997)
16. Georgescu-Roegen, N.: The Entropy Law and the Economic Process. Harvard University Press, Cambridge (1971)
17. Ayres, R.U.: Comments on Georgescu-Roegen. Ecol. Econ. **22**, 285–287 (1997)
18. Candeal, J.C., De Miguel, J.R., Induráin, E., Mehta, G.B.: Utility and entropy. Econ. Theory **17**, 233–238 (2001)
19. Costanza, R.: Embodied energy and economic valuation. Science **210**, 1219–1224 (1980)
20. Cleveland, C.J., Costanza, R., Hall, C.A.S., Kaufmann, R.: Energy and the U.S. economy: a biophysical perspective. Science **225**, 890–897 (1984)
21. Beaudreau, B.C.: Energy and Organization: Growth and Distribution Reexamined, 2nd edn. Greenwood Press, New York (2008)
22. Beaudreau, B.C., Pokrovskii, V.N.: On the energy content of a money unit. Phys. A: Stat. Mech. Appl. **389**, 2597–2606 (2010)
23. Sciubba, E.: On the possibility of establishing a univocal and direct correlation between monetary price and physical value: the concept of extended exergy accounting. In: Ulgiati, S. (ed.) Advances in Energy Studies Workshop: Exploring Supplies, Constraints, and Strategies (Porto Venere, Italy 2000), pp. 617–633. Servizi Grafici Editoriali, Padova (2001)
24. Valero, A.: Thermoeconomics as a conceptual basis for energy-ecological analysis. In: Ulgiati, S. (ed.) Advances in Energy Studies Workshop: Energy Flows in Ecology and Economy (Porto Venere, Italy 1998), pp. 415–444. MUSIS, Rome (1998)

# Chapter 12
# The Global Dynamics

**Abstract** The investigation of production activity of the humans on the Earth in this chapter is based on the Maddison's estimates of the gross world product and the aggregate numbers of the population. The results of the discussed theory of a social production allow one to describe changes in the technological use of humans efforts and external energy sources in the past centuries beginning with the second millennium of our era. It appears that the approach with two value-creating factors does not work properly in the earlier times, and an alternative description with the only value-creating factor was developed for population and production activity of humans on the Earth for the agricultural era. In conclusion, the future of the human population is discussed.

## 12.1 Reconstruction of the Past Production Activity

From prehistoric times, members of our species, *Homo sapiens*, built dwellings and made tools, weapons and other material objects to maintain their existence [1]. In contrast to other populations that inhabited the Earth, humans as a species live in an environment that is partially and increasingly self-created. All of the possessions constituting tangible and non-tangible wealth (clothes, buildings, home appliances, vehicles, infrastructure, utilities, communications networks and so on) are not natural things. Today, water, heat, electric power and even communications are delivered into our dwellings through pipes and wires. This entire artificial environment is created and maintained by the complex production system of the society, which reflects the generations of technological achievements. The increase of population and accompanying progress of productive activity are obviously interconnected. Better life determines growth of the population, while increase in population gives some stimulus for production. In this section, we attempt to segregate factors responsible for the growth of the global production system and draw a picture of development of the population.

The undertaken investigation is based on the empirical time series for the number of humans on the Earth $N$ and the gross world product (GWP) $Y$ in the period from the beginning of our era up to recent years; the numbers were estimated by Angus



V. N. Pokrovskii, *Econodynamics*, New Economic Windows,
https://doi.org/10.1007/978-3-319-72074-6_12

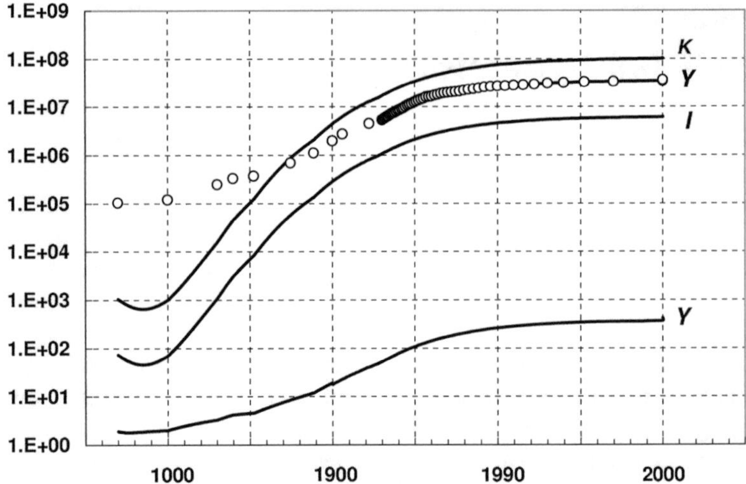


**Fig. 12.1** The Gross World Product. The empty circles show values of GWP in millions international (1990) dollars according to Maddison [2]. Solid lines nearby show values of investments $I$ and values of the basic production equipment $K$ also in millions international (1990) dollars. The bottom curve represents the values of GWP measured in energy scale, and the value $10^{16}$ J is accepted as a monetary unit

Maddison ([2], Appendix B) and presented in Figs. 12.1 and 12.2 by empty circles.[1] Our aim is to recognise and estimate the original sources of growth. The technological three-factor theory of economic growth determines some relations among macroeconomic characteristics, which allow us estimating these characteristics, being based only on the two time series given by Maddison. No arbitrary parameters are needed. The approach to the description of the global production, as a one-sector system, is rather rough, but hopefully acceptable as an initial approach to a problem.

### 12.1.1 Gross World Product

The overall economic activity of a society is characterised by the well-defined and quantifiable concepts of gross domestic product (GDP), which represents an assessment of results of all creative activity of the population in a unit of time. This is a measure of current achievements of the economy as the whole—a money measure of a multitude of things and services, created by society for unit of time (see discussion in Sect. 2.2.3). There are different methods for calculating GDP: it can be estimated as the results of production, that is, the sum of value of created products, or by the

[1]The accuracy of numbers in the earlier times could be hardly overestimated. In any case, there are and will be some corrections to the Maddison's estimates [3]. In this paper, we prefer to use the original Maddison's numbers.

account of the use of products, or by the contribution of separate components of the created value. The GWP is calculated as a total of the GDPs of all nations on the Earth. The estimates of the GWP in the 1990 international (1990) dollars are shown in Fig. 12.1, according to Maddison ([2], Appendix B). On the definition, the international Geary–Khamis dollar is a monetary unit of constant purchasing capacity.

Values $Y$ allow us to estimate the growth rate of the GWP; note that an average rate of growth of GWP from the beginning of our era till 1000 AC is equal to about 0.014% a year.

### 12.1.2 Working People and Consumption of Energy

According to the second line of Eq. (6.16), values of output (GWP in this case) $Y$ are determined by two production factors: labour $L$ and substitutive work $P$. These quantities are connected with the aggregate number of population $N$ and the total consumption of energy carriers $E$, correspondingly.

#### 12.1.2.1 Working People

There are some difficulties in estimation of labour force for the ancient times, when slaves, who certainly were a part of the human population, were used mainly as machines for performance of mechanical operations such as cultivation, harvesting, mining and propelling warships via oars [4]. We accept that the economically active part of the population through all the periods counts approximately half of the total number $N$ (a more detailed discussion can be found in Sect. 2.4.2 in Chap. 2). Numbers of population $N$ according to the estimates by Maddison [2] are shown in Fig. 12.2 with empty circles. The enlargement of human population is interconnected with the development of social production system.

The production efforts of humans can be estimated in energy units. The organism of every member of the population needs energy in amount of $\approx 4 \cdot 10^9$ J/year (see discussion in Sect. 2.4.1 in Chap. 2). The work on the production of useful things acquires an additional amount of energy; it can be approximated as $\approx 4 \cdot 10^8$ J/year. The entire work of the population (labour $L$ in power units) is depicted in Fig. 12.2.

The values of $L$ allow us to calculate the growth rate of labour as

$$\nu = \frac{1}{L}\frac{dL}{dt}. \tag{12.1}$$

By the said assumptions, this quantity, naturally, coincides with the growth rate of the population. According to Maddison's numbers [2], it is possible to find for the period from the beginning of our era till supposedly 1000 AC that the average growth rate of the population is equal to 0.017% a year, which is bigger than the growth rate of

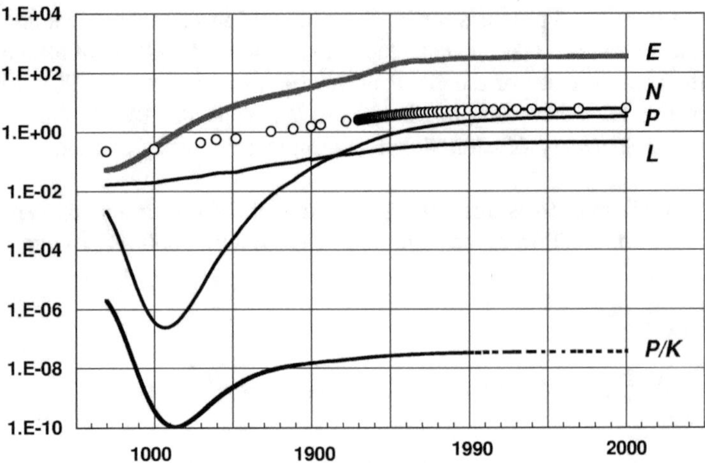


**Fig. 12.2** Population, energy and production factors. The empty circles represent values of the population according to Maddison [2] in billions of person. The top curve represents the total consumption of energy carriers according to the sources specified in the text. The bottom curves—production factors: labour $L$ (the continuous curve) and substitutive work $P$ (the dashed line), estimated by energy unit equal to $10^{18}$ J per year. The bottom-dotted curve represents an index of a level of technological progress $P/K$ in an arbitrary units

GWP at the same time. It means the decrease of labour productivity and appropriate deterioration of conditions of life of the population. However, after 1000 AC the growth rate of GWP exceeds the growth rate of the population, which shows the growth of labour productivity.

#### 12.1.2.2 Consumption of Energy

From the ancient times, economic activities of the humans were accompanied by attraction, in this or that way, natural energy sources for replacement of efforts of people in the production of useful things and services. It is interesting to follow the stumbling way of development and use of energy. The first deliberate use of fire as a source of useful heat seems to have occurred about one million years ago [5]. It was one of the first innovations in the utilisation of external sources of energy. As time went on, other important innovations in the exploitation of external sources of energy included the taming of animals (dogs for hunting; cattle, sheep and goats for work, milk, meat and wool; horses for riding) occurred. The invention of harnesses enabled teams of horses or a water buffalo to pull heavy ploughs that enabled human farmers to exploit grasslands and move away from the river valleys where agriculture began. Then came purely mechanical innovations for the use of power from wind (at first in sailing ships, later for grinding grain) and flowing water (also for grinding mills) (see also Sect. 11.1.2 in Chap. 11).

Assessments of the total consumption (use) of energy carriers $E$ are shown in Fig. 12.2 according to various sources: values $E$ for the recent decades (since 1965) are taken from a review [6], values of consumption of energy since 1820—from Vaclav Smil's work [7], assessments till 1820 are found on the assumption that the growth rate of consumption of energy increases from zero value in the beginning of our era up to value 0.0063 in 1820. It allows to estimate the world consumption of energy in the beginning of our era as $5.4 \cdot 10^{16}$ J per year or $2.4 \cdot 10^{8}$ J a year per a person.

The running of production equipment is obviously impossible without attraction of great amounts of energy to the production. However, the true substitutive work $P$, replacing efforts of the humans in processes of production, is only a small part of the total consumption of energy $E$. Values of substitutive work $P$ are shown in Fig. 12.2, though the discussion of the assessment of this quantity will be given later, in Sect. 12.1.4. One can see that the growth rate of substitutive work is, generally speaking, different from the growth rate of total consumption of energy carries. The difference is connected, first, with the fact that only one part of the total energy is directed to do work of production equipment; then, the amount of substitutive work depends on the efficiency of conversion, the growth rate of which is increasing with time.

A number of authors in the past and recent times have emphasised the importance of energy in economic activity. There were some hopes that introduction of the total consumption of energy $E$ into the theory helps to solve the problem of exogenous technical progress, and there were some attempts to introduce this quantity as a production factor in line with neoclassical factors $L$ and $K$, so that production function can be written in the form

$$Y = Y(K, L, E). \tag{12.2}$$

However, it has appeared that relation (12.2) does not reflect the proper role of capital and energy in production, and attempts to expand the neoclassical theory of economic growth by introducing 'consumed' energy as a production factor led to unresolved contradictions and were eventually unsuccessful. A principal step to the solution of the problem was made by Boudreau [8], who understood that the productivity of capital is, in fact, productivity of work that is provided with capital stock.

### *12.1.3 Capital Stock and Its Productivity*

Though true sources of value are labour and substitutive work, capital stock $K$, as value of collection of tools and equipment, is an essential variable in the theory, and output, according to Eq. (6.16), can also be connected with this quantity

$$Y = \xi K, \tag{12.3}$$

where $\xi$ is productivity of capital stock $K$ (the tools and equipment), which is the part of social wealth. It is difficult to compile the entire list of components of production equipment. Really, a dwelling, for example, can be considered as the appliance for creation of comfort, or just as a consuming thing. It is assumed that one can define the quantity $K$ in this or that way.

The growth rate of productivity of capital stock $\xi$ can be found according to the expression for the growth rate of output (6.21) that can be rewritten here as follows:

$$\frac{1}{\xi}\frac{d\xi}{dt} = \frac{1}{Y}\frac{dY}{dt} - \frac{\nu + (1 - \overline{\lambda})\mu}{\overline{\lambda}}. \tag{12.4}$$

One can see that, to estimate the growth rate of productivity of capital stock, one needs to know the growth rate of production of value, non-dimensional technological factor $\overline{\lambda}$, the growth rate of usage of labour $\nu$ and factor of depreciation $\mu$. The quantity $\overline{\lambda}$ is a measure of substitution of efforts of the humans with work of production equipment (see Chap. 5), which is believed to be known. If $\overline{\lambda} = 1$, variations in technology do not occur, labour productivity is constant and all increase in output is connected only with increase in number of workers. Human efforts are, certainly, the basic motive power, but, under condition, when $\overline{\lambda} < 1$, these efforts are partially replaced with work of the machines operating by the means of energy from external sources, and labour productivity increases. It is a general description of scientific and technological progress that enters naturally into a picture of the progress of mankind.

The growth rates of output and usage of labour can be estimated according to Maddison's numbers, which were discussed in the previous section, but values of coefficient of depreciation $\mu$ and measure of substitution $\overline{\lambda}$ should be set based on some speculations. Value of factor of depreciation is taken constant, $\mu = 0.07$ in all times. This value is chosen in such a way that in years close to our time value of productivity of capital stock comes close to known assessments.

Estimates of the measure of substitution $\overline{\lambda}$ could be found according to its expression (5.19) for steady-state situations

$$\overline{\lambda} = \frac{\nu + \mu}{\delta + \mu}.$$

In the beginning of considered period, the growth rates of labour $\nu$ and capital stock $\delta$ were small and less than coefficient of depreciation $\mu$, so that $\overline{\lambda} \approx 1$ in the first years of our era; the exact value is set $\overline{\lambda} = 1.02$, taking into account that some recession (the growth rate of GDP was less than the growth rate of number of population) was observed in the beginning of our era. After the year 1000 AC, one can observe the progress that requires lesser values of the quantity ($\overline{\lambda} < 1$), so that we assume the measure of substitution decreases down to modern value ($\overline{\lambda} \approx 0.8$), and substitutive work increases in due course in comparison with usage of labour.

To restore values of productivity $\xi$, one can use Eq. (12.4). Instead of specifying an initial value of the quantity for solution of the equation, we consider a final value of the quantity known; value $\xi \approx 0.5$ corresponds to values of productivity of

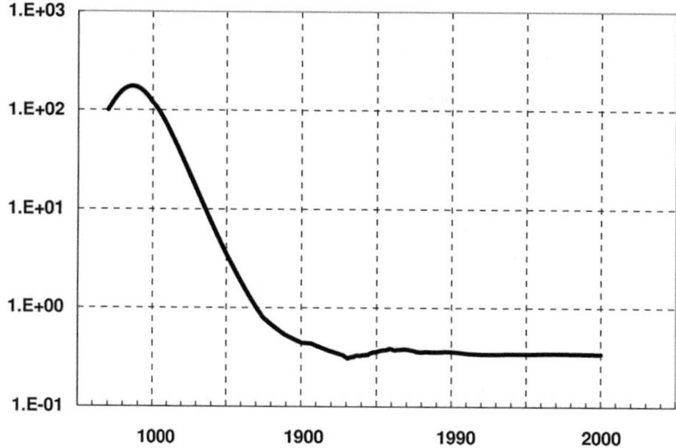


**Fig. 12.3** Productivity of capital stock

capital stock for a number of the countries in our time. This condition determines the value of productivity in the beginning of our era $\xi \approx 100$. The calculated values of productivity of production equipment are shown in Fig. 12.3. It is not surprising that in the beginning of our era, productivity of the capital stock appears big in comparison with modern values. The sources of value are labour and substitutive work, and in the epoch, when manual work prevailed and tools and appliances were rare, output per unit of capital stock (in value units) was big. Attraction of new energy sources (coal, oil) in seventeenth to nineteenth centuries demanded creation of bulky and expensive appliances that led to reduction of productivity of the production equipment.

When values of output and productivity of capital stock are known, the formula (12.3) allows us to restore values of capital stock that are shown in Fig. 12.1. Then, one can use the calculated values of the growth rate of capital stock $K$ and factor of depreciating, to restore, using a simple formula (5.6), the values of investments $I$ (production accumulation) shown also in Fig. 12.1. One can see that production accumulation (investment $I$) and capital stock were insignificant during an ancient epoch; the most part of the created products were used for personal and public consumption. Undoubtedly, the production itself was severely restricted by availability of labour and substitutive work, and the aims and character of production activity of humans in the beginning of our era were different from modern.

### 12.1.4 Substitutive Work

The simple method, which was described earlier in Sect. 7.1.2, allows us to calculate substitutive work $P$, if empirical time series of output $Y$, capital $K$ and labour $L$ are known. We do not have a value of substitutive work $P_0$ in the beginning of our era to proceed calculation, but we can judge about the modern values of substitutive work for the world, which, apparently, exceeds the value of substitutive work in the

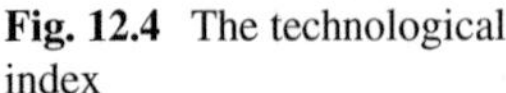

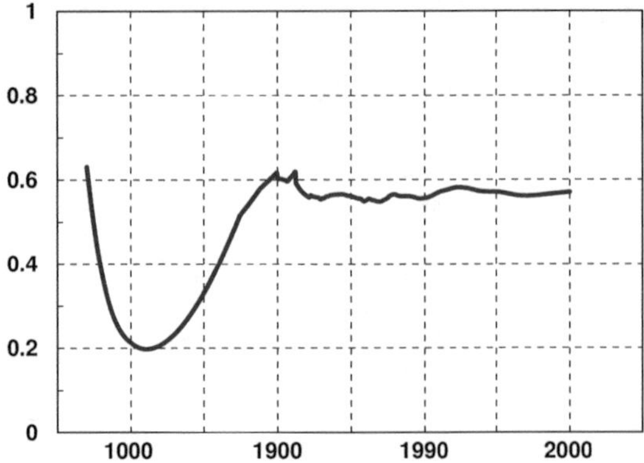


**Fig. 12.4** The technological index

USA economy in sometimes and was taken as $3 \cdot 10^{18}$ J in the year 2000. It allows us to make estimates of substitutive work $P$, moving during the course of calculations from the final values to initial. They found in this way values of substitutive work are shown in Fig. 12.2.

Simultaneously, values of the technological index $\alpha$, which are shown in Fig. 12.4, are estimated. These values have appeared to be small in the beginning of the era, which corresponds to the fact that the amount of production equipment is very small at that time. It is known (Chap. 6, Sect. 6.7) that the technological index $\alpha$ represents the fraction of expenses, necessary for maintenance of the production equipment, in the total expenses for the production factors. For comparison, it is possible to look at modern values of the technological index for the USA (see Fig. 7.1 in Chap. 7).

The amount of substitutive work $P$ is an essential quantity to describe the quality of the production system: according to Eq. (1.9), the ratio $P/K$, which is shown in Fig. 12.2, represents the neoclassical characteristic of technical progress. The other and better index of technical progress is the ratio of substitutive work to efforts (in energy units) of peoples, $P/L$. The quantity can be interpreted as the number of 'mechanical' workers, operating in the production processes, simultaneously with a 'live worker'. This quantity, which is shown in Fig. 12.5, is small in the beginning of our era, but by 2000 this ratio comes close to three.[2]

The results demonstrate that in the first millennium of our era the use of substitutive work and all indexes of the level of technological progress falls. It is desirable to have direct assessment of technological level in that epoch to judge, whether the first millennium has really been marked by technological recession, or such conclusion is only a consequence of a not precise assessment of a GWP and the number of population in the first years of our era.

[2]For comparison, we shall attract attention to Fig. 2.9: the number of 'mechanical workers' per a 'live worker', $P/L$, comes close to unit for Russia, while for the USA this quantity is more than ten.

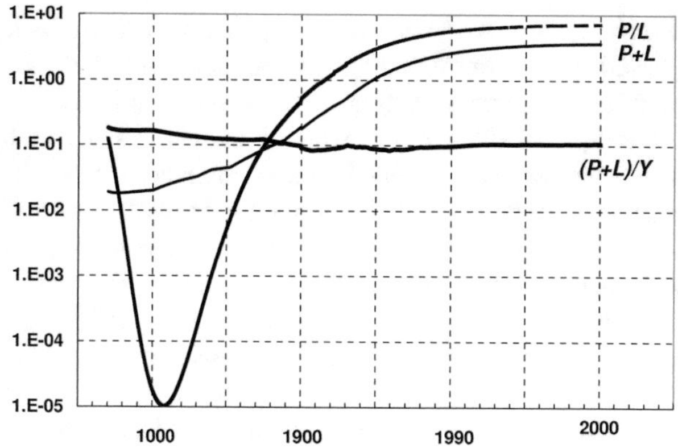


**Fig. 12.5** Substitutive work and efforts of humans. The sum of substitutive work and efforts of workers (in units of $10^{18}$ J) as well as their ratio are presented. The solid central line represents the energy content of international (1990) dollar (in units of $10^6$ J)

### *12.1.5 Universal Measure of Value*

Let us attract also attention to the important consequence of the approach: the theory allows introducing the absolute measure of value. In the Smith–Marx's theory, it is postulated that labour is the only source of all created wealth. When one includes the effect of substitution into the theory, one could expect that the total amount of work, as the sum of the properly accounted workers' efforts and true substitutive work of production equipment, could be the absolute measure of value (Chap. 1).

To confirm the validity of the statement, one has to investigate behaviour of the quantity $\epsilon$, which defines the work, needed for the creation of products valued of one monetary unit, or, in other words, 'energy content of monetary unit' (see Sect. 11.3). The quantity can be calculated as the ratio of the total work to the output

$$\epsilon = (P + L)/Y. \tag{12.5}$$

One cannot expect that this quantity is constant, when output $Y$ is measured with a special monetary unit of constant purchasing power; some dependencies can be easily found for the case. From the second line of relations (6.16) from Chap. 6, one gets that in the industrial epoch, when $P \gg L$,

$$\epsilon = \exp(\eta - \delta)t, \quad \eta - \delta > 0, \tag{12.6}$$

where $\eta$ and $\delta$ are the growth rates of substitutive work and capital stock, correspondingly. In the opposite case, when $L \gg P$,

$$\epsilon = \exp(\nu - \delta)t, \quad \nu - \delta < 0, \tag{12.7}$$

where $\nu$ is the growth rate of workers' efforts in production. The relations (12.6) and (12.7) can be tested for the considered case of the global economic development.

As calculated in the previous sections, results allow us to easily find the ratio of total work to the output $(P + L)/Y$, which is shown in Fig. 12.5 by the central solid line. The Maddison's data on GWP and all the macroeconomic quantities in the previous sections are estimated with international (1990) Geary–Khamis dollar, which is a monetary unit of constant purchasing power, and the ratio $(P + L)/Y$ defines the work, needed for creation of products, valued one international (1990) dollar, or, in other words, 'energy content of international (1990) dollar' for various epochs. This monetary unit does not present constant amount of value, considered as absolute quantity in the spirit of Marx's concept of value, and the 'energy content of monetary unit' can change during the times. In accordance with relations (12.6) and (12.7), the quantity decreases (from $\approx 2 \cdot 10^5$ J) at the beginning of the era during middle ages about thousand and half years, and increases slowly in the recent centuries. The estimated energy content of the international (1990) dollar for our epoch is $\approx 10^5$ J. This estimate corresponds to the assessments for the US dollar of 1996 (see Fig. 11.3 in Chap. 11).

The comparison of various assessments of an energy content of monetary units confirms that one can choose energy unit as a universal measure of value. Keeping in mind the U.S. dollar, it would be convenient apparently to choose $10^5$ J as an energy unit of value. Assessments of the GWP in energy scale, when the amount $10^{16}$ J is used as a unit of value, are shown in Fig. 12.1.

Note that the larger content of the monetary unit in the beginning of the era on the comparison with our days' content can be attributed to the overestimation of the efforts of people: the accumulation was not the aim of the early societies, and, for the satisfaction of the modest demand, it was enough to work not 6 h per day, but smaller quantity of hours than in modern times. Besides that, the possibilities of the human engine could be less in the beginning of our era, as it is possible to judge referring to the available research [9], so that it can be also a reason for the overestimating of calculated values.

### 12.1.6 Limit of Applicability of the Theory

The empirical figures for the GWP and number of population and the relations of the developed theory of economic growth allow one to make an assessment of the variables describing technological achievements of a population. It has been detected that the modern technological era starts not earlier that in the second millennium of our era; before this time the amount of the used substitutive work is too small to take it into account. The picture of technological development, given by the theory before the year 1000, seems to be unrealistic. Such picture results from Maddison's assessments, according to which the growth rate of GWP is less than the growth rate of the population in the first millennium of our era. Undoubtedly, objectives and character of production activity of a person in the beginning of our era differed from modern: during an antique epoch production, accumulation were insignificant (see Fig. 12.1), and the most part of the created products were used for personal and public consumption.

## 12.2 Constrained Growth of the Human Population

In an elementary approximation, a human population, however, as any other biological population, can be characterised by its number $N$; variation of this quantity is obeyed [10–12] a simple balance equation

$$\frac{dN}{dt} = (b - d)N, \tag{12.8}$$

where $b$ and $d$ are the coefficients of birth rate and death rate, accordingly.

Equation (12.8) is universal and applicable to any biological population. However, the quantities $b$ and $d$ are determined by circumstances of the life of individuals of the population. So, these quantities can depend on number of the same population $N$ or from some other variables; in case of a human population, it is necessary to enter characteristics of results of activity of production system. Further, we assume, following earlier work [13], that growth rate $b-d$ is essentially determined by quality of the condition of inhabitancy of the humans which, in addition to the natural environment, is possible to characterise, in the elementary approximation, by the social wealth per one individual. Unlike other populations, the humans live in the artificial environment that is an important attribute of human existence and should be accounted at any consideration.

### *12.2.1 The Estimates of Number of the Population*

Discovering and studying of remains of ancient people have given the basis for the statement that the humans have appeared in Africa about three millions years ago [14], and since that remote time has extended almost through the entire territory of the Earth. Prospective history of a stumbling movement of humans through marine passages and mountain ridges in time from two millions up to five hundred thousand years ago has been reproduced by means of computer model [15].

Some estimates of number of a population $N$ on the Earth are collected in Table 12.1 which is reproduced, in an essential part, from previous work [13]. Fast growth of a population occurred in some instants of history designated as the cultural revolution (nearby $10^6$ years ago), the agricultural revolution (nearby $10^4$ years ago) and the industrial revolution, which started in seventeenth to eighteenth centuries. During last four thousand years, the population of the Earth has increased from 30

**Table 12.1** The estimates of number of the population

| | $N$, $10^6$ man | $b-d$, year$^{-1}$ | $N/\overline{N}$ | $\overline{N}$, $10^6$ man | $a^* = r/\overline{N}$, man$^{-1}$year$^{-1}$ | $W/N$, \$/man |
|---|---|---|---|---|---|---|
| $10^5$ B.C. | 0.03 | | | 0.03 | $10^{-6}$ | 1 |
| 8000 B.C. | 10 | 0.0006 | 0.978 | 10.2 | $2.71 \times 10^{-9}$ | 10 |
| 1000 A.D. | 280 | 0.0004 | 0.986 | 284 | $9.75 \times 10^{-9}$ | 20 |
| 1650 | 516 | 0.003 | 0.892 | 579 | $4.79 \times 10^{-11}$ | 50 |
| 1850 | 1,171 | 0.007 | 0.747 | 1,567 | $1.77 \times 10^{-11}$ | 100 |
| 1900 | 1,668 | 0.007 | 0.747 | 2,232 | $1.24 \times 10^{-11}$ | 200 |
| 1920 | 1,968 | 0.008 | 0.711 | 2,767 | $1.10 \times 10^{-11}$ | 250 |
| 1960 | 3,308 | 0.019 | 0.314 | 10,530 | $2.63 \times 10^{-12}$ | 2500 |
| 1990 | 5,268 | 0.0166 | 0.401 | 13,146 | $2.11 \times 10^{-12}$ | 11000 |
| 1995 | 5,700 | 0.0141 | 0.454 | 12,577 | $2.20 \times 10^{-12}$ | 14000 |
| 2000 | 6,090 | 0.0126 | 0.545 | 11,172 | $2.48 \times 10^{-12}$ | 18000 |
| 2010 | 6,864 | 0.0112 | 0.596 | 11,523 | $2.40 \times 10^{-12}$ | 30000 |
| 2020 | 7,628 | 0.0096 | 0.653 | 11,674 | $2.37 \times 10^{-12}$ | 40000 |
| 2030 | 8,314 | 0.0076 | 0.726 | 11,458 | $2.42 \times 10^{-12}$ | 50000 |

The second and third columns give an assessment of an aggregate number and the growth rate of population according to Carr-Saunders [16], Clark [17] and Durand [18]. Values of these quantities for 1995–2030 are taken from the U.S. Census Bureau (https://www.census.gov/)

million up to almost seven billions. Dependence of number population in the last centuries can be approximated by the hyperbolic law:

$$N = c/(2030 - t),$$

where $t$ is the time (in years) of our chronology ([19], Sect. 25). Apparently, as Shklovsky marks, the specified approximation does not work for forecasting of the population growth in twenty-first century.

Last column of the table contains an assessment of the social wealth per one individual. The social wealth contains all artificial thing both tangible (buildings and constructions, cars, vehicles, networks of supply and the communications, house adaptations, clothes, etc.) and non-tangible ones (systems of knowledge, projects of the organisation, ethical rules, works of art, etc.), which ought to be estimated in terms of value. For the assessment of variation of value of the social wealth $W$, it is necessary to take advantage of the balance equation (6.28), which defines rate of change of the social wealth as a difference between result of productive activity of members of a society during some time unit (Gross domestic product) $Y$, discussed in Sect. 12.1.1, and disappearance of the social wealth during the same time interval, both as a result of direct consumption and because of ageing or deterioration of the social wealth.

### 12.2.2 Maltus Law—The Exponential Growth

In the simplest case, it is possible to assume that the growth rate of a population is constant

$$b - d = r.$$

Then, Eq. (12.8) gives the solution

$$N(t) = N(0) \exp\, rt. \tag{12.9}$$

In application to the human population, this law has been considered by Maltus [20], who has found that if nothing restrains the growth, the population is doubled approximately each 25 years,[3] that defines the growth rate

$$r \approx 0.0277\,\text{year}^{-1}, \tag{12.10}$$

so that $1/r \approx 36.1$ year appears to be the time for which the population increases in $e \approx 2.72$ time. The Maltus constant $r$ defines some timescale, that, probably, needs in specification. The value of the quantity is close to 42 years—the value of the constant introduced by S.P. Kapitsa [21]. This scale is close also to lifetime of one generation.

In application to the human population, the law (12.9) is valid, perhaps, for some populations during some periods of time, but does not describe observable growth of the human population for all its history. The growth rate of the human population increases, and we should discover the factors influencing the growth rate.

[3]Maltus writes in the sixth chapter of the first edition of the book [20]: 'It has been universally remarked that all new colonies settled in healthy countries, where there was plenty of room and food, have constantly increased with astonishing rapidity in their population. Some of the colonies from ancient Greece, in no very long period, more than equaled their parent states in numbers and strength. And not to dwell on remote instances, the European settlements in the new world bear ample testimony to the truth of a remark, which, indeed, has never, that I know of, been doubted. A plenty of rich land, to be had for little or nothing, is so powerful a cause of population as to overcome all other obstacles. ... The consequence of these favourable circumstances united was a rapidity of increase probably without parallel in history. Throughout all the northern colonies, the population was found to double itself in twenty-five years. ... In New Jersey the period of doubling appeared to be twenty-two years; and in Rhode island still less. In the back settlements, where the inhabitants applied themselves solely to. agriculture, and luxury was not known, they were found to double their own number in fifteen years, a most extraordinary instance of increase. Along the sea coast, which would naturally be first inhabited, the period of doubling was about thirty-five years; and in some of the maritime towns, the population was absolutely at a stand'.

### 12.2.3 Constrained Growth—Logistic Curve

From Malthus' times it becomes clear that availability of the resources limits the growth of human population, and investigating of this restriction has appeared central and fruitful in research of the problem. Limiting factors (food, dwellings, etc.) define that in each point in time there is some limiting value of the population number $\overline{N}(t)$. Growth of the population in this situation is described by known Verhulst–Pearl equation that was established first for biological populations [12, 22, 23]. Later, the equation was applied for description of the characteristic growth rate of a human population. The Verhulst–Pearl equation looks like

$$\frac{dN}{dt} = r\left(1 - \frac{N}{\overline{N}}\right)N, \tag{12.11}$$

where $r$ is the internal, biological growth rate of the population—Malthus' constant. In this case, the described growth rate of the population is not constant and has the form

$$b - d = r\left(1 - \frac{N}{\overline{N}}\right). \tag{12.12}$$

Equation (12.11) has a steady-state point $\overline{N}$ that is stable as $\frac{dN}{dt} < 0$ $N > \overline{N}$ and $\frac{dN}{dt} > 0$ $N < \overline{N}$. The obvious solution of Eq. (12.11) can be easily found if one records this equation in the form

$$\frac{dN}{N} - \frac{dN}{N - \overline{N}} = r\,dt.$$

Integration of this relation gives a solution—an equation of the logistic curve

$$N(t) = \frac{\overline{N}\,N(0)\exp\,rt}{\overline{N} - N(0) + N(0)\exp\,rt}. \tag{12.13}$$

The solution is valid both for $N(0) < \overline{N}$ and for $N(0) > \overline{N}$. At any initial value, the trajectory approach asymptotic line $N = \overline{N}$, as shown in Fig. 12.6.

At the constant value $\overline{N}$, the growth rate depends on the number of population, and Eq. (12.11) appears nonlinear. This nonlinearity is conjugated usually with effect of suppression of growth because of a mutual competition of individuals of the population. In this case, it is possible to introduce also the factor of mutual influence of individuals or the self-poisoning coefficient $a$ by a relation

$$\overline{N} = \frac{r}{a}. \tag{12.14}$$

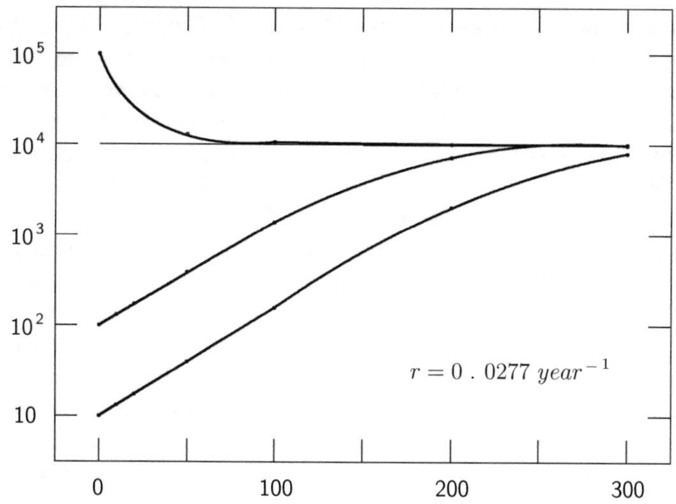


**Fig. 12.6** Evolution of the population. Trajectories of evolution approach the line $N = \overline{N}$ from top at $N > \overline{N}$ or from bottom at $N < \overline{N}$. Calculated according to Eq. (12.13) at $r = 0.0277\ \text{year}^{-1}$, $\overline{N} = 10^4$, and various initial values $N(0)$

It is possible to assume that Eq. (12.11) can be used for description of the human population growth at a changing value of $\overline{N}$ [13]. In this case, the parity (12.12) allows us, at known value of the Malthus' constant, to estimate empirical values of the limiting number $\overline{N}$ that are listed in Table 12.1. The possible values increase monotonically together with growth of the population number, reaching the maximum value at the end of the last century, and have in the beginning of our millennium value

$$\overline{N} \approx 1.15 \times 10^{10}. \tag{12.15}$$

The increase in the limiting number of a population can be connected with improvement in quality of the life of humans.

The corresponding values of the coefficient of mutual influence of individuals decrease from the natural (at the earliest stage of evolution of the humans) value

$$a \approx 10^{-6}\ \text{man}^{-1} \cdot \text{year}^{-1}. \tag{12.16}$$

### *12.2.4 The Changing Limit to Growth*

Addressing now specifically to the formulation of equation for description of growth of the number of individuals, we notice that the population on the Earth consist of a number of independently developing populations [24], that is, especially true for the initial stage of evolution of the mankind,[4] and consequently any equation describing growth of the population, should be covariant in relation to any breakage of the population, that is, should be valid both to separate independent populations and to the united population. For example, if for a compound population $N = N_1 + N_2$, the equations of growth both for $N$ and for $N_1$, $N_2$ should have the identical form. Equation (12.11) with constant value $\overline{N}$ does not satisfy this requirement, and, consequently, it is not suit for the description of growth of the population number, which, however, was discussed in the previous section. For the same reason, the nonlinear equations recorded by Kapitsa [21] do not suit for the aim also.

Nevertheless, Eq. (12.11) can serve a convenient starting point in our investigation and allows to interpret dynamics of population. According to this equation, the population number is trending to the possible value $\overline{N}$ that is determined by the reached level of technological and organisational development that allows individuals of the population to have convenience in supplying of food, an arrangement of dwelling, protection against illnesses and other advantages. The possible population number $\overline{N}(t)$ changes under its own law.

The assets, specifically influencing conditions of human life, includes all achievements of a population, both material and non-material, and is estimated in value units. A quantitative measure of the reached standard of living can be taken as amount of the social wealth per one individual $W/N$, and we assume that there is a function[5]

$$\overline{N} = Nf\left(\frac{W}{N}\right). \tag{12.17}$$

The modern numbers of value of social wealth can be found in national statistical collections. For the USA, for example, social wealth per one individual is approximately equal to 40 thousand dollars, but it is much less for the majority of countries of the world. It is difficult to estimate the social wealth in ancient times. What is the cost in modern prices of the primitive dwelling, several animal skins, primitive tools and primitive kitchen appliances, for example? Table 12.1 contains possible assess-

---

[4]S.P. Kapitsa [21] states opposite, for example, on p. 65: '... The coherence and the interdependence of the modern world caused the transport and information streams, unite all in a single entity and allow conclusive possibilities to consider the world today as a global system.... And in the remote past, when there were not so many humans, and the world was substantially divided, its populations slowly but surely cooperated'. It is possible to argue that for complex systems, which populations are, even the interoperability in some respects does not mean that there are no independent processes in the populations, which are, apparently, demographic processes. Anyway, the law of the composition of a set of dependent or independent populations should be discussed. Judging on some responses [25, 26], S.P. Kapitsa's concept is, at the best, ill-founded.

[5]For the description of features of growth of the population, Korotayev et al. [27] used two variables: the quantity that almost coincides with growth rate of well-being $W/N$ and *literacy*, as the factor influencing reduction of birth rate. However, literacy can be considered as a component of social wealth and separation of this part, without consideration of others, looks artificial. Apparently, it is possible to specify some other components of social wealth influencing decrease or increase in birth rate. However, so detailed consideration is inappropriate at our rough description, and we are limited to consideration of influence of one characteristic—well-being, that is, a quantity of the cumulative stored social wealth per one individual (material and non-material, that includes literacy).

ments which, undoubtedly, require to be corrected. However, the essential increase in social wealth from those remote times does not seem improbable.

The possible value of the quantity $\overline{N}$ increases as the social wealth $W$ increases. During ancient times, the amount of social wealth was very small, and the population number $N$ was close to the possible numbers $\overline{N}$ that are recorded in Table 12.1. The ratio $N/\overline{N}$ was close to unit in early years of evolution of the population, and deviation of this quantity from unit can be connected linearly with the value of assets per one individual

$$\frac{N}{\overline{N}(W/N)} \approx 1 - h\,\frac{W}{N}, \quad h\,\frac{W}{N} \ll 1, \tag{12.18}$$

where $h$ is a coefficient of influence of social wealth on the growth rate of the population. It is possible to address data in Table 12.1 to get the rough estimate of the coefficient of influence

$$h \approx 0.002 \text{ man} \cdot \text{dollar}^{-1}.$$

The possible value of the population number $\overline{N}(t)$ increases monotonically as the social wealth increases; however, the growth rate of the population can show non-monotonic dependence. A condition of transition from growth to falling (this event is defined as demographic transition) can be found as the condition of extremum of the growth rate

$$b - d = r\left(1 - \frac{N}{\overline{N}}\right). \tag{12.19}$$

Considering both actual and possible number of the human population, $N$ and $\overline{N}$ as functions of time, we find the extreme value of this function, which defines a condition of demographic transition as the condition of equality of the growth rates of functions $N$ and $\overline{N}$

$$\frac{1}{N}\frac{dN}{dt} = \frac{1}{\overline{N}}\frac{d\overline{N}}{dt}. \tag{12.20}$$

Before the point of demographic transition, the actual population closely follows an escaping limit. After a point of demographic transition speed of increase of number appears less than speed of increase of a limiting number.

### 12.2.5 Approximation of the Growth Rate of the Population

To describe the situation in more details, we shall consider coefficients of the birth and death rates separately, assuming that each one depends on well-being, characterised, in the simplest case, by only one quantity—the social wealth per one person $W/N$. We believe that influence of well-being leads to reduction of both the coefficients of birth rate and death rate in comparison with some uncertain natural values that are considered equal, assuming that in natural conditions the number of population does not change. In linear approximation,

$$b = r\left(a - h_b\,\frac{W}{N}\right), \quad d = r\left(a - h_d\,\frac{W}{N}\right), \quad \frac{W}{N} \ll 1. \tag{12.21}$$

The quantity $h$ in the ratio (12.18), according to the evidence, has to be positive, so that, from comparison of parities (12.18), (12.19) and (12.21) follows that the inequalities $h_d > h_b > 0$ are valid.

We believe further that the coefficients of birth rate and death rate continue to decrease monotonically at an increase of well-being down to some positive values, which in the simplest kind could be approximated by the functions

$$\begin{aligned} b &= r\left(a - \frac{\gamma x}{1+\beta\gamma x}\right), \quad \beta > 1 \\ d &= r\left(a - \frac{x}{1+\alpha x}\right), \quad \alpha > 1 \end{aligned} \tag{12.22}$$

Here, the dimensionless variables are introduced for convenience

$$x = h_d\,\frac{W}{N}, \quad h_b\,\frac{W}{N} = \gamma x, \quad \gamma = \frac{h_b}{h_d} < 1. \tag{12.23}$$

If the influence of well-being on the mortality-rate coefficient appears significantly stronger in comparison with the influence on the coefficient of birth rate, one has the relation $\gamma \ll 1$.

The approximation (12.22) defines the expression for the growth rate of population

$$b - d = r\left(\frac{x}{1+\alpha x} - \frac{\gamma x}{1+\beta\gamma x}\right). \tag{12.24}$$

The first term of the expansion of the function naturally returns us to the expressions (12.18), (12.19) and (12.21). At $x \to \infty$, the growth rate trends to a constant value

$$(b-d)_\infty = r\,\frac{\beta - \alpha}{\alpha\beta}. \tag{12.25}$$

The asymptotic value of the population, that is, the quantity $(b-d)_\infty$, is specific and should be estimated for each separated nation. It is possible to believe that the number of the population is stabilised, that is, becoming a constant; in this case, we should put $\beta = \alpha$.

Under the stated conditions, the function (12.24) possesses a maximum in a point which is defined as a point of demographic transition. Up to this point, the growth rate increases, and after a point of demographic transition, it falls. The situation is illustrated in Fig. 12.7.

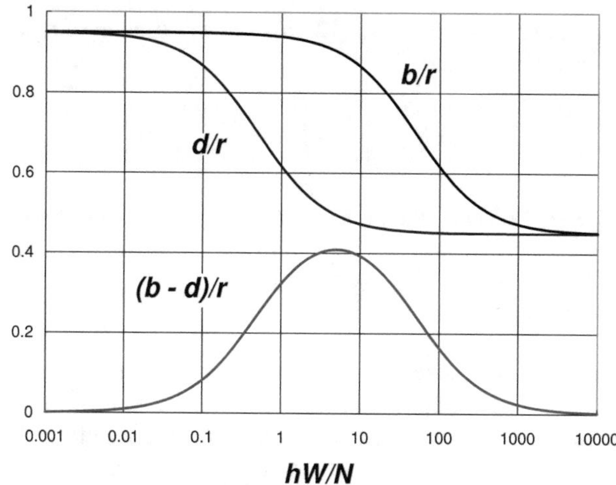


**Fig. 12.7** The demographic transition. The curves represent dependences (12.22) and (12.24) for values of parameters: $\alpha = 2$, $\beta = 2$, $\gamma = 0.01$

The behaviour of the growth rate reproduces a picture observable in the present period for the human populations as a whole and the population of the separate countries (see, for example, [21], p. 67, Fig. 3). However, actual dependences could appear more complex in comparison with discussed approximation: growth of a population could be accompanied not by one, as in the considered example, but several demographic transitions which could occur during earlier periods of evolution of mankind.

### *12.2.6 Description of Catastrophic Events*

The observable number of a human population is, as a rule, always less than possible value which is defined by value of well-being at present time

$$N < \overline{N}\left(\frac{W}{N}\right). \tag{12.26}$$

But what would occur, if there would be a catastrophe and suddenly all life-support systems would disappear? The social wealth decreases up to some value $W_k$, and a new value of the possible number falls to some level that could appear less or even much less than current value of the number of population

$$N \gg \overline{N}_k. \tag{12.27}$$

At earlier times of observation, Eq. (12.13) defines variation of number of a population in the form of linear law

$$N(t) = N(0)\left[1 + \left(1 - \frac{N(0)}{\overline{N}_k}\right) rt\right]. \qquad (12.28)$$

When the condition (12.27) is valid, the law of falling of the number can be recorded in the form

$$N = N(0)\left(1 - r\frac{N(0)}{\overline{N}_k}\, t\right). \qquad (12.29)$$

The above-written parity represents the law of reduction of the number of population in the case of a world catastrophe. This law can be also applied to cases when the elements of stored social wealth suddenly disappear, for example, in the case of the besieged city. Under what law the population of blocked Leningrad died out? For 2 years of blockade, the city lost, according to the different assessments, from 0.5 up to 1.5 million person at initial number of 5 million person. If we admit the linear law to be valid, formula (12.29) at $r = 0.0277$ defines that the initial population exceeded the possible value of number after accident $N_k$ about fourfold.

### *12.2.7 About Limits of Applicability of the Theory*

It is impossible, apparently, to explain, not referring to the social production system, the adaptation of the humans to conditions of existence, which gave increase in the human population from very small group one million years ago up to approximately seven billions now. Though based on some assumptions and simplifications, Eqs. (12.8), (12.12) and (12.24) can quantitatively describe huge growth of a human population in the last centuries and its reduction in cases, when the social wealth is disappearing.[6] The principle of the greatest attraction of energy (the maximal capacity) defines the general direction of progress via the simple mechanism of evolution: things and ideas which are necessary or useful to the survival of the humans are kept and collect.

The discussed parities are applicable both to the World's population as a whole and to separate populations, which are on a various level of evolution. Here, it is

[6]It could be said that increase in the human population is connected, eventually, with the fact that the humans have appeared to be capable to analyse their impressions from the world and to adapt the environment for their advantage. It has led to the invention of the appliances, allowing to attract the natural energy into production and to create abundance of consumer goods. Thus, the causal sequence is recorded in the form of a chain: *archive of knowledge—use of energy—social wealth—growth of the population*. Omitting the intermediate elements of the chain, B.M. Dolgonosov ([28], Chap. 1) has undertaken courageous research of dependence of the growth of population straightforwardly on the quantity of knowledge and has considered various scripts of progress. Unfortunately, it is unclear, what is the quantity of knowledge and whether this quantity is enough for characteristisation of the knowledge.

possible to notice that numbers of Table 12.1 show that the global growth rate of the population has started to decrease, beginning since about 1960, despite of the obvious increase of the social wealth. At first sight, it could mean that the improvement of the conditions of life now approaches to the limit, and the increase of social wealth in the future will not influence the growth rate of the population. But it is possible also another point of view: at the present stage of development of the social system, the part of national wealth $W$, considered in statistical reports, does not assist the survival of individuals of population, but, on the contrary, it is sometimes directly aimed at destruction of individuals. Besides, it is necessary to consider that the total social wealth is distributed non-uniformly, both inside a local population and in relation to various local populations: true value of well-being for the majority of individuals appears much smaller than average value of well-being $W/N$.

The approximation, according to which the human population and its production activity are considered by means of the variables, concerning the population as a whole, is very rough and can lead to doubtful consequences. To find out what variations of the theory are mandatory for more precise description, it is necessary to study, apparently, well-described dynamics of development of separate populations that would allow to present the future progress based on assumptions of the future behaviour of the humans. Probably, the modern hindrance of growth is similar to the hindrance after agrarian revolution, and some changes in the organisation of a social production–distribution scheme are necessary for the observable delay that would be replaced by a rise. It is possible to assume that the appropriate social mechanism of regulation of growth will be designed and carried out.

## 12.3 Production in the Times of Cattlemen and Farmers

Returning to discussion of the results of Sect. 12.1, we shall notice that the method used was based on concept of the two value-creating production factors, as was formalised by Eq. (6.16), and has allowed (at more or less reasonable assumptions) to describe the evolution of production activity of people in the last centuries, but, before the year 1000 of our era, the quantity of the substitutive work, one of the two factors, has appeared practically equal to zero (see Fig. 12.2), so that for consideration of production activity over the centuries before the year 1000, it is necessary to address a theory with the only one of the production factors; the theory is known as the labour theory of value (see Sect. 1.3.1). Our immediate task is to extrapolate the description of production activity of the population from 1000 of our era in the depth of centuries.

### 12.3.1 The System of Equations of a One-Factor Theory

The labour theory of value assumes (see Sect. 1.3.1) that the only value-creating factor is humans' work, which is estimated with work hours of one individual; it can be estimated by power units, as noted in Sect. 2.4.1. The theory determines the parity for the gross global output

$$Y = AL, \tag{12.30}$$

where $A$ is a labour productivity. Expenditures of labour $L$ are determined by the existing number of population

$$L = \chi N \tag{12.31}$$

with time-dependent factor of proportionality. Considering an initial stage of progress of mankind, we accept for the population $N$ (see Sect. 12.2.4) the law

$$\frac{dN}{dt} = rhK, \tag{12.32}$$

where $K$ is the value of existing production equipment (the basic production capital); this quantity is included in the quantity of social wealth $W$. We assume that the distinction between all social wealth $W$ and its part, participating in production as a fixed capital $K$, can be taken into account in Eq. (12.32) with time-dependent factor $h$. The Maltus' constant $r = 0,0277$ is considered as a characteristic of the population.

Dynamics of the production assets is determined by the equation

$$\frac{dK}{dt} = sY - \mu K, \tag{12.33}$$

where $s$ is the fraction of the total product invested in production activity; $\mu$ is the factor of deterioration of production funds (and also of the total social wealth).

The labour productivity $A$ in Eq. (12.30) changes in due course; to record the law of changing, we address the expression (5.16), which defines reduction of expenditures of labour at variation of production technology, and find

$$\frac{dA}{dt} = As(1 - \overline{\lambda})\frac{Y}{K}. \tag{12.34}$$

Technological coefficient $\overline{\lambda}$ is determined by a parity (5.19), that is, by the expression

$$\overline{\lambda} = \frac{\nu + \mu}{\delta + \mu}, \tag{12.35}$$

in which the growth rates of expenditures of labour and production capital are defined as

$$\nu = \frac{1}{L}\frac{dL}{dt}, \quad \delta = \frac{1}{K}\frac{dK}{dt}. \tag{12.36}$$

These quantities are calculated according to the above-recorded Eqs. (12.31)–(12.33).

The system of Eqs. (12.31)–(12.36) for five variables $Y$, $A$, $L$, $K$ and $N$ describes a trajectory of evolution, when the initial values of variables and four time-dependent parameters: $s(t)$, $h(t)$, $\chi(t)$ and $\mu(t)$, are given.

### 12.3.2 Reconstruction of the Dynamics of Growth

The formulated system of equations allows one to find out some laws of evolution before the epoch of the intensive use of energy in production. For the description of evolution of the world economy during the period before the year 1000 of our era, we assume the trajectory is given in the form of power (exponential) functions:

$$N = N_0 e^{\rho t}, \quad Y = Y_0 e^{\zeta t}, \quad L = L_0 e^{\nu t}, \quad K = K_0 e^{\delta t}, \quad A = A_0 e^{\psi t}. \tag{12.37}$$

The subscript *zero* designates values of variables in the year 1000 of our era. The growth rates of variables are defined by substitution of prospective functions (12.37) in Eqs. (12.31)–(12.36)

$$\delta = s_0 \frac{Y_0}{K_0} - \mu_0, \quad \psi = s_0(1-\overline{\lambda})\frac{Y_0}{K_0}, \quad \rho = rh_0 \frac{K_0}{N_0}, \quad \nu = \rho + \frac{1}{\chi}\frac{d\chi}{dt}, \quad \zeta = \psi + \nu. \tag{12.38}$$

The expressions for the growth rates include some parameters, whereas the deliberate form of time dependence of the parameters is required

$$s = s_0 e^{(\delta-\zeta)t}, \quad h = h_0 e^{(\rho-\delta)t}, \quad \chi = \chi_0 e^{(\zeta-\psi-\rho)t} = \chi_0 e^{(\nu-\rho)t}, \quad \mu = \mu_0. \tag{12.39}$$

We consider the year 1000 of our era as a reference point. The values of all variables as well as values of the parameters in this point are assumed to be given. The growth rates of parameters are determined by the existing relations; only the quantity $\nu - \rho$ should be set independently.

To find out the reference values of variables, we address, first of all, direct estimates. We use the known [2, 29] assessments of the population and the GWP that are defined for the year 1000 of our era: $N_0 = 267 \cdot 10^6$ person, $Y_0 = 121 \cdot 10^9$ international 1990 dollars. As a value of the basic production funds, we use the value established in Sect. 12.1, namely $K_0 = 1.02 \cdot 10^9$ international 1990 dollars. The work expenses for the year 1000 are estimated in Sect. 12.1 as $L_0 = 2.7 \cdot 10^{16}$ J in power units. This assessment is obtained on the assumption (see Sect. 12.1.2) that half of population is occupied in production, and everyone spends for performance of work $4 \cdot 10^8$ J/year. By this assumption, it was necessary to spend $2.21 \cdot 10^5$ J (see Sect. 12.1.2) for creating the product valued one dollar in the year 1000. Accepting that for one working hour the person spends $4.18 \cdot 10^5$ J, we find the quantity $L_0 = 64 \cdot 10^9$ man-h for the year 1000. Such a quantity could be provided by 22 million person (about eight percent of the population), working on eight hours each

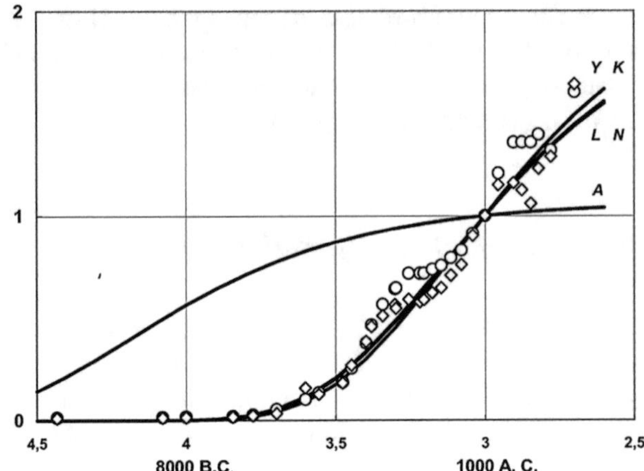


**Fig. 12.8** Trajectory of evolution. The values of the logarithm of remoteness of time from 2000 are drawn on the x-axes. The circles and rhombs present average values over three versions of the direct estimates, collected by Delong [29]. The empty circles present values of the population, the rhombs—the GWP. The curves represent the relative values of the variables specified by symbols at curves and calculated according to Eq. (12.37) with values of the growth rates determined in the text

day in a year. It is possible that capacity of human mechanism was less during ancient times (see Sect. 2.4.1) and assessments of expenditures of labour are underestimated. The labour productivity is determined as the ratio $A = Y/L$, and $A_0 = 4.18$ international 1990 dollars per one man-hour in the year 1000 of our era.

The available assessments of population [2, 29] allow us to establish, requiring the comprehensible conformity of points with exponential curve (see Fig. 12.8), value of the growth rate $\rho = 0.00073$ (here and later the growth rates are specified in terms of $year^{-1}$), which exceeds the value of the growth rate estimated in Sect. 12.1.2 $\rho = 0.00017$. By comparing with formula (12.38), we can also estimate the value of the parameter $h_0 = 0.007$ person for one dollar. The value of the parameter a few times exceeds the value estimated in Sect. 12.2.4, but it is necessary to take into account that the formula (12.18) contains value of production assets, unlike formulas of Sect. 12.2.4, where value of all social wealth is used. In a similar way, existing [2, 29] assessments of the GWP allow to establish specific value of the growth rate of total output as $\delta = 0.0008$, which coincides with the growth rate of the basic production assets, assuming constant share of investment product in the total output. According to Eq. (12.38), the quantity $\delta$ is determined by a fraction of investments in the output $s_0$ that cannot be too great because of limitation of the coefficient of depreciation $\mu_0 < 1$. The chosen value $s_0 = 0.004$, at which $\mu_0 = 0.48$, is close to the possible greatest value. When establishing the value of growth rate of expenditures of labour $\nu$, it is necessary to consider (see formula 12.38) that the technological coefficient $\overline{\lambda}$ cannot deviate too much from unit; the earlier found (in Sect. 12.1.3) value is $\overline{\lambda} = 0.97$. As a result, we find $\nu = 0.00074$, $\overline{\lambda} = 0.99987$.

The established values of the quantities allow us to calculate, using Eq. (12.38), the growth rate of labour productivity $\psi = 0.000064$ and the growth rate of a final output $\zeta = 0.0008$. Certainly, the values of the growth rates of variables are not defined unequivocally but the choice is very limited at the preset reference values of variables, so that we consider the trajectories of evolution represented in Fig. 12.8 to be highly probable. We shall notice that for the beginning of the new era, unlike Maddison's numbers (see Sect. 12.1), calculations define values: $N(1) = 130 \cdot 10^6$ person, $Y(1) = 56 \cdot 10^9$ international 1990 dollars. The trajectories of variables in Fig. 12.8 describe the slow evolution of a human population during millenniums after the beginning of agrarian revolution and before industrial revolution. We shall notice, however, that these curves do not reproduce the direct [29] assessment of values of GDP and population in 8000 and earlier years B.C. (see Fig. 12.8). In that, still more remote epoch—an epoch of hunters and gatherers—development occurred even more slowly.

The production activity of humans during the considered epoch was based on manual skills of farmers, cattlemen and handicraftsmen. The work was hard in some certain times, so as agriculture and cattle breeding had a seasonal nature, but, as for a year, work was not wearisome, anyway, less intense, than it is today, in the conditions of developed capitalism [30]. The objectives and character of production activity of humans during this epoch differed from modern ones, but, as well as presently, the production activity can be divided into three parts according to the character of the created products: very small fraction (no more than 1%) of all products was used for maintenance of the land and pastures, production and repairing of tools and appliances for manual work (the investment into the basic production assets). The another part contains the products, which were immediately consumed, and products, which can be considered as an investment in long-term projects: the weapon for protection and attacks, public constructions (palaces, temples), establishing the government system, creation of myths and works of art, scientific researches and so forth. The changes occurred very slowly, but permanently, and we collect the harvest of the great achievements of that epoch.

## 12.4 Are There Limits to Growth?

In all times, attempts to present a picture of the future development of a human population did not stop. Now we understand that the increase in the numbers of individuals is determined by possibilities of the population to utilise the greater and greater amounts of energy, which, as the result, leads to the growth of production and abundance of food and things. The demographic and economic progresses of the human population should be considered in common: on one hand, growth of the population brings to existence of additional working force, as it has been described in Sect. 2.4.2, determined by this an additional possibility to economic growth at given technology. On the other hand, the increase in population is connected with the best conditions of human existence, that is, finally, with results of economic activities, which, in turn, requires expenditures of labour and other factors of production. The described theory of economic growth connects output with technological usage of labour and energy. The equations of the theory allow us, more or less reasonable, to consider development of the population in the past, and one can try, on the base of the developed theory, to present a scheme of future development of the human population. In the simplest case, the global dynamics is considered for the entire Earth, without introduction of any structural variables (unstructured approximation), neglecting the existence of various states and various industries.

### *12.4.1 Asymptotic Relations*

The global production output $Y$, according to the second of Eq. (6.16), depends on expenditures of labour $L$ and substitutive work $P$ that are determined, correspondingly, by the population $N$ and the amount of total consumption of energy carriers $E$. The relationship between production factors $L$, $P$ and their primary sources $N$, $E$ appears, generally speaking, complex enough (see Sects. 2.4 and 2.5), but we can hope to find asymptotic relations between the output $Y$, production factors $L$ and $P$, and the specified quantities $N$ and $E$, assuming the future behaviour of variables.

One can note that increase in expenditures of labour is limited now by the increase in population, so that the growth rate of expenditures of labour coincides with the growth rate of the population,

$$L \sim N. \tag{12.40}$$

In the times of abundance of available energy (during the twentieth century and earlier), the growth rates of substitutive work $P$ exceed the growth rate of total consumption of energy $E$, and the latter exceeded the growth rate of the population $N$ (see Fig. 12.2), whereas the simple relation can be written as ([31], p. 120)

$$E \sim N^2. \tag{12.41}$$

As for the future, it can be expected that the amount of primary substitutive work will be approaching the total consumption of energy careers, so that the growth rate of substitutive work will not exceed the growth rate of the total consumption of energy carriers. In this situation, one can expect that the increase in substitutive work will be limited by the increase in the total consumption of energy carriers, so that the growth rates of substitutive work and total consumption of energy carriers will coincide, and the asymptotical relation can be written as

$$P \sim E. \tag{12.42}$$

Now, assuming recorded asymptotic parities (12.40) and (12.42) to be true, we can find the asymptotic behaviour of output in situations of the deficit of available energy. Following Akaev [31], we use the well-estimated quantities: the output and consumption of energy per capita, $Y/N$ and $E/N$, correspondingly. Then, the second of the relations (6.16) gives rise to asymptotic equation

$$\frac{Y}{N} \sim \left(\frac{E}{N}\right)^{\alpha}. \tag{12.43}$$

The coefficient of proportionality depends mainly on the ratio $P/E$ and trends asymptotically to constant. Value of the technological index $\alpha$ is situated between zero and unity; one can see in Fig. 12.4 that it is $\approx 0.6$ at the end of the last century. The technological index asymptotically trends to unity.

Equation (12.43) demonstrates that production output per capita in the situation of deficiency of energy carriers is determined eventually by the consumption of energy per capita. Could the humans provide the growth rate of energy consumption greater that the growth rate of the population? It follows from the equation that, for the decrease in the production output per capita not to be allowed, it is necessary to ensure that the growth rate of the energy consumption $E$ should be no less than the growth rate of the number of humans $N$.

One can expect that the ratio $Y/N$ should be approximately constant in the future. Then, it follows from Eq. (12.43) that the total consumption of energy $E$ appears to be proportional to the number of population $N$

$$E \sim N. \tag{12.44}$$

The crossover to the new paradigm of energy consumption from relation (12.41) to relation (12.44) was described by Akaev ([31], p. 124).

The peculiarity of the modern period of development is a transition from abundance of power resources, when the growth rate of substitutive work $P$ in industrialised countries exceeds the growth rate of total consumption of energy, to their deficiency. Really, we can see from two examples (the USA, Fig. 2.8 and Russia, Fig. 8.5) that, already to the middle of our century, the amount of substitutive work comes close to the total of energy available for use. The transition to the epoch of deficiency of energy carriers leads to variation of ways of use of energy carriers. The deficiency of energy carriers will lead also to an aggravation of struggle for available energy sources.

### 12.4.2 Forecast Dependences

Equations (12.41) and (12.44) establish some parity between number of population $N$ and the total consumption of energy $E$, but, to find out which of these two quantities is a leading component of progress, it is necessary to refer to a principle of evolution, which stated that the human population, similar to any other population or ecosystem, takes advantages to a survival, when consumption of energy increases, and the population, consequently, trends to absorb all available power resources (see

Sect. 11.1.3 of Chap. 11). This principle has been established by biophysicists [10, 33], and also it is known as a principle of the maximal power. According to this principle, the human population aspires to increase available energy, which allows the population increasing in the numbers. Consumption of energy appears a leading component of progress, and, for the discussion of the future of human population, it is important to have scripts of variation of available energy.

The future consumption of available energy was a subject of interest of many researchers, the majority of which supposed a monotonous increase of the consumed energy. For example, Belyaev et al. [32] have estimated requirement for global final energy in the year 2100 as $(4–6) \cdot 10^{20}$ J annually, that gives, accordingly, total consumption of primary energy carriers about $10^{21}$ J a year. Certainly, there are other possibilities: Plakitkin [34], for example, believes that the recession of consumption of energy carriers can already begin in the middle of our century. The stocks of mineral fuel will be disappearing, and they cannot be replaced by other energy sources in due time, which will lead to reduction of quantity of the consumed energy carriers.

Designing a long-term script of use of available energy, it is necessary to take into account that it is essential that the possibility of obtaining energy directly from the Sun is limited; prospective consumption of energy in 2100 is approximately one thousand times less than quantity of the energy received by the Earth from the Sun, which is equal to about $3 \cdot 10^{24}$ J annually. The human population will be compelled to address the use of nuclear energy and to aspire the development of new power sources. Possible scripts of growth of available energy are presented in Fig. 12.9.

The future numbers of population on the Earth were estimated independently. The majority of approaches give the values of 10–12 billion by the year 2100 [32], though some models assume reduction of a population from the middle of our century. Apparently, the growth of number of the population and the increase in use

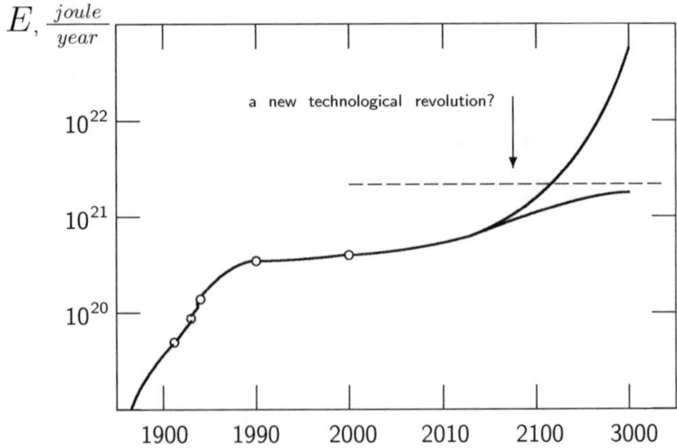


**Fig. 12.9** Availability of energy: the future. The dotted line presents the hypothetical limiting value of energy, which could be obtained from the Sun. The lower curve till the year 2100 corresponds to the estimates of Belyaev et al. [32]

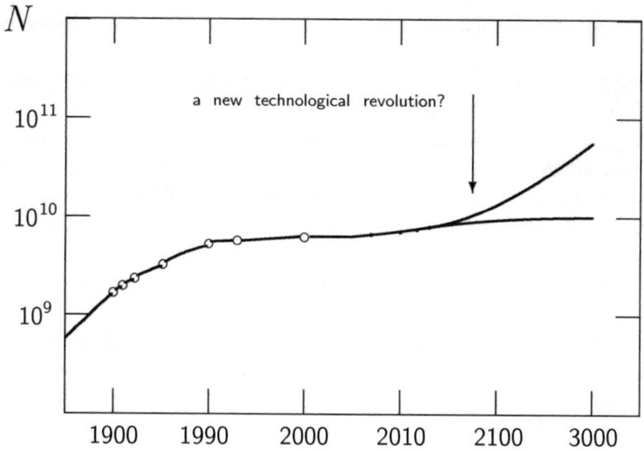


**Fig. 12.10** Growth of the population on the Earth: the future. The lower curve till the year 2100 corresponds to the known estimates [32]

of energy should be discussed in common. The numbers of the population on the Earth, according to the previous estimates in Sect. 12.2.1 and the described scripts of growth of available energy, are presented in Fig. 12.10.

In the case of constraint of availability of energy, number of the human population and global output will come close to constant values, and a case of the stable steady state (sustainable development) will be realised. In another case, when the nuclear energy and other energy sources will be widely used, growth will not stop and will have unpredictable consequences. However, it can appear that lack of available energy will bound arrogant projects of people.

## References

1. Masson, V.M.: Economica i sotzialnyi ctroi drevnikh obtshectv v svete dannykc arkheologii (Economy and Structure of Ancient Societies in a View of Archeological Data). Nauka, Leningrad (1976)
2. Maddison, A.: The World Economy: Volume 1: A Millennial Perspective and Volume 2: Historical Statistics. Appendix B. OECD Publishing (2006). http://www.ggdc.net/MADDISON/oriindex.htm. Accessed 17 Aug 2017
3. Lo, Cascio E., Malanima, P.: GDP in pre-modern agrarian economies (1–1820 A.D.): A revision of the estimates. Rivista di Storia Economica **25**, 391–419 (2009)
4. Wallon, H.: Histoire de l'esclavage dans l'antiquité. Deuxiéme édition. Paris, Tome premier (1879)
5. Berna, F., Goldberg, P., Howitzer, L.K., Brink, J., Holt, S., Bamford, M., Chazzan, M.: Microstratigraphic evidence of in situ fire in the Acheulean strata of Wonderwerk Cave, Northern Cape province, South Africa. Proc. Natl. Acad. Sci. U.S.A. **109**(20), E1215–20 (2012)
6. BP Statistical Review of World Energy. http://www.bp.com/statisticalreview. Accessed 17 Aug 2017

7. Smil, V.: Energy Transitions: History, Requirements and Prospects. Praeger, Santa Barbara (2010)
8. Beaudreau, B.C.: Energy and Organization: Growth and Distribution Reexamined, 2nd edn. Greenwood Press, New York (2008)
9. Fogel, R.W., Costa, D.L.: A theory of technophysio evolution, with some implications for forecasting population, health care costs, and pension costs. Demography **34**, 49–66 (1997)
10. Lotka, A.J.: Elements of Physical Biology. Williams and Wilkins, Baltimore (1925)
11. Volterra, V.: Lesons sur la mathematique de la lutte pour la vie. Marcel Brelot, Paris (1931)
12. Murray, J.D.: Mathematical Biology. Springer, Berlin (1989)
13. Pokrovski, V.N.: Physical Principles in the Theory of Economic Growth. Ashgate Publishing, Aldershot (1999)
14. Leakey, R.E.: The Making of Mankind. Elsevier-Dutton Publishing, New York (1981)
15. Mithen, S., Reed, M.: Stepping out: a computer simulation of hominid dispersal from Africa. J. Human Evol. **43**, 433–462 (2002)
16. Carr-Saunders, A.M.: World Population: Past Growth and Present Trends. Oxford University Press, London (1936)
17. Clark, C.: Population Growth and Land Use. Macmillan, London (1968)
18. Durand, J.D.: Historical estimates of World population: an evaluation. Popul. Dev. Rev. **3**(3), 253–269 (1977)
19. Shklovskiy I.S.: Vselennaya, zhizn, razum (The Universe, Life, Mind), 6th Enlarged Ed. In: Kardashev, N.S., Moroz, V.I. (eds.) Nauka: Gl. red. fiz.-mat. lit., Moscow (1987)
20. Maltus, T.R.: An Essay on the Principles of Population, as it Affects the Future Improvement of Society. J. Johnson, London (1798)
21. Kapitsa, S.P.: The phenomenological theory of world population growth. Physics-Uspekhi **39**(1), 57–71 (1996)
22. Verhulst, P.F.: Notice sur la loi que la population suit dans son accroissement. Corr. Math. et Phys. **10**, 113–121 (1838)
23. Pearl, R.: The growth of population. Quart. Rev. Biol. **2**, 532–548 (1927)
24. Sauvy, A.: General Theory of Population. Translated from the French edition (Paris, 1966) by Christophe Campos. Basic Books, New York (1969)
25. Tsirel S.V.: O fenomenologicheskoy teorii rosta naseleniya Zemli S.P. Kapitsy (On the phenomenological theory of population growth on the Earth by S.P. Kapitsa). Demoskop Weekly, pp. 139–140 (2003). http://www.demoscope.ru/weekly/2003/0139/analit02.php. Accessed 17 Aug 2017
26. Shishkov, YuV: Demograficheskie pokhozhdeniya fizika (The demographic adventures of a physicist). Obshhestvennye nauki i sovremennost **2**, 156–161 (2005). http://www.avmol51.narod.ru/Shishkov/d.htm. Accessed 17 Aug 2017
27. Korotaev A.V., Malkov A.S., Khalturina D.A.: Kompaktnaya matematicheskaya makromodel tekhniko-ekonomicheskogo i demograficheskogo razvitiya Mir-Sistemy (1-1973 gg.) (A compact mathematical macromodel of technical, economic and demographic development of the World-system (1-1973)). In: Malkov, S.Yu., Korotaev, A.V. (eds.) Istoriya i sinergetika: Matematicheskoe modelirovanie sotsialnoy dinamiki (History and Synergetrics: Mathematical Modelling of Social Dynamics), pp. 6–48. URSS: LIBROKOM, Moscow (2005)
28. Dolgonosov, B.M.: Nelineynaya dinamika ekologicheskikh i gidrologicheskikh protsessov (Nonlinear Dynamics of Ecological and Hydrological Processes). URSS: LIBROKOM, Moscow (2009)
29. DeLong J.B.: Estimating World GDP: One Million B.C. - Present (1999). http://www.j-bradford-delong.net/TCEH/2000/World_GDP/Estimating_World_GDP.html. Accessed 17 May 2017
30. Pietro, Basso: Modern Times, Ancient Hours: Working Lives in the Twenty-First Century. Verso, Paris (2003)
31. Akaev, A.A.: Ot epokhi velikoy divergentsii k epokhe velikoy konvergentsii: Matematicheskoe modelirovanie i prognozirovanie dolgosrochnogo tekhnologicheskogo i ekonomicheskogo razvitiya mirovoy dinamiki (From the Epoch of Great Divergence to the Epoch of

Great Convergence: Mathematical Modelling and Forecasting of Long-Term Technological and Technologic Progress of the Globe Dynamics). URSS: LENAND, Moscow (2014)
32. Belyaev, L.S., Marchenko, O.V., Filippov, S.P., Solomin, S.V., Stepanova, T.B., Kokorin, A.L.: World Energy and Transition to Sustainable Development. Kluwer Academic Publishers, Dordrecht (2002)
33. Odum, H.T.: Environmental Accounting. Emergy and environmental decision making. Wiley, New York (1996)
34. Plakitkin, J.A.: Mirovoe razvitie i zakonomernosti globalnoy energetiki. (The international development and global patterns of energy). Vestnik Rossiiskoi Akademii Estestvennykh Nauk **3**, 3–10 (2012)

# Chapter 13
# Principles of Organization of the National Economy

**Abstract** The organisation of production assumes the coordinated activity of huge groups of people, so that the modern (however, since prehistoric times) production has social character. On the other hand, consumption of created products has individual character and, consequently, there appears a problem of distribution: how to cut a social pie in such a way that everyone was happy. Laws of production, as they are described in the previous chapters, are identical for all countries and nations, the distinction of social systems is essentially connected with rules of distribution of the social product. These rules are formulated on the basis of theory that shows what production factors, in what degree, take part in the creation of the product. The formulated rules then are reformulated in laws, declared by the parliament and recorded in the constitution: rules of distribution of a social product are legitimised, and the life of the society (the social relations, including the production ones) will be organised in such a way to guarantee obedience to the recorded rules.

## 13.1 Production and Social Relations

The production processes transform the various substances and objects to the forms directly used by individuals, and, consequently, the analysis of production activity is impossible without taking into account the natural (ores, air, water...) and artificial (buildings, the industrial equipment...) environments that are an important *social resource*. The transformation of substances and rearrangement of objects appears possible only when humans' efforts (see Sect. 2.4.2) and substitutive work (see Sect. 2.5.5), which are also necessary to be considered as important *social resources*, are attracted to production. This activity brings to existence sets of *products* (the social product) that have value $Y$ that is connected in the universal way with an effective utilisation of the specified resources and can be presented in the following form (see the Eq. (11.15))

$$Y = \beta L + \gamma P. \tag{13.1}$$

The expenditures of human efforts $L$ and substitutive work $P$ are called production factors, which are true sources of value; the foundation of this statement can be



V. N. Pokrovskii, *Econodynamics*, New Economic Windows,
https://doi.org/10.1007/978-3-319-72074-6_13

found in the previous chapters. The marginal productivities $\beta$ and $\gamma$ are increasing functions of the ratio $P/L$; their properties are described in Sect. 6.3.3. The quantities of potential growth of human efforts and substitutive work are always in the centre of attention of a human society. The development of production system is determined, finally, by possibilities of attraction of workers and additional amount of external energy (see Sect. 11.1.3).

### *13.1.1 Social Resources*

The human being, as a main production resource, represents himself as a manager of each possible production projects. In the simplest case, a single individual appears to be an initiator, coordinator, executor and beneficiary of the project. When more complex projects are organised, the various functions are carried out, generally speaking, by various individuals; there appear some social production relationship of people. There exists some distribution of people according to their roles in production; it is convenient for the analysis to break all participants of production into two classes of people: businessmen (employers) and executors (hired workers). The relationship of these two classes was the object of special interest of Karl Marx and its followers. Certainly, now we refuse to accept the extreme concepts about businessmen (employers-capitalists) as parasites on a body of workers and should consider, following Schumpeter [1], their key role in production, as organisers and supervisers of the production processes.

The employer hires workers (executors) for a certain time (per hour, per day or for a year). The worker belongs to the employer only per certain operating time, during the rest of the time the person is left to himself. He can eat, spend time with family, care about health, both his and his relatives'. The situation can be described in such a manner that the employer rents an executor for certain time for certain payment, and all the other does not interest him [2]. The hired workers carry out the duties ascribed by instructions, and, for his action, the employer allocates a fraction of the created product to the workers in the form of wages or salaries.

The businessman (employer), as the organiser of the entire enterprise gets (or leases) production equipment—a set of appliances, which allows including production workers' efforts and substitutive work that are true sources of value. The installed equipment appears as a result of efforts of many generations of people and, in fact, is a public property, though it is used by employers, who are considered as proprietors of the production equipment. In fact, it is necessary to consider, that the production equipment is in everlasting rent by the businessman-employer, but the equipment to run, the businessman-employer must attract and pay off social resources: workers and substitutive work.

A value assessment of the production equipment is called basic production capital, and the capacity to increase value of the created products is attributed to the capital. However, not the capital, but capability of the production equipment to use working force and to replace efforts of humans with natural energy in the production of services and things, is the real cause of the increasing value. This capacity is the main property of the fixed capital (see Sect. 7.2.3).

In the widespread understanding, the capital is everything that provides profit. If one has production shares, he gets dividends, if one's money lay in bank, the person receives percent. The shares and money are capital in wider understanding. However, money and the shares are only symbols that bring nothing without huge work on production of value within the framework of the capitalist organisation of the national economy. The mystical force of the capital brings profit results from the rules of distribution of the social product created by workers and substitutive work. Only efforts of humans (in view of a substitution effect as we know from the previous chapters) lead to increase of the capital.

### *13.1.2 Inevitability of the Central Supervision*

The Gross Domestic Product as the result of social production activity, is a value assessment of the created products, which differ from each other by a number of characteristics, including the period that proceeds from consumption of a product till the occurrence of appreciable effect from its usage. The set of the created products includes the products bringing immediate satisfaction to a person (a feed, clothes, heating...), so the products that will be useful in production in a few years (investments into production), and products which will bring effect probably in tens and hundreds of years (education, the science, protection of borders and so forth). The last group includes products which, in the last decades, are discussed as the investment into 'the human capital' and the scientific and technical progress. All the listed groups of products are necessary for reproduction and progress of the human society (see also Sect. 2.3).

The businessmen are organisers of the production processes. They plan the business on their own risk, whereas the greater the time of getting the effect, the more the risk and the less the businessman's desire to be engaged in those projects. It leads to the fact that entrepreneurial business concentrates in the activity of the organisation of production with the least, as possible, waiting time for the effect. Businessmen, as a rule, have short-term targets, disregarding the projects that are not bringing immediate benefit. Thus, a care of some set of social needs is compelled to take up by the central body (the government, the state), because it not on the forces to the separate businessmen or because it does not bring them any immediate income. This set includes power supply, transport, communication, protection against epidemics and natural disasters, roads, post service, education, information service, internal and external security, social insurance, care of old men and invalids, as well as cares of defence of borders and maintenance of the areas of habitat. In total, the modern national state monitors the general situation and are organising the national projects, creating the favourable conditions for people to live. Besides, the central body is compelled to interfere in the relationships between people on the basis of vague and not always adequate concepts about justice.

The central body, to activate and perform the national projects, needs resources and for this purpose, in this or that way, the fraction of the social product created by workers' efforts and substitutive work should be allocated. The centre accepts the duties and gets the appropriate rights. Certainly, the government can have its own short-term or long-term production projects, even it should have them in the areas concerning all national economy, but the main duty of the centre in economic area is the maintenance of social welfare and influence on economic processes to create favourable conditions for activity and development.

## 13.2 Distribution of the Social Product

Thus, in the elementary approximation, it is necessary to consider the production–distribution relations between the three aggregate economic subjects: group of businessmen (employers), group of hired workers (executors) and the central body (government) that, on definition, represents all national interests. Each of the economic subjects takes part in the social production and aspires to increase its share of the limited social product. Economic subjects are compelled to cooperate with each other to establish the to-all-understandable distribution.

### *13.2.1 Fundamental Principle of Distribution*

There is an amazing harmony in declaration of the principle of distribution: all investigators—and Marx [3], and Clark [4], and many others adhere to consensus—the distribution should be fair. To each participant—a share of the product (reward), fitting his contribution to the creation of this product. The fair, natural, as Clark emphasises, principle of distribution is proclaimed, and now remains only to establish contribution of each participant to creation of the product to set the rules and to write appropriate laws.

Though everyone agrees that distribution should be done on fairness, the rules of distribution are formulated variously, depending on concepts of the investigator. So, we know, that Marx and its followers believed that the total produced value is created by the efforts of workers and, consequently, all social products belong to those who specifically work physically. But, probably, in such direct form these concepts are already sent to historical legends, so as production of value cannot be explained only by expenditures of labour, and, by the way, the role of businessmen should be appreciated.

The used in the 'market' countries 'fair' rules of distribution of the social product have been stated in the end of the nineteenth century on the base of the concept about productive force of capital [4, 5]. Clark in his work was based on the statement, that production is determined by the four basic factors: the capital money, the capital-means of production, entrepreneur capacities and efforts of hired workers. The social product, Clark stated, is distributed according to productivity of the production factors: the owner of the monetary capital receives interest, the owner of the production capital—the rent, the businessman—enterprise profit, and the hired worker— wage.

Clark explains the rules, on which the social product can be divided into unequal but 'fair' shares. There is no indispensability to discuss details of rules of distribution: the concept that the capital (in the widest sense, including money and assessments of property) possesses productive force, is a myth, probably a sincere error, or a mystification, that was thought up for the justification 'fair' distributions, at which the lion's share gets the owners of 'capital'.

From the point of view of the theory of real production, the owners of the paper and monetary capital cannot participate in the primary sharing of a social product (though they can obtain compensation for services later) and consequently, we begin discussion with consideration of a simple situation, when the social product is distributed between employers and employees, which was a subject of special interest of Karl Marx, who has originated concepts surplus product and exploitation of hired workers by employers.

### *13.2.2 Surplus Product*

The expected result of production activity appears as a consequence of the usage of production factors: expenditures of labour $L$ and substitutive work $P$. Naturally, expenses on restoration of production factors, that is workforce and the productive energy should be compensated from the output, if, certainly, the organisers of production do not set up an aim to corrupt the social production system. The businessman pays operational expenditures: to the hired worker—in the form of wages $w$; compensation of cost of the depleted equipment is interpreted as a payment for use of substitutive work $p$ (see the discussion and expression (2.35) in Sect. 2.5.3). When the current production expenditures are subtracted from the social product (13.1), there remains still some quantity named (according to Marx) the surplus product

$$\Delta Y = (\beta - w)L + (\gamma - p)P. \tag{13.2}$$

The proprietor-tenant of the production equipment keeps traditionally in possession the surplus product, and it is necessary to agree with it, as the businessman is an organiser and coordinator of production processes. However, the businessman tries to maximise the profit and, accordingly, is compelled to minimise the expenses, so that salary of a working person $w$ for the specifically performed work, as a rule, does not compensate all expenses (education, healthcare...) on restoration of the human population and even, as it is spoken, the workforce. In a similar way, the operational expenditure of the businessman on maintenance of the equipment $p$ does not compensate the social expenses on research and design of the appliances which allow usage of energy in creative aims.[1] The investments in demographic and research projects give effects over the tens years and, as a rule, do not attract private business. These circumstances confirm an indispensability of existence of the centralised social fund for organisation of national projects.

The expression (13.2) defines the fraction of a social product, which could be received by businessmen (employers), if they did not pay taxes. Nevertheless, there is an agreement that the centre, anyhow, represents national interests and to transfer some money into the social fund is necessary. Businessmen oppose, but are compelled to agree with withdrawal of a part of profit[2] and further we consider rules of withdrawal of a part of the created surplus product and formation of social fund.

### *13.2.3 Conventional Scheme of Taxation*

On the established tradition, the taxes represent (to say nothing of the details) deductions from profit of the enterprises and the income of separate persons [8, 9]. In terms of Eq. (13.2), some fraction of surplus value $\Delta Y$ and some fraction of wages of hired workers $w$ is deducted into the social fund, so that the distribution of the social product between businessmen, hired workers and the government looks like

$$\begin{aligned} \Pi &= [\beta - (1+\theta_{\mathrm{L}})w]L + (\gamma - p)P - \theta \Delta Y, \\ N &= (1-\theta_w)wL, \\ G &= (\theta_w + \theta_{\mathrm{L}})wL + \theta \Delta Y, \end{aligned} \tag{13.3}$$

where $\theta$ is a norm of deductions from profit of the enterprise, $\theta_{\mathrm{L}}$ is a norm of the social tax (paid by the enterprises), $\theta_w$ is a norm of the tax on the income of hired workers. There is also an important payment for the production usage of the natural environment, as one of the major social resources.

There are also taxes of other types, which have auxiliary character and bring the smaller contributions to social fund, but, eventually, they can be reduced to deductions from incomes and surplus value.

In fact, taxes represent the necessary payment for possibility of usage of social resources (the workforce, productive energy and the natural environment) by the proprietor-tenant. This payment provides expenses for education, professional training, healthcare and so on, i.e provides restoration of the population and methods of production. The payment provides expenses for the searches of new energy sources,

---

[1] Such expenses are compensated partially by payment for use of inventions and know-how in production through the various patent and licence schemes.

[2] Researchers of the problem of the property rights [6, 7] state, that the ownership of the production equipment does not assume the ownership of a productive opportunity and the created product. David Ellerman [7] names as 'a fundamental myth' the widespread belief, that the proprietor of the enterprise has the right to appropriate the product created by this enterprise.

research of a possibility of use of energy in production aims. Anyhow, two production factors: work of humans and capacity of energy to participate in production have to be supported by the central body.

The conventional scheme of taxation according to the income and profits does not look fair and causes some questions: why the money that I have earned by fair work, belongs not only me, but also to that man who did nothing. The progressive scale of taxes is unfair: a person, who earns more, has to pay more. The businessmen try to avoid taxes: there are some offshore schemes, the shadow enterprises, black salaries and other inventions. The right to withdrawal of a part of a product is challenged; the businessmen are convinced, that they have earned the profit fairly. The existing mechanism of distribution of the social product appears imperfect, though, in fact, it is admitted that the surplus product belongs not only to the proprietor-tenant of the means of production, but also to the entire society.

### 13.2.4 Optional Scheme of Taxation

Is it possible to propose a better way of formation of the social fund for organisation of the economic life of the society? Readers of the previous chapters of the book had a possibility to make sure, that only workers' efforts and substitutive work create value; the businessman rents workers and equipment to use efforts of the workers and outer energy in production. There is no other source, apart from the surplus product for the used social resources to be paid off in view of their expanded reproduction. It would be fair to establish deductions from the surplus product according to the use of workers' efforts and substitutive work, and, thus, the distribution of the social product between businessmen, hired workers and the government, accordingly, should look like

$$
\begin{aligned}
\Pi &= (\beta - w - d_{\mathrm{L}})L + (\gamma - p - d_{\mathrm{P}})P, \qquad (13.4)\\
N &= wL,\\
G &= d_{\mathrm{L}}L + d_{\mathrm{P}}\,P,
\end{aligned}
$$

where $d_{\mathrm{L}}$ is a norm of payment for use of work force (the social tax paid by the enterprises), $d_{\mathrm{P}}$ is a norm of payment for use of substitutive work. The quantities $d_{\mathrm{L}}$ and $d_{\mathrm{P}}$ are limited: on the one side, the income of businessmen $\Pi$ should be positive in reasonable amount, so norms cannot be too great; on the other side, payments should compensate an exhaustion of social resources and, on possibility, provide their development, so that the quantities $d_{\mathrm{L}}$ and $d_{\mathrm{P}}$ cannot be too small.

Apart from the payments for the use of work force and substitutive work, there should also be an important payment for the production usage of the natural environment, as one of the major social resources. The method of calculation of the prices of natural products are discussed in Sect. 2.6.

The relationship between the norms of payment per units of expenditures of labour and substitutive work $d_L$ and $d_P$ are being established according to average values of the quantities: the marginal productivities $\beta$ and $\gamma$, the direct compensation for the factors of production $w$ and $p$, and the fraction of the income of businessmen in the total final output $\pi = \Pi/Y$, and are calculated according to the relations

$$d_L + w = \beta(1 - \pi), \quad d_P + p = \gamma(1 - \pi). \tag{13.5}$$

The proposed scheme of loading of the social purse by payment for the use of social resources (each production enterprise of any ownership: state, private or cooperative) conforms to the principle of fairness, and, besides, it can appear convenient in the practical attitude, so as to supervise workers and substitutive work is easier, than to track the profits and incomes. There appears to be disclosed a wonderful property of distribution (13.4) to stimulate increase of efficiency of the use of social resources. The matter is that the marginal productivities of a separate enterprise, $\tilde{\beta}$ and $\tilde{\gamma}$, differ, certainly, from their average values over the entire system. In the case when the local values of marginal productivities exceed their average values, the owner of the enterprise receives additional profit—*entrepreneur premium*, and in this case, according to Schumpeter [1], we can name him a neo-entrepreneur. The enterprise gets advantage to a survival. In the case when local values of marginal productivities appears to be less than their average values of the entire system, the owner of the enterprise will feel difficulties and should take measures to improve the production processes, for example, to replace manual work by mechanical one (this increases the ratio $P/L$ and, in the turn, increases the marginal productivities, as it is seen in Sect. 6.3.3). Otherwise, the enterprise will be eliminated. Thus, distribution (13.4) without any additional slogans stimulates an increase of efficiency of the use of the social resources. Let us notice, that distribution (13.3) does not possess such property. Moreover, the conventional taxation oppresses production activity and stimulates concealment of profits and incomes.

## 13.3 Capitalism as Economic System

The capitalist system is based on the rules of distribution, according to which the social surplus product is considered to be a property of owners of 'capital', under which is understood as the direct means of production (the physical capital), so various certifications on the property rights (money, shares and other papers). The owners of 'capital' try to increase the surplus value in any way, that leads to a polarisation of a society [10]. Tools of redistribution of the surplus value in a capitalist society, such as shares and taxes, are adjusted on supporting of the existing rules of distribution [11].

### 13.3.1 The Tendencies of Capitalistic Production

The capitalist system of economy is very distant from the splendid picture representing an empire of fairness and freedom with the mythical 'invisible hand', which directs processes of production towards satisfaction of wishes of people-consumers of products. The essence of a capitalist way of production was described by Marx in Chap. XV, third part, third volume of 'Capital': '... production of the surplus value... is the direct objective and crucial motive of capitalist production. Therefore, it is impossible to represent capitalist production by what it is not actually, such production that has an objective consumption or manufacturing of consumer goods...'. Since Marx's times the principles of capitalist system have not changed, as, for example, a known businessman and a public person Soros testifies ([12], Chap. 6).

The primary activity of a businessman in the capitalist system is the pursuit of the surplus product, which is got in the process of production and this compels the businessman to stimulate demand. The businessmen operate by the principle: trust the 'invisible hand' but keep your powder dry. They encourage consumption in any way, unnecessary and even harmful products are imposed. The society has turned into a cooperation of consumers focusing on consumption not only fundamental products (food, clothes and shelter), but also on consumption of what is super necessary. There are parasitic enterprises, the enterprises serving to satisfy the super demands of the upper layers of the society, and so on. The capitalist system objectively encourages the actions and production, making huge profit, even they are causing harm doubtlessly. Any absurd statements, like that wars in general are useful, military expenses lead to growth of general welfare, are propagating.

The pursuit of profit are forcing businessmen to expand production and to encourage consumption, to struggle for commodity markets, so that capitalism trends to propagate the production—distribution relationship as widely as possible [13]. The trends of the capitalist system to expansion lead to the borders between the national states to be washed away, and the production system to cover over the Globe, forming the global production system including interethnic corporations, enterprises, banks... The globalisation of economy is supported by the non-economic ways down to application of force.

### 13.3.2 The World Financial System

The way of extraction of the surplus value at production activity (the productive capitalism) is supplemented with a powerful method of extraction of profit directly in the process of money circulation (the usurious capitalism, perfectly described by Katasonov [14]). To realise this way of getting profit, there is no indispensability to have or rent the expensive production equipment, it is enough to have the possibility to supply money on loan. The grateful borrower should return money with percent: money generate money, money has been turning to financial capital.

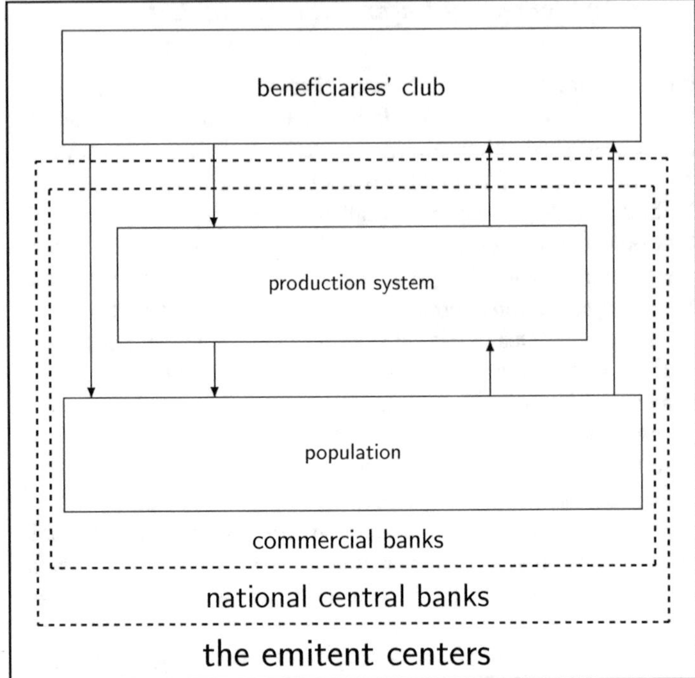


**Fig. 13.1** The scheme of the global monetary system. The international organisations, national central banks and commercial banks form the monetary environment for activity of economic subjects. The production system is presented by the international corporations and national producers. The national production–distribution systems are not closed and cooperate with each other and with the international organisations. It is supposed, that there is a group of organisers of the system—the club of beneficiaries

The modern financial system covers all the World, and it looks as it rises over the real production, though the financial sector has no sense of its existence without the sector of real production. The scheme of the global financial system as it can be presented, according to the descriptions of Katasonov [14] and Yakunin with co-authors [15], is shown on Fig. 13.1. The role of the informal over-national government plays the group of beneficiaries, who are presented by such political structures as Bilderberg group, Advice on International Relations, Trilateral Commission and by the supported financial organisations: the International Monetary Fund, Bank for International Settlements (BIS) in Basel, the World Bank and the World Trade Organization. The role of the Central World bank is playing by the Federal Reserve System in the United States of America which issues out world money in the form of the 'U.S. dollars'. The listed top-level organisations cooperate with the regional and national central banks that are emitting 'national' money which in that or this way are compelled to be adhered to dollar. At the very bottom of bank hierarchy, there is a huge quantity of commercial banks, cooperating directly with producers and consumers. There appears a huge quantity of non-cash credit money on this level. Katasonov ([14], p. 171) assimilates the global financial hierarchy to the church organisation with the priests-bankers and considers faith in mystical force of money as a cult of money.

The system was created for the organisation of streams of money (value) from the bottom upwards; it cannot exist without the rigid discipline and management (anyway, signs of operating processes are noticeable, as Yakunin with co-authors [15] have noted). The functioning of the system is supported by the various means, down to application of military force. The centres of disobedience are being brutally suppressed [16].

### 13.3.3 What a Capitalist Makes Money for?

Capitalism, or as now it is accepted to speak, 'market economy' is developing spontaneously, and, probably, it is wrong to ask, for what purposes it evolves. But the changes are being initiated by separate persons, and motivations of owners of 'capital' appear to dominate; therefore, their intentions are interesting to understand. We shall follow Maxim Gorky, who still in the beginning of the last century was curious to ask an imaginary millionaire, what he does make his money for [17]. The millionaire '... has raised a little his shoulders, his eyes move slightly in the orbits, and he answered: - I am making other money by them. - What for? - To make more money... - What for? - I have repeated. He has bent to me, resting the elbows on the handles of his armchair, and has asked with a tone of some curiosity: - Are you a madman?'

Nevertheless to the end of the appointment, there appears a phantom of an answer the question set by Maxim Gorky: ' - And I wander, do you have spare kings in Europe? - he has asked slowly. - It seems to me, all of them are spare ones! - I have answered. He has spitted to the right and told: - I think to employ a pair of good kings for myself, ah? - For what do you need them? - Funny, you know. I would order them to fight here... He has pointed a platform before the house and has added by a tone of a question: - From one hour up to half past one every day, ah? After a breakfast, it is amusing to bestow half an hour to the art... Well.'

And so, that is the matter! Money gives a satisfaction to all desires, whatever ludicrous they would not be. Money are unlimited possibilities: let us imagine, what pleasure it is to sit on a gold toilet bowl, or to depart from the distant Siberia on his own airplane to a winter resort, to France in a close circle of 20 beauties. Money is the power [18]: let us think, what pleasure is to reshuffle the government or to organise a fight not only between two spare kings, but also a battle between various states. Oh you, the poor creatures!

### *13.3.4 Labour Productivity and 'Unused' People*

The capitalist production system is based on the mechanised work, on attraction of huge quantity of external energy to the production. So, for example, in the USA, as it was already marked in Sect. 11.3, the more than ten mechanical workers is acting per an alive worker. The freedom of business and personal initiative are assuring the progress of social productive forces, introduction of innovations and growth of labour productivity. This is what the capitalist system is fairly proud of ([19], part I).

As productivity of work increases and no new workplaces appear, some people, who earlier performed the entire work, now, being replaced with work of mechanisms, appear to be free. There appears a tendency of increasing the number of people, who do not work (or work insufficiently), but, nevertheless, as a members of the society, obtain the public welfare. The number of unemployed and the rentiers increases, as well as the number of unused people. The growth of labour productivity objectively brings to existence the 'spare' people.

Even under more uniform distribution of labour efforts, any simplification of work, reduction of duration of the working day, there remains the question: what to do with 'extra' people who do not work but only eat. From the point of view of logic of the capitalist relations, there is no place for the 'extra' people on the planet. As the invented-by-Gorky millionaire has perfectly stated: '... In the country there should be so many people, to buy everything that I wish to sell. There should be so many workers that I did not require them. But - no one extra!'

How many people, thus, should live on the Globe? Possibilities of the modern production determine, that it is enough to have one billion individuals to make and consume everything that is made. Certainly, not all people from this 'gold billion' will appear to be equal: the best people will receive privileges and benefit from the public welfare, and the others will have a possibility to work. As for Russia is concerned, it is counted that enough to have, well... thirty millions person to ensure necessary for the 'gold billion' fuel. Also it is not necessary to hesitate in application of methods of optimization of a population: all means—wars, epidemics, sterilisation—are good. There, where one worships money, the price of a human life appears insignificant. Business is business, nothing more.

## 13.4 How Socialism was Being Built

Though the capitalism organisation of the national economies has widely extended on the Globe, it is not the best and unique one, and from time to time there are plans and attempts of realisation of other rules of distribution of the social product [16, 20]. Unlike the capitalistic, alternative approaches assume, that the surplus product is a national property.

Over a long time it was believed, that it is necessary to expel money, private property and elements of the market to reach the fair distribution of social product.

As it is known, a practical realisation of the communist project have started in Russia after the Great October Revolution of 1917. The objectives of the project were the general welfare, a society without exploiters, a society of universal equality. The analysis of movement to the objectives and the description of reached results can be found in numerous publications, from some of which [16, 21, 22] we extract the further statements.

### 13.4.1 Money, Budget and Financial System

During the revolutionary euphoria of 1918, they tried to eliminate money, but eventually had refused from this idea; money had become an effective tool of control and management of the national economy. The financial system of the Soviet Union completely was in the hands of the state; the social fund was formed, mainly, due to the tax on the turnover of the production enterprises. There were two systems of prices (wholesale and retail, see Sect. 10.5.2) for each product, and independently two types of money were circulating in the system: cash money—for households and credit money—for production enterprises ([12, Chap. 12], [22]). The financial system of the Soviet Union was crashed in 1988, when a law 'About the state enterprise (association)' had appeared, and free transformation of credit money into cash was allowed by the law. The system was made open to the external world (the monopoly of foreign trade was cancelled), after that the financial system cooperated with the World financial system that has immediately started the organisation of operations on extraction of created in Russia value [14].

### 13.4.2 Personal and Private Ownership

Over the centuries the private ownership on the means of production was considered as an absolute evil by all progressive thinkers. The famous 'Manifest of Communist Party' by Marx and Engels contains the programme requirement of elimination of a private ownership, so that one of the first actions of the Bolsheviks after revolution of 1917 was elimination of the right of private ownership on the means of production. But the right to have personal property was remained, the private production activity was allowed, it was authorised to have a horse, the tools and a personal plot. The borders between the private and personal property remain unclear.

The nationalisation of the basic production assets, on the assumption, should lead to harmonisation of the social production and individual consumption. But the actual situation in the country has appeared essentially distinct from the expected one: the underground (shadow) private activity, despite of huge risks for businessmen (down to execution), flourished under 'the developed socialism', and the state could not defend against it [23]. The reform of the year 1965 has provided some independence in decision-making and allowed the state enterprises to allocate the profit. The reform validates the right to consider profit as a business property, and this was the beginning of process of gradual return to the principle of private ownership, which has ended with the total legitimising of private ownership in the year 1990.

### *13.4.3 Planning and Control*

The nationalisation is only the first step on transformation of a private property into social one. For the successful functioning of the national economy, which, as a national property, should be operated as a united system, similar to a huge factory, it was necessary to organise an efficient management. For these aims, the central state body of the USSR (Gosplan) has been created in year 1921 on the basis of commission GOELRO. The Gosplan of the USSR worked out the plans of development firstly for a year, and since 1925 for 5 years; it existed till the year 1991.

The directive planning in the USSR prescribe the output of a product and its price that encouraged the enterprise to lower the net cost of products. This mechanism worked until the reform of 1965 which legitimises self-financing of the enterprises and, in fact, has destroyed planning and has made a national economy uncontrollable by the previous methods. With the aims to overcome uncontrollability, by the end of an epoch of 'the developed socialism' there appear many industrial ministries, but the management was not effective enough (see details in the Chap. 8).

With an aim to overcome the crisis of management, the theory of optimum performance of socialist economy was being developed by Soviet economists (see, for example, [24]). With the advent of high-speed computers and cybernetics, there were hopes to defeat crisis of management, however, the efforts of economists have been directed to creation of market methods of management, and attempt to design of the nationwide automated centralised control system has not crowned success [25]. Not denying an indispensability of use of computers for the purposes of indicative planning, nevertheless, we shall notice, that such projects cannot totally resolve the problem: it is impossible to supervise specific activity of millions of persons, and it is not necessary to try to do it. Under the attempts to establish the control, the controlled ones invent roundabout ways. For example, in the last years of the Soviet Union, instructions have been issued for use of the copiers: each work ought to be registered in a special book, and the sample of the work has to be remained. As a result, the cunning people disconnected blocking and duplicated everything what they wish. To foresee all the nuances of behaviour of a person appears to be impossible.

### *13.4.4 The Reforms and the Results*

Over the all history of the Soviet Union, the contradictions between a social production and personal consumption, between discipline of directive planning and personal free entrepreneurship were remained. These contradictions existed objectively and did not find the due solution. The restrictions on the initiative activity, the excessive control and planning of production and distribution did not allow to prosper talents and initiative of people, became a brake of progress. The repeatedly undertaken attempts to improve performance of the system had, as a sample, the capitalistic organisation of production, and it is natural, that, eventually, the society has returned to the 'sacred' law of private ownership on the means of production and to the 'sacred' rule of appropriation of surplus value by the owners of 'capital' with all consequences of the act.

Studying of the results of reforms does not leave doubts, that ill luck came to Russia in the period of reorganisation: the national economy has degraded and the quality of life has worsened, which in detail is described and analysed in numerous publications [26–30]. The researcher of the Russian national economy V.M.Simchera[3] dryly summarises in Sect. 11.5 of his book [27]: 'The country, instead of the persistently outlined doubling of the growth for 1986–2000 and not less than one and a half-fold growth for 1990–2000, had suddenly fallen down practically in all set of leading parameters. As a result, instead of achievement the leading world indexes by the end of XX century, which would exceed the values of the majority of key parameters for the beginning of the XX century in many times, the country, due to the unsuccessfully conducted reforms, has lowered down to the minimal marks, three times below possible, having lost practically everything, that the whole century has hardly saved up and carefully protected. The standard of living in the country in these conditions, naturally, could not rise. On the contrary, it, as never earlier in the XX century, has quickly gone down, hardly attaining marks of poor 60th years of the last century'.

However, the inspirers and executors of reforms [22, 31] estimate the results differently. So, for example, E.G.Yasin[4] does not lose optimism and in the final section of his work [22] writes: 'At the first stage the minimum is made – the market economy, instead of the planned one, is created. Now it is necessary to make a following step: to do the economy effective... ' and in the foreword to Gaydar and Chubays's book [31] adds: 'The flourishing of the Russian market economy is still ahead, I trust in it.'

## 13.5 The First Step to Effective Production and Fair Future

The activity on production and distribution of the social product is obeying to certain laws, which are understood usually as some parities between various quantities. As an example of the economic law, one can refer to the law of production of value (13.1), establishing a relationship between output and the used social resources.

[3]Vasily Mihajlovich Simchera for many years (2000–2012) supervised over Scientific Research Institute of Statistics of Federal Service of the State Statistics.

[4]From the information [22] 'About the author': 'In the years of reforms in Russia, E.G.Yasin was one of the basic developers of some governmental and independent programs of social and economic progress of the country, including the first full-scale system of measures of consecutive transformation of the planned economy into a market one.'

This relationship is determined by the existing state of the production system, but not wishes of people. But, apart from the laws, the economic life of a society is determined by rules that are established by people, for example, by the rules of distribution of the created social product (13.3) or (13.4).

As for another example of economic rule, one can refer to the rule, according to which the social product belongs to the proprietor of the production equipment. During the centuries, this rule has been seeming to be fundamental and solid, that is why the question about the ownership of the means of production appeared to be the main issue of all social transformations. In the year 1917 in Russia, the Bolsheviks struggled for the means of production to have been nationalised; the meaning of transformation of the ninetieth year of the last century in Russia—the struggle for transfer of the means of production to private hands. As in one, so in the other case, the change of the social situation has not led to stable social relations, and it forces to give a thought on the reasons for the existing problem. It is possible to think, that the matter is not in that persons own production actives, but in that they use the social resources, not caring about their maintenance. The principle of assignment of surplus value without any constraining forces destroys any society. This is a problem for each national society; as an example, it can be considered for Russia.

### *13.5.1 Problems of the Modern Russia*

On the official assessments ([32], Table 5.1), there were 71,391 thousand persons (including, accepting Khanin's estimate [33], about 3000 thousand businessmen) in the national economy of Russia in 2013, who, with common efforts, have created a social product (Gross Domestic Product) valued as $66755 \cdot 10^9$ rubles in current prices ([32], Table 12.4). According to Rosstat ([32], Tables 12.4 and 12.7) the greater fraction of GDP ($34399 \cdot 10^9$ rubles) was get by hired workers in the form of wages and salaries (including insurance payments) and was consumed by householders, the other part represents profit of the enterprises before payment of taxes, valued, according to Rosstat ([32], Table 12.4), as $32499 \cdot 10^9$ rubles. The pure income of hired workers, after a deduction of a payment of social insurances and the tax on the individuals ($4694 \cdot 10^9$ and $2499 \cdot 10^9$ rubles, accordingly, according to Rosstat [32], Table 23.3), appears to be $27435 \cdot 10^9$ rubles. The profit of the enterprises, after payment of taxes on production ($12987 \cdot 10^9$ rubles a year, according to Rosstat, [32], Table 12.4), makes up $19519 \cdot 10^9$ rubles a year, and is spent for accumulation (investments) and personal consumption of businessmen. Notice, that active 4% of participants (businessmen) appropriate about 22% of the created social product.

As compared with the amount of the per capita Gross Domestic Product, Russia essentially concedes to countries of the developed capitalism, which is connected, naturally, with the lower labour productivity in Russia in comparison with those countries (see, for an example, a footnote in Sect. 8.10). And, we shall note that the lower general productivity testify inefficiency of the organisation of production, inefficiencies of the use of social resources, first of all, the possibilities of people, in comparison with the countries of the developed capitalism. It is a fundamental problem that Russia has inherited from the Soviet Union; the labour productivity is necessary to be considered as the basic criterion of success of economic system. Eventually, the socialism in Russia has given up the place to capitalism only because has not surpassed it in labour productivity.

The accompanying problem, which is the eternal feature of capitalistic society [10], is the growing stratification of the population of Russia, first of all in the income and wealth (see, for example, discussion in the monograph of Kashin and Abramov ([34], p. 82 and following). With the above assessments of the social product, it is possible to find, that average value of income of a hired worker made up about 32 thousand rubles a month that is comparable with specified by Rosstat average value (29792 rubles a month). At the same time the average income of the businessman, which was spent for consumption and accumulation, made up 542 thousand rubles a month. The average figures do not show the all depth of inequality: according to Rosstat ([32], Table 6.26) in 2013 15,5 million person had the income below a minimum living standard (less than 7306 rubles a month), at the same time there were many people, whose monthly income exceeded hundred thousand rubles. By official assessments in 2013 more than 30 thousand persons 'have earned' more than 10 million rubles. The inequality of incomes is connected with rules of distribution that are set up in the interests of proprietors of 'the capital'.

The stability of any capitalist society is supported by the central bodies: the state takes measures on the reduction of non-uniformity of distribution of incomes, redistributing the social product via public fund in favour of the deprived layers of the population, and suppresses possible protests against unfair distribution. In the countries of the developed capitalism, under effective production, some measures, in particular, introduction of the progressive-scale tax on income and profit of businessmen, appears sufficient to satisfy the greater part of the population. The situation appears different in Russia: when the production is disorganised, the possibility of the government to offer any measures of redistribution are limited: businessmen blackmail with an imaginary or real danger of oppression of the development of production, if the deductions in the social fund is being increased. The social system of Russia is in an unstable state and various scenarios of process of the social transformations, discussed, for example, by Khanin [33], are possible.

### *13.5.2 Social Fund of Russia*

The existing system of taxation in Russia, as Kashin and Abramov [34] state, appears inefficient; they evidence that the Russian tax system is complex, confused and controversial. It contains many (more than 15) positions, but, basically, the scheme of the taxation (in the year 2013) can be reduced to the rules of distribution (13.3), whereas the norm of deductions from the profit of enterprise is $\theta = 0.20$, the norm of a payment for social insurance (paid by the enterprises) is $\theta_L = 0.25$, the norm of the taxation of the income of hired workers $\theta_w = 0.13$.

In Russia, in 2013, from $66755 \cdot 10^9$ rubles of the Gross Domestic Product, $24443 \cdot 10^9$ rubles or 36.6% of GDP had come into hands of the government and was used in interests of all people ([32], Table 23.3). The specified volume of social fund was comprised with the direct payments of hired workers in the form of taxes and payments for social insurance, that makes up $7193 \cdot 10^9$ rubles a year, and businessmen as taxes on profit $12987 \cdot 10^9$ rubles a year (including payments for the usage of the natural resources $2598 \cdot 10^9$ rubles). Other deliveries in the budget in quantity $4262 \cdot 10^9$ rubles are carried by the value-added tax, excise and other additional taxes to both hired workers, and businessmen.

Thus, about 10% of the social fund are delivered by the enterprises as profit tax; indirect taxes from the enterprises provide 50–70% of social fund. The resulted figures show, that the society does not consider the surplus product as a property of businessmen only; the businessmen, in their turn, allocating these fractions from the profit, consider all the deductions unfair. The social fund is distributed over welfare actions (education, culture, cinematography, mass media, healthcare, physical training and sports and social policy)—about 60%; support of the national economy—about 13%; the internal and external national security—about 15%.

As we see, the combination of principles of taxation is used, in fact, when the social fund of Russia is formed: about 20% is delivered according to the rule of taxation on profit of the enterprises and income of physical bodies and nearby 25%—as payment for the using of the social resources (the workers and the nature), so that transition to the uniform principle of compensation of social resources does not look wild and unexpected.

### 13.5.3 The Reform of Taxation

The used now in Russia and in many other countries the principle of taxation on income and profit does not look fair and does not encourage the growth of labour productivity; at the best, the taxation does not influence the labour productivity in any way. Moreover, the practical realisation of this principle is connected with the significant difficulties [34]. The system of taxation in Russia requires modification, and transition to the new scheme of taxation (13.4), instead of the used now rules (13.3), can provide the first step to resolve some essential problems of Russia.

The advantage of the proposed transition to the principle of payment for social resources is, first, the fairness of distribution of payments on the enterprises, second, the simplification of the supervision of gathering of payments in the social fund (so as it is easier to control the used resources, than profit and income), and, the last, though, perhaps, the most important, such system includes the automatic mechanism of revealing and exception of inefficient proprietors, that will lead to an increase of the labour productivity.

The necessary variations for transition to the rules (13.4) could be seen as follows: the profit taxes on the enterprises and incomes of hired workers are being cancelled, step by step, and three basic payments for the usage of the social resources are established.

*Payment for the use of hired workers.* If an enterprise (of any type: private, state or cooperative) has hired workers (receiving salaries and wages), the social payment is paid for use of everyone who is working. This payment corresponds to the social tax that is paid by businessmen now. The norm of the payment is established annually on the basis of the average quantities. Any taxes to the income of individuals are cancelled.

*Payment for the use of services of production equipment.* The services of production equipment are estimated by substitutive work. If the enterprise uses energy-driven equipment, the tax is paid for use of each unit of primary productive energy (see the discussion of energy, as a production factor, in Sect. 2.5). The norm of taxation is established annually on the basis of the average quantities. This payment corresponds to the existing tax on profit of enterprises and replaces it.

*Payment for the use of natural resources.* Problems of degradation of a natural environment (exhaustion of natural resources and environmental pollution) are connected with development of the system of production. The usage of the national natural resources in production ought to be compensated. The norm of the resources tax is being established and reconsidered annually.

These payments do not exclude the existence of any minor payments and fees, for the production system to be more precisely adjust.

For the illustration, we shall demonstrate assessments of possible norms of payment for Russia in 2013, assuming, that the government wishes to generate the social fund in volume $24443 \cdot 10^9$ rubles, as it has been really collected. Apart from the payment for the natural resources in the amount of $2598 \cdot 10^9$ rubles, the government had collected $21845 \cdot 10^9$ rubles a year, as payment for use of the production factors. Adhering to the previous assessments, we believe, that the use of the labour is charged $9324 \cdot 10^9$ rubles a year and for the use of substitutive work $12529 \cdot 10^9$ rubles a year. So as the aggregate number of workers is 71391 thousand person, which have fulfilled $124 \cdot 10^9$ man-hours in the year, and the general consumption of substitutive work is $127042 \cdot 10^{18}$ J a year, we receive the estimates of norms: $d_L = 75$ rubles for a working hour of one person, $d_P = 986 \cdot 10^{-13}$ rubles for a joule of the substitutive work. At the fixed volume of social fund, distribution of payments over enterprises, in comparison with existing one, can change; the quantity of payments are counted up on the basis of norms of payment for the use of social resources.

### 13.5.4 Who is for the Reform? Who is Against?

The possible transformation of the taxation system are necessary in interests of the whole society. It ought to be stated that the all advantages that bourgeois gained at restoration of capitalism to Russia, including the right of a private property on the means of production, ought to be kept. The created product is to be considered as property of a businessman, it can be consumed, sold, destroyed, used in production, make everything, that the businessman wishes, while it is supposed, that he *completely* compensates the used social resources. This is the only addition to the rules of distribution that are in operation now. The new rules can work effectively only at freedom of business, when each member of the society can organise a business according to his own wishes and initiatives. Any difficulties at registration of the enterprise should not be provided.

The talent of the businessman is uncovered in an effective utilisation of social resources; the most successful businessmen, who, according to Schumpeter [1] can be called neo-entrepreneur (see Sect. 5.5), deserves to get *entrepreneur premium*, the others ought to be satisfied with average value of profit. The victims of transformation could be a group of unfair and inefficient businessmen, who, naturally, will undertake all necessary steps for resistance to modification of the system of taxation and will make great noise.

By using the new principle of taxation, the government receives the tool for an assessment of efficiency of use of social resources, which will allow to supervise (to operate) growth of labour productivity and to regulate the distribution of social product. New rules are directed to stimulation of effective development, to the creation, not to any destruction. The growth of labour productivity leads to improvement of well-being of everybody, which smoothes contradictions and softens customs. It would be desirable to hope that the government will manage to use this tool in social interests.

## References

1. Schumpeter, J.A.: Theorie der wirtschaftlichen Entwicklung: Eine Untersuchung über Unternehmergewinn, Kapital, Kredit, Zins und den Kojunkturzyklus. Dunker und Humblot, Berlin (1911)
2. Ellerman, D.: On the renting of persons: the neo-abolitionist case against today's peculiar institution. Econ. Thought **4**(1), 1–20 (2015). http://www.worldeconomicsassociation.org/files/journals/economicthought/WEA-ET-4-1-Ellerman.pdf. Accessed on 17 Aug 2017
3. Marx, K.: Capital. Encyclopaedia Britannica, Chicago (1952). English translation of Karl Marx, Das Kapital. Kritik der politischen Oekonomie. Otto Meissner, Hamburg (1867)
4. Clark, J.B.: The Distribution of Wealth: A Theory of Wages, Interest and Profits. Macmillan, New York (1899)
5. Wicksteed, Ph.H.: Essay on the Coordination of the Laws of Distribution. MacMillan, London (1894)
6. Kamenetskiy, V.A., Patrikeev, V.P.: Sobstvennost v XXI stoletii (The Ownership in XXI Century). Ekonomica, Moscow (2004)
7. Ellerman, D.: On the role of capital in "capitalist" and in labor-managed firms. Rev. Radic. Polit. Econ. **39**(1), 5–26 (2007)
8. Kutsin, N.A., Kalyuzhnyy, V.V., Mozenkov, O.V., Balykin, V.D.: Sovremennaya teoriya i praktika nalogooblozheniya (The Modern Theory and Practice of Taxation). Prapor, Kharkov (2001)
9. Mayburov, I.A., Sokolovskaya, A.M.: Teoriya nalogooblozheniya: Prodvinutyy kurs (The Theory of Taxation: Advanced Course). Yuniti-Dana, Moscow (2011)

10. Piketty, T.: Capital in the Twenty-First Century. Harvard University Press, Cambridge (2014)
11. Kelso, L.O., Kelso, P.H.: Democracy and Economic Power: Extending the ESOP Revolution Through Binary Economics. Ballinger Publishing, Cambridge (1986). Reprinted by University Press of America, Lanham (1991)
12. Soros, G.: The Crisis of the Global Capitalism: Open Society Endangered. Public Affairs, New York (1998)
13. Egishyants, S.A.: Tupiki globalizatsii: torzhestvo progressa ili igry satanistov (The Deadlocks of Globalization: A Triumph of Progress or Game of Satanists?). Veche, Moscow (2004)
14. Katasonov, V.Yu.: Kapitalizm. Istoriya i ideologiya "denezhnoy tsivilizatsii" (Capitalism. History and ideology of 'the monetary civilization'). Institut russkoy tsivilizatsii (Institute of the Russian Civilization), Moscow (2013)
15. Yakunin, V.I., Sulakshin S.S., Averkova, N.A., Bagdasaryan, V.E., Bogdan, I.V., Vershinin, A.A., Genyush, S.V., Deeva, M.V., Korobkova, A.Yu., Kuropatkina, O.V., Orlov, I.B., Safonova, YU.A., Sulakshina, A.S., Shestopalova A.V.: Politicheskoe izmerenie mirovykh finansovykh krizisov: Fenomenologiya, teoriya, ustranenie (The Political Measurement of World Financial Crises: Phenomenology, Theory, Elimination). Nauchnyy Ekspert, Moscow (2012)
16. Antonov, M.F.: Ot kapitalizma – k totalitarizmu! Mir v XXI veke i sudby Rossii (From Capitalism – to Totalitarianism! The World in the XXI Century and the Destiny of Russia). Moscow (2008). http://mantonov.chat.ru/from_to/index.htm. Accessed on 17 Aug 2017
17. Gorky, M.: Odin iz koroley respubliki (One of the Kings of the Republics). Sobranie sochineniy v 30 tomakh (The complete works in 30 volumes), 1949–1956, vol. 7. GIKHL, Moscow (1906)
18. Bichler, S., Nitzan, J.: Capital as power: toward a new cosmology of capitalism. Real-World Econ. Rev. (61), 65–84 (2012). http://www.paecon.net/PAEReview/issue61/BichlerNitzan61.pdf. Accessed on 17 Aug 2017
19. Kornai, Y.A.: Razmyshleniya o kapitalizme (The Thoughts about Capitalism). Trans. from Hungarian. Izdatelstvo Instituta Gaydara, Moscow (2012)
20. Skoblikov, E.A.: Tretiy put - cherez revolyutsiyu, perevorot ili transformatsiyu obshhestva (The Third Way - through Revolution, Mutiny or Transformation of the Society)?, NEFORMAT. Accent Graphics Communications, Montreal (2014)
21. Polterovich, V.M.: Ekonomicheskoje ravnovesije i khozjaistvennyi mekanizm (The Economic Equilibrium and Mechanism of Management). Nauka, Moscow (1990)
22. Yasin, E.G.: Rossiyskaya ekonomika: Istoki i panorama rynochnykh reform. Kurs lektsiy (The Russian Economy: The Sources and Panorama of Market Reforms. A Course of Lectures). GU VSHE, Moscow (2002)
23. Bogdanov, S.V.: Khozyaystvenno-korystnaya prestupnost' v SSSR 1945–1990 gg.: faktory vosproizvodstva, osnovnye pokazateli, osobennosti gosudarstvennogo protivodeystviya (Economic-Mercenary Criminality in the USSR 1945–1990: The Factors of Reproduction, the Basic Parameters, Peculiarities of the State Counteraction). Dissertatsiya ... doktora istoricheskikh nauk: 07.00.02 (The Dissertation... Doctors of History: 07.00.02), Kursk (2010)
24. Fedorenko, N.P. (ed.): Vvedenie v teoriyu i metodologiyu sistemy optimalnogo funktsionirovaniya sotsialisticheskoy ekonomiki (Introduction to the Theory and Methodology of the system of Optimum Performance of Socialistic Economy). Nauka, Moscow (1983)
25. Pikhorovich, V.V.: Ocherki istorii kibernetiki v SSSR: Na puti k razvenchaniyu mifov (Sketches of the History of Cybernetics in the USSR: On a Way to Discern the Myths). URSS, Moscow (2016)
26. Glazyev, S.Yu., Kara-Murza, S.G., Batchikov, S.A.: Belaya kniga: Ekonomicheskie reformy v Rossii 1991–2001 gg (The White Book: Economic Reforms in Russia 1991–2001). Eksmo, Moscow (2003)
27. Simchera, V.M.: Razvitie ekonomiki Rossii za 100 let: Istoricheskie ryady, vekovye trendy, institutsional'nye tsikly (Development of the Economy of Russia over 100 years: Historical Series, Century Trends, Institutional Cycles). Ekonomika, Moscow (2007)
28. Simonyan, R.Kh.: Bez gneva i pristrastiya: Ekonomicheskie reformy 1990-kh godov i ikh posledstviya dlya Rossii (Without Rage and Predilection: Economic Reforms of 1990th years and their Consequences for Russia). Ekonomika, Moscow (2010)

29. Akhmetov, K.: Assimetrichnaya ekonomika (The Assymetric Economy). LEM (Lem), Almaty (2012). See also: Kurman Akhmetov. Paradoksal'naya finansovaya sistema SSSR (A Paradoxic Financial System of the USSR). http://via-midgard.info/news/copyright/13661-paradoksalnaya-finansovaya-sistema-sssr.html. Accessed on 17 Aug 2017
30. Kolganov, A.I. (ed.): Politekonomiya provala: Priroda i posledstviya rynochnykh 'reform' v Rossii (Political Economy of a Devastation: The Nature and Consequences of Market 'Reforms' in Russia). URSS, Moscow (2013)
31. Gaydar, E., Chubays, A.: Razvilki noveyshey istorii Rossii (The Fork Points of the Newest Yistory of Russia). OGI, Moscow (2011)
32. Rosstat: Rossiyskiy statisticheskiy ezhegodnik 2013 (The Russian Statistical Year-Book 2013). Rosstat, Moscow (2013)
33. Khanin, G.I.: Sovremennaya rossiyskaya burzhuaziya – opyt ekonomicheskogo eskiza (The Modern Russian Bourgeois - an Endeavor of Economic Sketch). Terra Economicus **11**(1), 10–29 (2013)
34. Kashin, V.A., Abramov, M.D.: Promyshlennaya politika i nalogovoe regulirovanie (The Production Policy and Tax Regulation). IPR RAN, Moscow (2015)

# Appendices

## A. The Sector Balance in the Russian Economy

The input-output analysis of the national economy of Russia for year 2003, completed by Goskomstat (2006), is based on the division of the economy in 22 sectors; the nomenclature table on the next pages contains the list of titles and codes of products/sectors, according to the Russian Classification System of Sectors (Goskomstat, 2000) that was used by Goskomstat of Russia in that time, also as their numbers in the summary balance table (Table A.1) that reproduces Table 8.1 of Goskomstat (2006) (with correction of an obvious mistake of the original table in the line of product 17). The first column of the balance table contains the numbers of products in the nomenclature table, and the first line—numbers of appropriating sectors, creating these products. The numbers in the table are expenses in rubles, specified at the identical total output of each sector equal to 1000 rubles, so that the coefficient of direct input $a_{ij}$ is the figure in the table, located on intersection of line $i$ and the column $j$ and divided by 1000. For the compactness of presentation, the symbols are used in Table A.1:

$\Sigma_b$ – value of intermediate productive consumption of all products used by the selected sector;

$\Sigma_p$ – value of a part of final output of the selected sector distributed over all sectors for intermediate productive consumption;

$Z$ – value added within the selected sector;

$Y$ – value of final product of the selected sector.

## References


1. Goskomstat: Sistema tablits 'Zatraty - Vypusk' Rossii za 2003 god (The System of Tables 'Input-Output' for Russia in 2003). Rosstat, Moscow (2006)
2. Goskomstat: The Russian Classification System (RCS) 'The Sectors of a National Economy'. Edition 15.02.2000. The document has become invalid, see 'The Russian Classification System of Titles and Codes of Economic Activities (RCSEA) 029–2007'. https://normativ.kontur.ru/. Accessed on 26 Aug 2017






V. N. Pokrovskii, *Econodynamics*, New Economic Windows,
https://doi.org/10.1007/978-3-319-72074-6

**Table A.1** The Input–output balance for the Russian economy in year 2003

| | 1 | 2 | 3 | 4 | 5 | 6 | 7 | 8 | 9 | 10 | 11 | 12 | 13 | 14 | 15 | 16 | 17 | 18 | 19 | 20 | 21 | 22 | $\Sigma p$ | Y |
|---|---|---|---|---|---|---|---|---|---|---|---|---|---|---|---|---|---|---|---|---|---|---|---|---|
| 1 | 131.1 | 39.33 | 64.1 | 30.89 | 52.33 | 51 | 96.69 | 36.52 | 45.88 | 70.18 | 43.43 | 14.84 | 25.01 | 14 | 14.44 | 39.05 | 7.61 | 2.54 | 126.5 | 36.76 | 15.69 | 21.7 | 980.6 | 19.39 |
| 2 | 100.9 | 226.5 | 14.2 | 89.46 | 27.74 | 9.34 | 70.05 | 17.36 | 29.89 | 49.58 | 5.51 | 13.69 | 6.43 | 23.45 | 21 | 39.03 | 8.88 | 5.56 | 33.71 | 8.58 | 20.01 | 12.2 | 835.1 | 164.9 |
| 3 | 21.99 | 0.27 | 231.9 | 2.49 | 24.97 | 0.49 | 4.56 | 2.36 | 4.16 | 4.14 | 2.26 | 1.02 | 0.88 | 1.33 | 0.78 | 0.95 | 0.51 | 0.11 | 4.72 | 3.48 | 0.42 | 1.1 | 317.9 | 682.1 |
| 4 | 0.4 | 0.02 | 0.05 | 29.25 | 0.03 | 0.02 | 0.01 | 0.04 | 0.12 | 0.14 | 0.02 | 0.01 | 0 | 0.23 | 0.16 | 0 | 0 | 0 | 0.01 | 0.05 | 0.01 | 0.0 | 34.6 | 965.4 |
| 5 | 3.22 | 7.54 | 19.66 | 4.28 | 269.6 | 9.46 | 26.27 | 111.9 | 12.11 | 69.58 | 2.55 | 2.69 | 7.75 | 50.46 | 0.25 | 10.56 | 1.7 | 0.96 | 12.67 | 0.75 | 10.52 | 0.0 | 639.5 | 360.5 |
| 6 | 5.65 | 0.42 | 0.09 | 0 | 44.86 | 377.1 | 14.17 | 59.14 | 9.92 | 11.47 | 0 | 2.24 | 189.1 | 3.66 | 0 | 0.43 | 0.16 | 0.05 | 0.43 | 0.18 | 7.01 | 0.0 | 732.1 | 267.9 |
| 7 | 4.7 | 11.01 | 41.01 | 0 | 5.57 | 8.38 | 259.4 | 34.53 | 37.22 | 29.05 | 67.13 | 9.87 | 40.19 | 17.79 | 15.02 | 11.33 | 13.34 | 4.93 | 10.65 | 35.92 | 31.14 | 4.3 | 699.5 | 300.5 |
| 8 | 15.89 | 14.82 | 71.26 | 40.58 | 17.5 | 19.56 | 21.53 | 220.3 | 44.01 | 25.03 | 7.61 | 14.54 | 20.26 | 72.54 | 28.24 | 74.98 | 21.88 | 43.86 | 22.84 | 14.8 | 59.73 | 38.0 | 917.8 | 82.2 |
| 9 | 0 | 0.1 | 11.36 | 0.04 | 2.42 | 0.58 | 21.85 | 6.67 | 245.5 | 10.51 | 1.7 | 14.51 | 93 | 24.59 | 0.6 | 5.72 | 9.63 | 38.66 | 3.8 | 4.09 | 4.72 | 3.3 | 512.3 | 487.7 |
| 10 | 1.26 | 0.8 | 4.95 | 0.6 | 1.77 | 0.35 | 4.85 | 3.17 | 4.11 | 135.4 | 0.49 | 3.81 | 1.39 | 117.7 | 1.39 | 5.7 | 4.43 | 0.9 | 12.85 | 4.26 | 3.78 | 0.2 | 324.1 | 675.9 |
| 11 | 0.23 | 0.29 | 4.46 | 0.52 | 1.85 | 0.18 | 5.52 | 3.08 | 10.93 | 3.23 | 401.7 | 2.16 | 9.65 | 1.25 | 0.78 | 2.58 | 3.86 | 4.86 | 2.67 | 8.89 | 2.62 | 7.9 | 490.2 | 509.9 |
| 12 | 0.11 | 0.18 | 1.08 | 0.05 | 0.61 | 0.14 | 7.4 | 0.94 | 0.88 | 0.23 | 0.62 | 268.6 | 16.08 | 0.31 | 15.71 | 1.81 | 16.94 | 2.2 | 1.7 | 60.36 | 2.9 | 18.2 | 429.0 | 571.0 |
| 13 | 2.6 | 1.33 | 4.96 | 0.22 | 2.81 | 0.74 | 5.11 | 4.15 | 2.26 | 2.39 | 2.2 | 7.12 | 48.24 | 1.04 | 34.17 | 4.64 | 3.76 | 114.2 | 9.83 | 6.62 | 5.78 | 6.6 | 283.7 | 716.3 |
| 14 | 25.9 | 16.66 | 8.25 | 10.86 | 13.14 | 12.11 | 16.73 | 13.57 | 7.28 | 10.49 | 8.07 | 10.03 | 9.22 | 3.94 | 5.33 | 31.05 | 4.31 | 2.45 | 60.15 | 10.37 | 14.49 | 17.2 | 325.6 | 674.4 |
| 15 | 0 | 0.15 | 0 | 0 | 0.4 | 0 | 0 | 0.07 | 0 | 0 | 25.49 | 229.3 | 64.1 | 0 | 217.3 | 0 | 5.87 | 0 | 0.33 | 9.8 | 1.88 | 10.4 | 580.2 | 419.8 |
| 16 | 55.4 | 49.29 | 72.35 | 53.53 | 61.1 | 18.43 | 53.59 | 31.33 | 52.21 | 75.53 | 20.34 | 23.66 | 23.92 | 54.16 | 21.53 | 49.75 | 83.93 | 27.09 | 24.83 | 19.99 | 22.66 | 59.2 | 969.9 | 30.1 |
| 17 | 95.8 | 45.31 | 30.73 | 32.19 | 59.70 | 27.26 | 54 | 40.59 | 55.94 | 60.92 | 40.81 | 46.41 | 32.36 | 45.12 | 28.87 | 72.8 | 34.25 | 26.39 | 39.17 | 31.81 | 25.91 | 28.9 | 972.2 | 27.8 |
| 18 | 6.36 | 4.69 | 5.24 | 2.48 | 3.66 | 2.55 | 5.59 | 5.07 | 4.11 | 4.66 | 4.18 | 3.71 | 7.69 | 5.6 | 0.87 | 8.83 | 12.59 | 38.94 | 4.56 | 9.82 | 6.25 | 7.1 | 172.5 | 827.5 |
| 19 | 12.03 | 2.73 | 12.61 | 11.89 | 3.17 | 3.48 | 5.17 | 8.13 | 5.94 | 6.74 | 6.98 | 4.91 | 6.11 | 5.01 | 2.61 | 11.02 | 3.34 | 4.6 | 18.32 | 29.18 | 12.21 | 37.3 | 232.5 | 767.5 |
| 20 | 0.94 | 0.42 | 0.97 | 0.94 | 0.55 | 0.48 | 0.59 | 0.63 | 0.36 | 0.53 | 0.22 | 0.34 | 0.31 | 0.62 | 0.07 | 2.1 | 0.32 | 0.87 | 0.99 | 17.31 | 0.71 | 0.9 | 51.2 | 948.8 |
| 21 | 4 | 11.74 | 2.69 | - | 3.52 | 10.25 | 4.05 | 24.11 | 1.31 | 0.35 | 0 | 0.34 | 0.74 | 0.87 | 0.1 | 2.03 | 0.71 | 1.83 | 0.33 | 0.11 | 224.7 | 54.0 | 368.7 | 631.3 |
| 22 | 13.25 | 11.04 | 11.81 | 15.11 | 13.83 | 9.86 | 5 | 8.23 | 7.67 | 4.7 | 3.13 | 4.17 | 5.37 | 3.99 | 2.63 | 15.47 | 4.84 | 8.46 | 6.26 | 6.1 | 9.89 | 38.1 | 230.9 | 769.1 |
| $\Sigma_b$ | 501.7 | 444.6 | 613.7 | 325.4 | 611.1 | 561.8 | 682.2 | 631.9 | 581.8 | 574.8 | 644.4 | 678.0 | 607.8 | 447.6 | 411.9 | 389.8 | 242.8 | 329.5 | 397.3 | 319.2 | 483.0 | 366.5 | 10,847 | |
| Z | 498.3 | 555.4 | 386.3 | 674.6 | 388.9 | 438.2 | 317.8 | 368.1 | 418.2 | 425.1 | 355.6 | 321.9 | 392.2 | 552.4 | 588.1 | 610.2 | 757.1 | 670.5 | 602.7 | 680.8 | 517.0 | 633.5 | | 11,153 |

## The List of Products/Sectors and their Codes

| Number in Table A.1 | Title of product/sector | Title of the sector in the RCS | The RCS code |
|---|---|---|---|
| 1 | Electricity and heat energy | Electric power industry | 11100 |
| 2 | Oil and Gas extraction | Oil and Gas extraction | 11210 |
| | | Petroleum-refining industry | 11220 |
| | | Gas industry | 11230 |
| 3 | Coal | Coal industry | 11300 |
| 4 | Combustible slates and peat | Slate industry | 11410 |
| | | Peat industry | 11610 |
| 5 | Ferrous metals | Ferrous metallurgy | 12100 |
| 6 | Nonferrous metals | Nonferrous metallurgy | 12200 |
| 7 | Products of the chemical and petrochemical industry | Chemical and petrochemical industry (without apharmaceuticalindustry) | 13000 |
| | | Chemi - farmaceutical industry | 19310 |
| 8 | Mechanisms and the equipment, products of metal working | Mechanical engineering and metal working (without the industry of iatrotechnics) | 14000 |
| | | Industry of iatrotechnics | 19320 |
| 9 | Wood product, pulp and paper | Wood product, pulp and paper manufacturing | 15000 |
| 10 | Construction materials (including products of the glass and porcelain-faience manufacturing) | Construction materials manufacturing | 16100 |
| | | Glass and porcelain-faience manufacturing (without the enterprises on manufacture medical from - from glass and porcelain) | 16500 |
| | | Manufacture of medical products from glass, porcelain and plastic | 19330 |
| 11 | Products of non-heavy industries | Non-heavy industries | 17000 |
| 12 | food | Food manufacturing | 18000 |
| | | Flour and groats manufacturing | 19210 |
| | | Fishing and other water bioresources | 21500 |
| 13 | Other industrial products | Microbiological industry | 19100 |
| | | Animal food industry | 19220 |
| | | Printing industries | 19400 |
| | | Other industrial productions | 19700 |
| | | State control production, state supervision over the standards and means of measurements | 19800 |
| | | Management of Companies and Enterprises | 19900 |
| 14 | Production of construction | Construction | 60000 |
| 15 | Crop, Animals, Forestry products and Related services | Agriculture | 20000 |
| | | Forestry | 30000 |

(continued)

(continued)

| Number in Table A.1 | Title of product/sector | Title of the sector in the RCS | The RCS code |
|---|---|---|---|
| 16 | Transportation and communication | Transportation and communication | 50000 |
| 17 | Trading-intermediary services (including services of public catering) | Commerce and public catering | 70000 |
| | | Logistics and selling | 80000 |
| | | Warehousing | 81000 |
| | | Operations with real estate | 83000 |
| | | General activity on maintenance of functioning of the market | 84000 |
| | | Gathering scrap and breakage | 87300 |
| 18 | Products of other activities | Information and Data Processing | 82000 |
| | | Publishing industries | 87100 |
| | | Private gardening | 87400 |
| | | Other goods industries | 87500 |
| 19 | Services of house facilities and non-productive types of consumer services of the population | Housing and communal services | 90000 |
| | | Non-productive types household the population | 90300 |
| 20 | Services of healthcare, physical training and social security, formation, culture and art | Healthcare, physical training and social security | 91000 |
| | | Educational Services | 92000 |
| | | Culture and Arts | 93000 |
| 21 | Science, geology and investigation of bowels, geodetic and hydro-meteorological services | Scientific service | 95000 |
| | | Geology and investigation of bowels, geodetic both hydro-meteorological services | 85000 |
| 22 | Services of financial intermediary, insurance, management and public associations | Finance, credit, insurance and provision of pensions | 96000 |
| | | Public Administration | 97000 |
| | | Public Associations | 98000 |

## B. Macrocharacteristics of the U.S. Economy

Time series on an annual basis provide the basic empirical data for the test of every theory of economic growth. National statistical compilations contain information about gross national product, labour, energy and other quantities. High art and hard work are needed to elaborate time series for economic quantities. Sometimes, it needs in courage as well.

**Table A.2** Macrocharacteristics of the U.S. economy

| Year | Population $N \cdot 10^{-3}$ | GNP $Y \cdot 10^{-6}$ $ (1996) | Investment $I \cdot 10^{-6}$ $ (1996) | Capital $K \cdot 10^{-6}$ $ (1996) | Labour $L \cdot 10^{-6}$ man-h | Substitutive work $P$ quad | Primary energy $E$ quad |
|---|---|---|---|---|---|---|---|
| 1891 | – | 214,663 | 47,059 | 967,288 | 39,927 | 0.0028 | 7.04 |
| 1892 | – | 236,357 | 60,567 | 1,047,614 | 41,640 | 0.0033 | 7.33 |
| 1893 | – | 222,488 | 45,865 | 1,123,236 | 41,071 | 0.0037 | 7.63 |
| 1894 | – | 216,891 | 43,545 | 1,172,219 | 39,234 | 0.0053 | 7.93 |
| 1895 | – | 246,560 | 54,564 | 1,231,646 | 42,384 | 0.0054 | 8.23 |
| 1896 | – | 241,260 | 44,133 | 1,287,255 | 42,398 | 0.0065 | 8.62 |
| 1897 | – | 267,115 | 53,596 | 1,331,494 | 44,119 | 0.0066 | 9.02 |
| 1898 | – | 272,514 | 50,050 | 1,382,140 | 44,483 | 0.0076 | 9.42 |
| 1899 | – | 305,550 | 62,951 | 1,432,732 | 48,624 | 0.0077 | 9.91 |
| 1900 | 76,090 | 308,472 | 71,130 | 1,489,105 | 49,404 | 0.0088 | 10.70 |
| 1901 | 77,580 | 343,044 | 79,186 | 1,543,855 | 52,181 | 0.0089 | 11.69 |
| 1902 | 79,160 | 340,172 | 82,638 | 1,601,086 | 55,141 | 0.0090 | 13.38 |
| 1903 | 80,630 | 355,526 | 81,633 | 1,666,279 | 57,160 | 0.0096 | 15.06 |
| 1904 | 82,170 | 348,691 | 67,877 | 1,716,012 | 56,213 | 0.0111 | 16.06 |
| 1905 | 83,820 | 378,409 | 73,291 | 1,774,784 | 59,734 | 0.0113 | 16.35 |
| 1906 | 85,450 | 431,208 | 98,273 | 1,858,641 | 62,715 | 0.0114 | 13.78 |
| 1907 | 87,010 | 434,922 | 96,940 | 1,947,528 | 64,516 | 0.0127 | 14.67 |
| 1908 | 88,710 | 387,770 | 60,558 | 2,018,638 | 61,265 | 0.0144 | 16.06 |
| 1909 | 90,490 | 443,541 | 96,205 | 2,071,670 | 65,471 | 0.0145 | 15.46 |
| 1910 | 92,410 | 442,748 | 95,114 | 2,137,790 | 67,694 | 0.0149 | 16.35 |
| 1911 | 93,860 | 457,558 | 83,589 | 2,199,302 | 69,182 | 0.0159 | 17.94 |
| 1912 | 95,360 | 476,230 | 98,175 | 2,251,571 | 71,857 | 0.0160 | 18.43 |
| 1913 | 97,230 | 495,300 | 109,430 | 2,326,498 | 72,900 | 0.0180 | 17.74 |
| 1914 | 99,110 | 446,562 | 58,553 | 2,398,262 | 71,077 | 0.0205 | 17.54 |
| 1915 | 100,550 | 460,975 | 56,161 | 2,445,923 | 70,822 | 0.0229 | 17.94 |
| 1916 | 101,960 | 537,895 | 94,979 | 2,494,075 | 77,346 | 0.0231 | 20.12 |
| 1917 | 103,270 | 513,626 | 75,380 | 2,548,021 | 79,096 | 0.0236 | 20.91 |
| 1918 | 103,210 | 543,839 | 69,314 | 2,596,922 | 78,440 | 0.0264 | 21.70 |
| 1919 | 104,510 | 575,043 | 116,507 | 2,643,806 | 75,961 | 0.0287 | 21.70 |
| 1920 | 106,460 | 587,921 | 157,481 | 2,707,472 | 77,128 | 0.0305 | 19.62 |
| 1921 | 108,540 | 552,754 | 65,451 | 2,763,599 | 69,459 | 0.0348 | 20.12 |
| 1922 | 110,050 | 581,482 | 82,070 | 2,814,941 | 75,597 | 0.0350 | 18.63 |
| 1923 | 111,950 | 672,617 | 135,455 | 2,917,966 | 82,960 | 0.0353 | 19.23 |
| 1924 | 114,110 | 693,915 | 86,915 | 3,039,886 | 80,919 | 0.0397 | 21.31 |
| 1925 | 115,830 | 704,811 | 127,210 | 3,161,810 | 84,418 | 0.0403 | 20.70 |
| 1926 | 117,400 | 755,827 | 134,742 | 3,288,677 | 88,209 | 0.0415 | 20.42 |
| 1927 | 119,040 | 760,285 | 131,413 | 3,405,460 | 88,573 | 0.0458 | 21.61 |
| 1928 | 120,510 | 816,254 | 133,016 | 3,515,442 | 89,667 | 0.0504 | 21.61 |
| 1929 | 121,770 | 822,198 | 141,193 | 3,635,742 | 92,218 | 0.0532 | 22.10 |
| 1930 | 123,080 | 751,500 | 119,147 | 3,702,810 | 85,803 | 0.0577 | 22.10 |

(continued)

**Table A.2** (continued)

| Year | Population $N \cdot 10^{-3}$ | GNP $Y \cdot 10^{-6}$ \$ (1996) | Investment $I \cdot 10^{-6}$ \$ (1996) | Capital $K \cdot 10^{-6}$ \$ (1996) | Labour $L \cdot 10^{-6}$ man-h | Substitutive work $P$ quad | Primary energy $E$ quad |
|---|---|---|---|---|---|---|---|
| 1931 | 124,040 | 703,600 | 95,211 | 3,719,694 | 77,930 | 0.0616 | 19.62 |
| 1932 | 124,840 | 611,800 | 69,100 | 3,685,926 | 68,577 | 0.0641 | 18.43 |
| 1933 | 125,580 | 603,300 | 63,036 | 3,642,074 | 68,744 | 0.0627 | 18.43 |
| 1934 | 126,370 | 668,300 | 72,261 | 3,628,941 | 69,262 | 0.0619 | 18.63 |
| 1935 | 127,250 | 728,300 | 84,404 | 3,639,025 | 73,774 | 0.0615 | 18.93 |
| 1936 | 128,050 | 822,500 | 108,340 | 3,695,540 | 81,575 | 0.0618 | 19.62 |
| 1937 | 128,820 | 865,800 | 116,353 | 3,762,608 | 87,334 | 0.0620 | 18.83 |
| 1938 | 129,820 | 835,600 | 102,012 | 3,797,784 | 80,190 | 0.0655 | 19.62 |
| 1939 | 130,880 | 903,500 | 120,585 | 3,862,507 | 85,803 | 0.0657 | 20.76 |
| 1940 | 132,120 | 980,700 | 136,303 | 3,945,521 | 91,125 | 0.0661 | 23.69 |
| 1941 | 133,400 | 1,148,800 | 183,970 | 4,096,776 | 103,518 | 0.0670 | 25.47 |
| 1942 | 134,860 | 1,360,000 | 229,151 | 4,329,638 | 113,359 | 0.0681 | 27.85 |
| 1943 | 136,740 | 1,583,700 | 269,298 | 4,591,813 | 120,357 | 0.0692 | 29.14 |
| 1944 | 138,400 | 1,714,100 | 276,264 | 4,804,742 | 119,483 | 0.0757 | 30.13 |
| 1945 | 139,930 | 1,693,300 | 242,855 | 4,896,667 | 113,651 | 0.0804 | 31.22 |
| 1946 | 141,390 | 1,505,500 | 215,735 | 4,929,967 | 116,640 | 0.0794 | 30.62 |
| 1947 | 144,130 | 1,495,100 | 257,875 | 5,019,781 | 120,868 | 0.0792 | 30.13 |
| 1948 | 146,630 | 1,560,000 | 291,303 | 5,140,316 | 122,909 | 0.0827 | 30.62 |
| 1949 | 149,190 | 1,550,900 | 294,899 | 5,285,005 | 117,809 | 0.0894 | 32.01 |
| 1950 | 152,270 | 1,686,600 | 345,504 | 5,494,417 | 122,919 | 0.0948 | 33.30 |
| 1951 | 154,880 | 1,815,100 | 356,476 | 5,716,491 | 124,413 | 0.1023 | 33.30 |
| 1952 | 157,550 | 1,887,300 | 367,447 | 5,937,394 | 125,663 | 0.1116 | 34.29 |
| 1953 | 160,180 | 1,973,900 | 392,164 | 6,188,547 | 127,280 | 0.1221 | 37.56 |
| 1954 | 163,030 | 1,960,500 | 391,654 | 6,414,843 | 122,270 | 0.1329 | 39.35 |
| 1955 | 165,930 | 2,099,500 | 433,277 | 6,684,991 | 127,953 | 0.1381 | 38.85 |
| 1956 | 168,900 | 2,141,100 | 430,935 | 6,924,888 | 131,174 | 0.1473 | 38.85 |
| 1957 | 171,980 | 2,183,900 | 435,866 | 7,157,984 | 130,269 | 0.1598 | 39.84 |
| 1958 | 174,880 | 2,162,800 | 418,833 | 7,340,662 | 126,309 | 0.1720 | 41.72 |
| 1959 | 177,830 | 2,319,000 | 463,953 | 7,594,864 | 132,272 | 0.1775 | 43.51 |
| 1960 | 180,670 | 2,376,700 | 466,357 | 7,836,872 | 133,578 | 0.1905 | 43.71 |
| 1961 | 183,690 | 2,432,000 | 470,568 | 8,072,548 | 133,225 | 0.2055 | 44.30 |
| 1962 | 186,540 | 2,578,900 | 514,763 | 8,354,421 | 134,940 | 0.2191 | 47.83 |
| 1963 | 189,240 | 2,690,400 | 546,897 | 8,663,731 | 137,276 | 0.2340 | 49.65 |
| 1964 | 191,890 | 2,846,500 | 587,743 | 9,012,907 | 140,097 | 0.2513 | 51.83 |
| 1965 | 194,300 | 3,028,500 | 645,272 | 9,411,797 | 144,126 | 0.2691 | 54.02 |
| 1966 | 196,560 | 3,227,500 | 683,714 | 9,835,779 | 146,916 | 0.2917 | 57.02 |
| 1967 | 198,710 | 3,308,300 | 693,556 | 10,231,152 | 147,620 | 0.3183 | 58.91 |
| 1968 | 200,710 | 3,466,100 | 747,120 | 10,653,728 | 150,607 | 0.3441 | 62.41 |
| 1969 | 202,680 | 3,571,400 | 760,064 | 11,071,613 | 153,556 | 0.3663 | 65.63 |
| 1970 | 205,050 | 3,578,000 | 739,764 | 11,424,775 | 152,147 | 0.3968 | 67.86 |

(continued)

**Table A.2** (continued)

| Year | Population $N \cdot 10^{-3}$ | GNP $Y \cdot 10^{-6}$ \$ (1996) | Investment $I \cdot 10^{-6}$ \$ (1996) | Capital $K \cdot 10^{-6}$ \$ (1996) | Labour $L \cdot 10^{-6}$ man-h | Substitutive work $P$ quad | Primary energy $E$ quad |
|---|---|---|---|---|---|---|---|
| 1971 | 207,660 | 3,697,700 | 783,096 | 11,804,436 | 151,990 | 0.4284 | 69.31 |
| 1972 | 209,900 | 3,898,400 | 863,164 | 12,258,669 | 158,838 | 0.4459 | 72.76 |
| 1973 | 211,910 | 4,123,400 | 935,651 | 12,756,051 | 164,395 | 0.4700 | 75.81 |
| 1974 | 213,850 | 4,099,000 | 894,004 | 13,155,176 | 165,966 | 0.5004 | 74.08 |
| 1975 | 215,970 | 4,084,400 | 840,255 | 13,466,597 | 161,152 | 0.5382 | 72.04 |
| 1976 | 218,040 | 4,311,700 | 907,749 | 13,838,519 | 167,209 | 0.5485 | 76.07 |
| 1977 | 220,240 | 4,511,800 | 1,007,912 | 14,290,877 | 173,140 | 0.5661 | 78.12 |
| 1978 | 222,590 | 4,760,600 | 1,104,643 | 14,814,054 | 180,386 | 0.5864 | 80.12 |
| 1979 | 225,060 | 4,912,100 | 1,150,502 | 15,341,921 | 184,244 | 0.6162 | 81.04 |
| 1980 | 227,760 | 4,900,900 | 1,080,070 | 15,752,771 | 182,185 | 0.6630 | 78.44 |
| 1981 | 229,940 | 5,021,000 | 1,097,206 | 16,147,910 | 184,644 | 0.6970 | 76.57 |
| 1982 | 232,170 | 4,919,300 | 1,041,772 | 16,447,136 | 181,172 | 0.7458 | 73.44 |
| 1983 | 234,300 | 5,132,300 | 1,138,894 | 16,824,687 | 184,208 | 0.7683 | 73.32 |
| 1984 | 236,370 | 5,505,200 | 1,317,892 | 17,359,824 | 194,388 | 0.7752 | 76.97 |
| 1985 | 238,490 | 5,717,100 | 1,430,937 | 17,943,972 | 194,400 | 0.8266 | 76.78 |
| 1986 | 240,680 | 5,912,400 | 1,498,944 | 18,534,452 | 199,432 | 0.8701 | 77.07 |
| 1987 | 242,840 | 6,113,300 | 1,523,250 | 19,087,645 | 204,292 | 0.9017 | 79.63 |
| 1988 | 245,060 | 6,368,400 | 1,554,809 | 19,636,149 | 208,570 | 0.9402 | 83.07 |
| 1989 | 247,340 | 6,591,800 | 1,618,728 | 20,167,768 | 212,477 | 0.9790 | 84.72 |
| 1990 | 249,910 | 6,707,900 | 1,602,374 | 20,650,376 | 214,686 | 1.0840 | 84.34 |
| 1991 | 252,640 | 6,676,400 | 1,515,484 | 20,931,321 | 211,162 | 1.1560 | 84.52 |
| 1992 | 255,420 | 6,879,529 | 1,606,935 | 21,299,101 | 213,181 | 1.1949 | 85.87 |
| 1993 | 258,140 | 7,063,412 | 1,719,528 | 21,748,774 | 216,989 | 1.2295 | 87.58 |
| 1994 | 260,680 | 7,347,348 | 1,817,471 | 22,246,928 | 223,330 | 1.2505 | 89.25 |
| 1995 | 262,803 | 7,531,325 | 1,916,976 | 22,787,433 | 225,363 | 1.3137 | 91.22 |
| 1996 | 265,229 | 7,810,009 | 2,054,615 | 23,410,780 | 227,962 | 1.3851 | 94.22 |
| 1997 | 267,784 | 8,161,271 | 2,232,978 | 24,089,180 | 234,445 | 1.4373 | 94.73 |
| 1998 | 270,248 | 8,502,032 | 2,469,906 | 24,873,020 | 237,892 | 1.5234 | 95.15 |
| 1999 | 272,691 | 8,880,300 | 2,674,882 | 25,723,677 | 240,859 | 1.6329 | 96.77 |
| 2000 | 282,172 | 9,205,400 | 2,607,368 | 27,655,636 | 245,567 | 1.7130 | 98.91 |
| 2001 | 285,040 | 9,274,509 | 2,608,868 | 28,468,773 | 243,494 | 1.8540 | 96.38 |
| 2002 | 287,727 | 9,422,759 | 2,608,868 | 29,216,898 | 241,983 | 1.9884 | 98.03 |
| 2003 | 290,211 | 9,659,247 | 2,715,579 | 29,983,768 | 242,761 | 2.1063 | 98.16 |
| 2004 | 292,892 | 10,010,697 | 2,876,488 | 30,792,361 | 245,433 | 2.2113 | 100.35 |
| 2005 | 295,561 | 10,304,854 | 3,029,802 | 31,585,756 | 250,541 | 2.2839 | 100.51 |
| 2006 | 298,363 | 10,591,133 | 3,110,069 | 32,444,607 | 256,064 | 2.3627 | 99.86 |
| 2007 | 301,290 | 10,805,961 | 3,106,975 | 33,214,996 | 258,936 | 2.4608 | 101.60 |
| 2008 | 304,060 | 10,926,080 | 3,020,894 | | 255,441 | | 99.40 |

The population estimates were found on a website of the U.S. Census Bureau (https://www.census.gov/). Values of gross national product $Y$, gross investment $I$ and capital $K$ are available on the website of the U.S. Bureau of Economic Analysis (https://www.bea.gov). The dollar (1996) values for $Y$ from year 1929 and for $I$ and $K$ from year 1925 are reproduced in the Table. The investment $I$ is understood, in terms of the U.S. Bureau of Economic Analysis, as a sum of investments in private fixed assets, in government fixed assets and in consumer durable goods, which make up capital $K$. The time series for labour $L$ for the latest decades (from year 1948) are found on the website of the U.S. Bureau of Labour Statistics (https://stats.bls.gov/). The series of quantities compiled by different researchers are used to restore absolute values of quantities $Y$, $I$, $K$ and $L$ for the earlier years, whereas there is no need to discuss the discrepancies between series from different sources here, for we use the series not for analysis of economic growth but only for illustration of methods of analysis. Data for total consumption of energy $E$ are taken from the website of the U.S. Department of Energy (https://eia.gov/) for years from 1949 and from Historical Statistics (Historical Statistics of the United States: Colonial Times to 1970, Parts 1 and 2. U.S. Department of Commerce, Washington, 1975) for the earlier years. The Table contains also values of substitutive work $P$ estimated in Chap. 7.

## C. Macrocharacteristics of the Russian Economy

The difficulties of collecting macroeconomic time series for Russia are connected with our turbulent development, which did not support the objective establishment of facts. Unfortunately, as writes Bessonov (Bessonov, 2005, p. 169), 'Up to date there is no practice of the publication of the greatest possible long time series of the parameters of economic dynamics in Russia. So, till now there are no publications in Russia (as before in the Soviet Union) that could be similar to *Historical Statistics of the United States* and would cover, as full as possible, if not centuries, the last decades of the existence of the state (that, we shall note, has arisen much more centuries earlier, than the USA)'.

One can think, that the situation has been changing. The values of GDP are estimated by the Russian Federal Service of the State Statistics (Rosstat) since 1989 on the basis of methodological approaches of the system of the national accounts accepted today in the majority of the countries of the world. Figures for the last years can be found in collections of the Rosstat (http://www.gks.ru/), but the values of quantities for the period before year 1991 should be restored. Recently Ponomarenko (2002) has undertaken reconstruction of some parameters of macroeconomic statistics for Russia for the period from year 1961 to 1990 and, as a result of a great and tedious work, has estimated the values of GDP and some other quantities. Ponomarenko has overcome also additional difficulties arising at separation of data for Russia from all-union statistics of the USSR. The collection of empirical data for the Russian Federation, which are discussed and used in Chap. 8, is presented in Table A.3.

**Table A.3** Macrocharacteristics of the Russian economy

| Year | Population $N \cdot 10^{-3}$ | GDP $Y \cdot 10^{-9}$ ruble per year | GDP $Y \cdot 10^{-9}$ R(2000) per year | Investment $I \cdot 10^{-9}$ R(2000) per year | Basic capital $K \cdot 10^{-9}$ R(2000) | Humans efforts $L \cdot 10^{-6}$ man-h per year | Substitut. work $P$ quad per year | Primary energy $E$ quad per year |
|---|---|---|---|---|---|---|---|---|
| 1961 | 120,766 | 123 | 3,687 | 731 | 8,448 | 111,569 | 0.0007 | 12.47 |
| 1962 | 122,407 | 133 | 3,960 | 803 | 8,841 | 113,085 | 0.0009 | 13.03 |
| 1963 | 123,848 | 138 | 4,113 | 708 | 9,290 | 114,417 | 0.0011 | 14.38 |
| 1964 | 125,179 | 146 | 4,387 | 964 | 9,627 | 115,646 | 0.0013 | 15.17 |
| 1965 | 126,309 | 154 | 4,628 | 932 | 10,205 | 116,691 | 0.0016 | 16.16 |
| 1966 | 127,189 | 167 | 4,956 | 1,153 | 10,729 | 117,502 | 0.0019 | 17.12 |
| 1967 | 128,026 | 181 | 5,262 | 1,219 | 11,453 | 118,276 | 0.0024 | 17.97 |
| 1968 | 128,696 | 195 | 5,590 | 1,282 | 12,214 | 118,895 | 0.0029 | 18.88 |
| 1969 | 129,379 | 205 | 5,733 | 1,216 | 13,007 | 119,526 | 0.0036 | 19.77 |
| 1970 | 130,079 | 228 | 6,225 | 1,474 | 13,702 | 120,172 | 0.0044 | 20.12 |
| 1971 | 130,704 | 240 | 6,487 | 1,516 | 14,628 | 120,751 | 0.0055 | 21.20 |
| 1972 | 131,446 | 246 | 6,673 | 1,563 | 15,559 | 121,435 | 0.0068 | 22.52 |
| 1973 | 132,210 | 268 | 7,253 | 1,897 | 16,499 | 122,142 | 0.0084 | 23.53 |
| 1974 | 132,940 | 281 | 7,571 | 1,895 | 17,737 | 122,817 | 0.0108 | 24.56 |
| 1975 | 133,775 | 290 | 7,943 | 2,129 | 18,923 | 123,588 | 0.0137 | 25.42 |
| 1976 | 134,690 | 304 | 8,249 | 2,237 | 20,294 | 124,433 | 0.0177 | 26.94 |
| 1977 | 135,645 | 322 | 8,577 | 2,283 | 21,720 | 125,316 | 0.0228 | 27.39 |
| 1978 | 136,596 | 339 | 8,807 | 2,463 | 23,134 | 126,195 | 0.0291 | 28.67 |
| 1979 | 137,551 | 350 | 8,960 | 2,496 | 24,671 | 127,076 | 0.0376 | 29.33 |
| 1980 | 138,291 | 369 | 9,343 | 2,563 | 26,181 | 127,760 | 0.0485 | 29.74 |
| 1981 | 139,028 | 385 | 9,529 | 2,597 | 27,173 | 128,441 | 0.0569 | 30.46 |
| 1982 | 139,816 | 418 | 9,671 | 2,826 | 28,140 | 129,169 | 0.0662 | 31.23 |
| 1983 | 140,766 | 434 | 9,934 | 2,972 | 29,278 | 130,046 | 0.0786 | 31.93 |
| 1984 | 141,842 | 455 | 10,054 | 2,914 | 30,493 | 131,041 | 0.0937 | 33.31 |
| 1985 | 142,823 | 476 | 10,284 | 2,998 | 31,273 | 131,946 | 0.1042 | 33.49 |
| 1986 | 143,835 | 490 | 10,634 | 2,975 | 32,082 | 132,881 | 0.1159 | 32.82 |
| 1987 | 145,115 | 504 | 10,699 | 3,012 | 32,811 | 134,064 | 0.1259 | 35.21 |
| 1988 | 146,343 | 530 | 10,962 | 3,277 | 33,198 | 135,198 | 0.1294 | 35.74 |
| 1989 | 147,401 | 568 | 11,126 | 3,409 | 33,819 | 137,282 | 0.1337 | 35.55 |
| 1990 | 148,041 | 607 | 10,940 | 3,192 | 34,523 | 136,776 | 0.1517 | 35.64 |
| 1991 | 148,543 | | 10,393 | 3,192 | 34,124 | 134,181 | 0.1554 | 35.00 |
| 1992 | 148,704 | | 8,887 | 2,554 | 30,355 | 130,951 | 0.0694 | 34.10 |
| 1993 | 148,673 | | 8,113 | 2,618 | 26,716 | 128,734 | 0.0242 | 31.97 |
| 1994 | 148,366 | 611 | 7,083 | 2,554 | 23,884 | 124,436 | 0.0141 | 29.23 |
| 1995 | 148,306 | 1,429 | 6,757 | 1,542 | 21,565 | 120,727 | 0.0087 | 27.94 |
| 1996 | 147,976 | 2,008 | 6,514 | 1,215 | 19,668 | 119,824 | 0.0054 | 27.36 |
| 1997 | 147,502 | 2,522 | 6,604 | 1,119 | 18,926 | 117,442 | 0.0049 | 25.77 |
| 1998 | 147,105 | 2,685 | 6,253 | 980 | 18,166 | 115,636 | 0.0039 | 25.96 |
| 1999 | 147,539 | 4,546 | 6,650 | 1,043 | 17,650 | 116,224 | 0.0034 | 27.01 |
| 2000 | 146,890 | 7,303 | 7,312 | 1,232 | 17,500 | 118,796 | 0.0032 | 27.46 |
| 2001 | 146,304 | 8,944 | 7,688 | 1,294 | 18,479 | 118,523 | 0.0034 | 27.70 |

(continued)

**Table A.3** (continued)

| Year | Population $N \cdot 10^{-3}$ | GDP $Y \cdot 10^{-9}$ ruble per year | GDP $Y \cdot 10^{-9}$ R(2000) per year | Investment $I \cdot 10^{-9}$ R(2000) per year | Basic capital $K \cdot 10^{-9}$ R(2000) | Humans efforts $L \cdot 10^{-6}$ man-h per year | Substitut. work $P$ quad per year | Primary energy $E$ quad per year |
|---|---|---|---|---|---|---|---|---|
| 2002 | 145,167 | 10,819 | 8,057 | 1,312 | 19,611 | 121,319 | 0.0036 | 27.93 |
| 2003 | 144,964 | 13,208 | 8,645 | 1,431 | 21,058 | 120,906 | 0.0038 | 28.76 |
| 2004 | 144,168 | 17,027 | 9,265 | 1,559 | 18,976 | 122,440 | 0.0039 | 28.74 |
| 2005 | 143,474 | 21,610 | 9,856 | 1,647 | 18,925 | 147,703 | 0.0041 | 28.73 |
| 2006 | 142,754 | 26,917 | 10,660 | 1,873 | 18,807 | 148,613 | 0.0043 | 29.74 |
| 2007 | 142,115 | 33,248 | 11,570 | 2,337 | 21,015 | 150,503 | 0.0045 | 30.14 |
| 2008 | 142,009 | 41,277 | 12,177 | 2,591 | 21,960 | 151,173 | 0.0049 | 30.83 |
| 2009 | 141,900 | 38,807 | 11,224 | 2,307 | 23,805 | 147,081 | 0.0054 | 29.86 |
| 2010 | 142,800 | 46,309 | 11,730 | 2,318 | 23,604 | 148,977 | 0.0056 | 31.61 |
| 2011 | 142,900 | 59,698 | 12,238 | 2,262 | 22,140 | 149,742 | 0.0054 | 32.73 |
| 2012 | 143,000 | 66,927 | 12,669 | 2,382 | 22,955 | 150,288 | 0.0055 | 33.24 |
| 2013 | 143,300 | 71,017 | 12,831 | 2,430 | 24,124 | 149,499 | 0.0056 | 32.69 |
| 2014 | 143,700 | 79,200 | 12,925 | 2,746 | 24,059 | 149,223 | 0.0058 | 33.29 |
| 2015 | 146,300 | 83,233 | 12,559 | 2,605 | 24,252 | 150,859 | 0.0059 | 32.45 |
| 2016 | 146,804 | 86,044 | 12,531 | 2,638 | 25,015 | 151,003 | 0.0060 | |

The values of GDP for 1961–1990 in current and comparable prices are due to the assessment by Ponomarenko (2002). For the extension of the series, we used values of the relative index of GDP specified in a table of the work of Bessonov (2002) in comparison with the values of Rosstat, so that the estimates of the GDP for the last years (1995–2015) are those given by Rosstat.

The values of investments for the Russian national economy in 1961–1990, which were put in the basis of time series, are estimated by Ponomarenko (2002) in the current and comparable prices. The extension of the series for the upcoming years is based on estimates of the investments issued by Rosstat. Let us note, that the Bessonov and Voskoboynikov's recent research (2006) on the methods of assessment of investments and basic production assets in the Russian transitive economy shows, that, probably, estimates of Rosstat for 1990–2000 require clarification, and there are alternative assessments of investments, which are different from the figures, established by Rosstat. In particular, values of investments for 1990 in the prices of that year (177.2 billion rubles), calculated by Ponomarenko, have appeared to be less in comparison with data by Rosstat (249.1 billion rubles). After some meditations, the values of investments in the current and comparable prices since 1995 are shown according to publication of Rosstat, and for the intermediate years the values are reconstructed due to Bessonov and Voskoboynikov's assessments (2006).

The values of fixed capital $K$ since 2000 are taken as the estimates of Rosstat. The values of capital in the previous years are restored with use of Eq. (8.3) at the known investments and given coefficient of depreciation $\mu = 0.04$. The initial value of basic production assets in year 1960 was taken equal to two thirds of official value of production funds in statistics of the USSR.

The values of expenditures of workers' efforts $L$ in man-hours since 2005 are found in the publication of Rosstat. The estimates of efforts in the previous years, which are found in the way, described in Sect. 8.3, are brought into accord with the assessments by Rosstat. The Table contains also the numbers of total population $N$.

The method of estimating values of substitutive work $P$ is described in Sect. 8.6. The table contains also total consumption of energy carriers $E$ according to statistical service of the Russian Federation. Values of consumption of energy carriers $E$ and substitutive work $P$ are estimated in *quads* ($1\,\text{quad} = 10^{15}\,\text{BTU} \approx 10^{18}\,\text{J}$) per year.

Let us notice once more, that we tried to be close to the official figures by Rosstat, as far as possible, and did not have an objective to discuss accuracy of assessments of the considered quantities. The time series of the quantities are used by us for the illustration of methods of analysis and forecast of economic activity.

## References

1. Bessonov, V.A.: Problemy postroeniya proizvodstvennykh funktsiy v rossiyskoy perekhodnoy ekonomike (Problems of formulation of the production function for the Russian transitive economy). In: Bessonov, V.A., Tsukhlo, S.V. (eds.) Analiz dinamiki rossiyskoy perekhodnoy ekonomiki (The Analysis of Dynamics of the Russian Transitive Economy), S.5–89. Institut ekonomiki perekhodnogo perioda, Moscow (2002)
2. Bessonov, V.A.: Problemy analiza rossiyskoy makroekonomicheskoy dinamiki perekhodnogo perioda (Problems of the analysis of Russian macroeconomic dynamics of the transition period). Institut ekonomiki perekhodnogo perioda, Moscow (2005)
3. Bessonov, V.A., Voskoboynikov, I.B.: O dinamike osnovnykh fondov i investitsiy v rossiyskoy perekhodnoy ekonomike (On dynamics of the fixed capital and investments in the Russian transitive economy). Ekonomicheskiy zhurnal VSHE (Econ. J. HSE). **10**(2), pp. 193–228 (2006)
4. Ponomarenko, A.N.: Retrospektivnye natsional'nye scheta Rossii: 1961–1990 (The Retrospective National Accounts of Russia: 1961–1990). Finansy i statistika, Moscow (2002)

# Index




V. N. Pokrovskii, *Econodynamics*, New Economic Windows,
https://doi.org/10.1007/978-3-319-72074-6

Printed in the United States
By Bookmasters